CO_2 Capture and Sequestration

CO_2 Capture and Sequestration

Editor

Diganta Bhusan Das

Basel • Beijing • Wuhan • Barcelona • Belgrade • Novi Sad • Cluj • Manchester

Editor
Diganta Bhusan Das
Loughborough University
Loughborough
UK

Editorial Office
MDPI
St. Alban-Anlage 66
4052 Basel, Switzerland

This is a reprint of articles from the Special Issue published online in the open access journal *Clean Technologies* (ISSN 2571-8797) (available at: https://www.mdpi.com/journal/cleantechnol/special_issues/CO2_capture_sequestration).

For citation purposes, cite each article independently as indicated on the article page online and as indicated below:

Lastname, A.A.; Lastname, B.B. Article Title. *Journal Name* **Year**, *Volume Number*, Page Range.

ISBN 978-3-7258-1131-1 (Hbk)
ISBN 978-3-7258-1132-8 (PDF)
doi.org/10.3390/books978-3-7258-1132-8

© 2024 by the authors. Articles in this book are Open Access and distributed under the Creative Commons Attribution (CC BY) license. The book as a whole is distributed by MDPI under the terms and conditions of the Creative Commons Attribution-NonCommercial-NoDerivs (CC BY-NC-ND) license.

Contents

About the Editor vii

Preface ix

Diganta Bhusan Das
CO_2 Capture and Sequestration
Reprinted from: *Clean Technol.* **2024**, 6, 25, doi:10.3390/cleantechnol6020025 1

Jonathan Paul Marshall
A Social Exploration of the West Australian Gorgon Gas, Carbon Capture and Storage Project
Reprinted from: *Clean Technol.* **2022**, 4, 6, doi:10.3390/cleantechnol4010006 4

Fernanda M. L. Veloso, Isaline Gravaud, Frédéric A. Mathurin and Sabrine Ben Rhouma
Planning a Notable CCS Pilot-Scale Project: A Case Study in France, Paris Basin—Ile-de-France
Reprinted from: *Clean Technol.* **2022**, 4, 28, doi:10.3390/cleantechnol4020028 28

Anja Pfennig and Axel Kranzmann
Understanding the Anomalous Corrosion Behaviour of 17% Chromium Martensitic Stainless Steel in Laboratory CCS-Environment—A Descriptive Approach
Reprinted from: *Clean Technol.* **2022**, 4, 14, doi:10.3390/cleantechnol4020014 47

Luqman Kolawole Abidoye and Diganta B. Das
Carbon Storage in Portland Cement Mortar: Influences of Hydration Stage, Carbonation Time and Aggregate Characteristics [†]
Reprinted from: *Clean Technol.* **2021**, 3, 34, doi:10.3390/cleantechnol3030034 66

Thomas Quaid and M. Toufiq Reza
Carbon Capture from Biogas by Deep Eutectic Solvents: A COSMO Study to Evaluate the Effect of Impurities on Solubility and Selectivity
Reprinted from: *Clean Technol.* **2021**, 3, 29, doi:10.3390/cleantechnol3020029 84

Santosh Khokarale, Ganesh Shelke and Jyri-Pekka Mikkola
Integrated and Metal Free Synthesis of Dimethyl Carbonate and Glycidol from Glycerol Derived 1,3-Dichloro-2-propanol via CO_2 Capture
Reprinted from: *Clean Technol.* **2021**, 3, 41, doi:10.3390/cleantechnol3040041 97

Thomas Deschamps, Mohamed Kanniche, Laurent Grandjean and Olivier Authier
Modeling of Vacuum Temperature Swing Adsorption for Direct Air Capture Using Aspen Adsorption
Reprinted from: *Clean Technol.* **2022**, 4, 15, doi:10.3390/cleantechnol4020015 111

Kamal Jawher Khudaida and Diganta Bhusan Das
A Numerical Analysis of the Effects of Supercritical CO_2 Injection on CO_2 Storage Capacities of Geological Formations
Reprinted from: *Clean Technol.* **2020**, 2, 21, doi:10.3390/cleantechnol2030021 129

Tryfonas Pieri and Athanasios Angelis-Dimakis
Model Development for Carbon Capture Cost Estimation
Reprinted from: *Clean Technol.* **2021**, 3, 46, doi:10.3390/cleantechnol3040046 161

Szabolcs Szima, Carlos Arnaiz del Pozo, Schalk Cloete, Szabolcs Fogarasi, Ángel Jiménez Álvaro, Ana-Maria Cormos, et al.
Techno-Economic Assessment of IGCC Power Plants Using Gas Switching Technology to Minimize the Energy Penalty of CO_2 Capture
Reprinted from: *Clean Technol.* **2021**, 3, 36, doi:10.3390/cleantechnol3030036 . . . . **178**

Jennifer Reeve, Oliver Grasham, Tariq Mahmud and Valerie Dupont
Advanced Steam Reforming of Bio-Oil with Carbon Capture: A Techno-Economic and CO_2 Emissions Analysis
Reprinted from: *Clean Technol.* **2022**, 4, 18, doi:10.3390/cleantechnol4020018 . . . . **202**

About the Editor

Diganta Bhusan Das

Diganta Bhusan Das is a *Reader in Porous Media* within the Chemical Engineering Department of the School of Aeronautical, Automotive, Chemical and Materials Engineering (SAACME) at Loughborough University (LU). His research relates to fluid flow, mass transport, and reactions in porous media, spanning between water and process systems engineering to bioengineering. He is motivated to develop smart porous media technologies supported by cutting-edge fabrication and digital technologies to address some of the most critical societal challenges: NetZero, water security, sustainability, and material circularity. The underlying concept is that they leverage porous materials with interconnected void spaces (pores) and particles/fibres of different sizes and shapes to achieve specific engineering goals or functions. Therefore, his research vision is to integrate porous media theories, process engineering, and system autonomy to develop innovative sustainable porous media technologies, enhancing their performance, reducing the need for constant human oversight, and enabling their adoption to solve the above societal challenges. In 2021, Diganta was elected as a Fellow of the Royal Society of Chemistry (FRSC) and Royal Society of Biology (FRSB) for leading contributions in solute transport and reactions in porous materials and tissue engineering, respectively. Diganta has successfully mentored seven postdoctoral and twenty-five PhD researchers across different institutions. He has authored/co-authored >150 peer-reviewed journal publications. In addition, he has co-authored six books and co-edited three books related to his research interests, which serve both as references for research and books for research-informed teaching.

Preface

CO_2 capture and sequestration (CCS) technologies aim to capture carbon dioxide (CO_2) from CO_2 sources (e.g., fossil fuel power plants), separate the CO_2, and store it in suitable media. CO_2 can be captured using various technologies, including absorption, adsorption, cryogenic processes, and membrane gas separation. Therefore, accurate selection, design, modelling, and optimisation of the processes for CO_2 capture and the tuning of the material properties are essential. There are different methods for CO_2 sequestration, e.g., (i) geological sequestration, which injects different phases of CO_2 into the subsurface; (ii) oceanic storage, which dissolves CO_2 into an ocean at different depths; (iii) the solid-phase reaction of CO_2 with metal oxides to produce stable carbonates with no risk of CO_2 release to the atmosphere; etc. The flow, transport, and reaction of CO_2 during CCS and other related matters, such as monitoring critical parameters, are also essential. To address these points, a Special Issue (SI) of *Clean Technologies* and an e-book with all published papers have been organized which highlight the recent trends and innovative developments in CCS. In particular, the published papers in the SI and the e-book highlight the following issues:

- Socio-political issues related to CCS project development and deployment.
- Fundamental technical issues concerning the development and deployment of CCS projects.
- The synthesis of value-added chemicals using captured CO_2 in CCS projects.
- Applications of mathematical modelling for the development of CCS projects.
- The development of techno-economic costing models for CCS projects.

Overall, the SI and e-book cover a diverse range of topics, including some of the most pressing concerns for the future growth and development of CCS projects.

Diganta Bhusan Das
Editor

Editorial

CO$_2$ Capture and Sequestration

Diganta Bhusan Das

Department of Chemical Engineering, Loughborough University, Loughborough LE11 3TU, Leicestershire, UK; d.b.das@lboro.ac.uk; Tel.: +44-1509222509

Citation: Das, D.B. CO$_2$ Capture and Sequestration. *Clean Technol.* **2024**, *6*, 494–496. https://doi.org/10.3390/cleantechnol6020025

Received: 27 December 2023
Accepted: 12 March 2024
Published: 16 April 2024

Copyright: © 2024 by the author. Licensee MDPI, Basel, Switzerland. This article is an open access article distributed under the terms and conditions of the Creative Commons Attribution (CC BY) license (https://creativecommons.org/licenses/by/4.0/).

CO$_2$ capture and sequestration (CCS) aims to capture carbon dioxide (CO$_2$) from CO$_2$ sources (e.g., fossil fuel power plants), separate the CO$_2$, and store it in suitable media. CO$_2$ can be captured using various technologies, including absorption, adsorption, cryogenic processes, and membrane gas separation [1]. Therefore, accurate selection, design, modelling and optimisation of the processes for CO$_2$ capture and the tuning of the material properties are essential. There are different methods used for CO$_2$ sequestration, e.g., (i) geological sequestration that injects different phases of CO$_2$ into the subsurface [2], (ii) oceanic storage that dissolves CO$_2$ into an ocean at different depths [3], (iii) the solid-phase reaction of CO$_2$ with metal oxides to produce stable carbonates with no risk of CO$_2$ release to the atmosphere [4], and others. The flow, transport, and reaction of CO$_2$ during CCS and other related matters, such as monitoring critical parameters, are also essential [5].

To address these points, a Special Issue (SI) of *Clean Technologies* has been organised to highlight the recent trends and innovative developments in CCS [6]. Thirteen (13) submissions were received, which underwent a rigorous peer review process. Two papers were declined at the peer review stage, and the remaining eleven papers [7–17] have now been published [6]. The published papers are also being compiled as an edited e-book to be published by MDPI. The papers [7–17] highlight several common and important issues.

Issues related to CCS project development and deployment have been considered by Marshall [7] and Veloso et al. [8]. Marshall [7] has identified that although CCS projects are essential to lower gas emissions, they have not achieved their desired objectives in Australia. To investigate the reasons for this failure, Marshall [7] undertook a historical and social study of the Gorgon gas project in Western Australia, considered one of the world's most significant CCS projects. The study has rightly concluded that CCS's social dynamics must be included in CCS project projections to enhance the accuracy of their expectations, without which the project projections are likely to miss their targets. Veloso et al. [8] emphasised that there are few commercial-scale CCS projects worldwide, and almost all are in the USA and China. Despite the many CCS pilot-scale projects planned in Europe, only two commercial-scale projects operate today. To help improve this situation, the authors have proposed a 'multicriteria regional-scale approach' that can help select the most promising locations in France to deploy CCS pilot-scale projects. Subsequently, the authors have assessed different aspects of CCS technology at the regional scale, including the key economic performance indicators of the CCS project. The authors have rightly concluded that the CCS projects should be located strategically close to potential CO$_2$ sources in case of the confirmation of proven resources.

Several fundamental issues concerning CCS have also been addressed in the SI. Pfennig and Kranzmann [9] considered cases where CO$_2$ is compressed to sequestrate it into deep geological formations. In this process, the corrosion of injection steel pipes can occur due to the contact of the metal with CO$_2$ and saline water in the geological formation. The published work is supported by the authors' laboratory experiments, which have evaluated corrosion kinetics on stainless steels X$_{35}$CrMo$_{17}$ and X$_5$CrNiCuNb$_{16-4}$ with approximately 17% Cr. The relationship between the corrosion rate and ionic species diffusion into the metal has been studied to determine the longevity of the chosen steels in a CCS environment. In the paper by Abidoye and Das [10], the effects of particle size, carbonation time, curing time and pressure on the efficiency of carbon storage in Portland cement mortar as the media for CCS have been investigated. The authors have shown how carbonation efficiency increases with decreased particle size using data generated in pressure chamber experiments. Overall, these authors show that carbonation efficiency increases with smaller-sized particles or higher-surface areas, carbonation time and higher pressure, but it decreases with hydration/curing time. Quaid and Reza [11] analysed deep eutectic solvents (DESs) for their carbon capture and biogas upgrade applications. In particular, they analysed how the presence of contaminants in biogas may affect the carbon capture by DESs. The behaviour of DESs under different temperatures, pressures, and influences from pollutants has been studied, which suggests that a complex interplay of variables must be understood when choosing DESs for CO_2 absorption for biogas uplifting.

This Special Issue also highlights how the captured CO_2 may be further used to synthesise value-added chemicals. Khokarale et al. [12] demonstrate that industrially important solvents, namely, dimethyl carbonate (DMC) and glycidol, could be synthesised in a combined process using glycerol-derived 1,3-dichloro-2-propanol and captured CO_2 via a metal-free reaction route under mild conditions.

The mathematical modelling applications in CCS have been demonstrated by Deschamps et al. [13] and Khudaida and Das [14]. Deschamps et al. [13] used conservation of mass and energy principles and equations of states to evaluate the performance of a vacuum temperature swing adsorption (VTSA) process for direct CO_2 capture from the air at an industrial scale. A parametric study on the effects of the main operating conditions has been undertaken to assess the performance and energy consumption of the VSTA. The developed approach considers how the lab-scale process could be upscaled to a larger industrial scale. In contrast to lab- or industrial-scale processes, Khudaida and Das [14] attempted to conduct a numerical study on the significance of injecting CO_2 into deep saline aquifers at the scale of geological formations. Several CO_2 injection scenarios and aquifer characteristics have been investigated to enhance current knowledge on the effects of the residual and solubility trapping of CO_2 on the sequestration mechanisms. For example, it was shown how the extent of subsurface heterogeneity increases the residual trapping of CO_2 in geological formations.

Finally, this Special Issue highlighted the critical issues relating to the techno-economic costing of CCS projects. Pieri and Angelis-Dimakis [15] reviewed the current approaches used to quantify CO_2 capture costs. It has been shown that with the existing knowledge in the literature, one can estimate capture costs based on the amount of CO_2 captured and the technologies used in CO_2 capture technology. In the paper by Szima et al. [16], it has been pointed out that increased levelized electricity costs within CCS projects are associated with significant energy penalties involved in CO_2 capture. Consequently, Szima et al. evaluated three CCS approaches that rely on integrated gasification combined cycles: (i) gas switching combustion (GSC), (ii) GSC with added natural gas firing to increase the turbine inlet temperature, and (iii) oxygen production pre-combustion that replaces the air separation unit with more efficient gas switching oxygen production reactors. This comparison has enabled the authors to identify the most promising solution for further development and exploitation in CCS. Reeve et al. [17] carried out a techno-economic analysis of three processes for hydrogen production from advanced steam reforming (SR) of bio-oil as an alternative route to hydrogen with bioenergy with carbon capture and storage (BECCS): conventional steam reforming (C-SR), C-SR with CO_2 capture (C-SR-CCS), and sorption-enhanced chemical looping (SE-CLSR). The analysis concluded that SE-CLSR is comparable to C-SR-CCS in terms of the levelized cost of hydrogen (LCOH).

Overall, it is evident that this Special Issue and the forthcoming e-book cover a diverse range of topics, including some of the most pressing concerns for CCS. I envisage that the authors of the published papers and I, as the guest editor of the SI, can motivate future directions and progress in CCS.

I appreciate the efforts of the authors and referees of all the accepted and declined papers. These contributions have made this Special Issue a true success. Finally, I acknowledge the Editorial Office for supporting this Special Issue and the edited e-book.

Conflicts of Interest: The author declares no conflict of interest.

References

1. Chowdhury, S.; Kumar, Y.; Shrivastava, S.; Patel, S.K.; Sangwai, J.S. A Review on the Recent Scientific and Commercial Progress on the Direct Air Capture Technology to Manage Atmospheric CO_2 Concentrations and Future Perspectives. *Energy Fuels* **2023**, *37*, 10733–10757. [CrossRef]
2. Abidoye, L.K.; Khudaida, K.J.; Das, D.B. Geological Carbon Sequestration in the Context of Two-Phase Flow in Porous Media: A Review. *Crit. Rev. Environ. Sci. Technol.* **2015**, *45*, 1105–1147. [CrossRef]
3. Ho, H.-J.; Iizuka, A. Mineral carbonation using seawater for CO_2 sequestration and utilization: A review. *Sep. Purif. Technol.* **2023**, *307*, 122855. [CrossRef]
4. Tyagi, P.; Singh, S.; Malik, N.; Kumar, S.; Malik, R.S. Metal catalyst for CO_2 capture and conversion into cyclic carbonate: Progress and challenges. *Mater. Today* **2023**, *65*, 133–165. [CrossRef]
5. Arellano, Y.; Tjugum, S.-A.; Pedersen, O.B.; Breivik, M.; Jukes, E.; Marstein, M. Measurement technologies for pipeline transport of carbon dioxide-rich mixtures for CCS. *Flow Meas. Instrum.* **2024**, *95*, 102515. [CrossRef]
6. Available online: https://www.mdpi.com/journal/cleantechnol/special_issues/CO2_capture_sequestration (accessed on 19 January 2024).
7. Marshall, J.P. A Social Exploration of the West Australian Gorgon Gas, Carbon Capture and Storage Project. *Clean Technol.* **2022**, *4*, 6. [CrossRef]
8. Veloso, F.M.L.; Gravaud, I.; Mathurin, F.A.; Rhouma, S.B. Planning a Notable CCS Pilot-Scale Project: A Case Study in France, Paris Basin—Ile-de-France. *Clean Technol.* **2022**, *4*, 28. [CrossRef]
9. Pfennig, A.; Kranzmann, A. Understanding the Anomalous Corrosion Behaviour of 17% Chromium Martensitic Stainless Steel in Laboratory CCS-Environment—A Descriptive Approach. *Clean Technol.* **2022**, *4*, 14. [CrossRef]
10. Abidoye, L.K.; Das, D.B. Carbon Storage in Portland Cement Mortar: Influences of Hydration Stage, Carbonation Time and Aggregate Characteristics. *Clean Technol.* **2021**, *3*, 34. [CrossRef]
11. Quaid, T.; Reza, M.T. Carbon Capture from Biogas by Deep Eutectic Solvents: A COSMO Study to Evaluate the Effect of Impurities on Solubility and Selectivity. *Clean Technol.* **2021**, *3*, 29. [CrossRef]
12. Khokarale, S.; Shelke, G.; Mikkola, J.P. Integrated and Metal Free Synthesis of Dimethyl Carbonate and Glycidol from Glycerol Derived 1,3-Dichloro-2-propanol via CO_2 Capture. *Clean Technol.* **2021**, *3*, 41. [CrossRef]
13. Deschamps, T.; Kanniche, M.; Grandjean, L.; Authier, O. Modeling of Vacuum Temperature Swing Adsorption for Direct Air Capture Using Aspen Adsorption. *Clean Technol.* **2022**, *4*, 15. [CrossRef]
14. Khudaida, K.J.; Das, D.B. A Numerical Analysis of the Effects of Supercritical CO_2 Injection on CO_2 Storage Capacities of Geological Formations. *Clean Technol.* **2020**, *2*, 21. [CrossRef]
15. Pieri, T.; Angelis-Dimakis, A. Model Development for Carbon Capture Cost Estimation. *Clean Technol.* **2021**, *3*, 46. [CrossRef]
16. Szima, S.; del Pozo, C.A.; Cloete, S.; Fogarasi, S.; Álvaro, Á.J.; Cormos, A.; Cormos, C.; Amini, S. Techno-Economic Assessment of IGCC Power Plants Using Gas Switching Technology to Minimize the Energy Penalty of CO_2 Capture. *Clean Technol.* **2021**, *3*, 36. [CrossRef]
17. Reeve, J.; Grasham, O.; Mahmud, T.; Dupont, V. Advanced Steam Reforming of Bio-Oil with Carbon Capture: A Techno-Economic and CO_2 Emissions Analysis. *Clean Technol.* **2022**, *4*, 309–328. [CrossRef]

Disclaimer/Publisher's Note: The statements, opinions and data contained in all publications are solely those of the individual author(s) and contributor(s) and not of MDPI and/or the editor(s). MDPI and/or the editor(s) disclaim responsibility for any injury to people or property resulting from any ideas, methods, instructions or products referred to in the content.

Article

A Social Exploration of the West Australian Gorgon Gas, Carbon Capture and Storage Project

Jonathan Paul Marshall

Social and Political Sciences, University of Technology Sydney, Sydney 2007, Australia; jonathan.marshall@uts.edu.au

Abstract: Carbon capture and storage (CCS) appears to be essential for lowering emissions during the necessary energy transition. However, in Australia, it has not delivered this result, at any useful scale, and this needs explanation. To investigate the reasons for this failure, the paper undertakes a historical and social case study of the Gorgon gas project in Western Australia, which is often declared to be one of the biggest CCS projects in the world. The Gorgon project could be expected to succeed, as it has the backing of government, a practical and economic reason for removing CO_2, a history of previous exploration, nearby storage sites, experienced operators and managers, and long-term taxpayer liability for problems. However, it has run late, failed to meet its targets, and not lowered net emissions. The paper explores the social factors which seem to be disrupting the process. These factors include the commercial imperatives of the operation, the lack of incentives, the complexity of the process, the presence of ignored routine problems, geological issues (even in a well-explored area), technical failures, regulatory threats even if minor, tax issues, and the project increasing emissions and consuming carbon budgets despite claims otherwise. The results of this case study suggest that CCS may work in theory, but not well enough under some contemporary forms of social organisation, and the possibilities of CCS cannot be separated from its social background. Social dynamics should be included in CCS projections to enhance the accuracy of expectations.

Keywords: CCS; carbon capture; socio-technical; energy transitions; disorder

Citation: Marshall, J.P. A Social Exploration of the West Australian Gorgon Gas, Carbon Capture and Storage Project. *Clean Technol.* 2022, 4, 67–90. https://doi.org/10.3390/cleantechnol4010006

Academic Editor: Diganta B. Das

Received: 5 December 2021
Accepted: 30 January 2022
Published: 9 February 2022

Publisher's Note: MDPI stays neutral with regard to jurisdictional claims in published maps and institutional affiliations.

Copyright: © 2022 by the author. Licensee MDPI, Basel, Switzerland. This article is an open access article distributed under the terms and conditions of the Creative Commons Attribution (CC BY) license (https://creativecommons.org/licenses/by/4.0/).

1. Introduction

Through a case study, this paper explores the social, organisational, and ecological contexts of carbon capture and storage (CCS), as displayed by the Chevron Gorgon gas project in West Australia, and suggests explanations for its apparent failure. The prime suggestion is that technology is a social venture, which cannot be separated from its complex social background.

In social studies of science and technology, it is standard to assert that technology is invented, understood, developed, used, promoted, managed, installed, regulated, designed, financed, and sold in differing social, economic, and power relations and that these factors have consequences. Technologies may be driven by these relations, take them for granted, or be designed to reinforce them, although technologies frequently have disruptive unintended consequences. Technologies can work in theory but be found socially impractical, be hindered by social practices (intentionally or unintentionally), or have less success than supposedly technically inferior inventions. Some good introductions to this subject include [1–3]. However, this paper requires no specialist knowledge.

Technologies can also involve compelling 'social imaginaries', especially those technologies which exist in theory or fail to work the way they are intended. These imaginings may then function as a rhetoric to persuade people of an existing, or forthcoming, "beneficial reality" [4]. Consequently, technologies can be used politically, or to avoid facing disturbing problems. In illustration, this paper explores the unintentional social and technical undermining of carbon capture as a working solution for greenhouse gas (GHG) emission problems.

It seems important to understand that societies are a subset of complex interactive systems [5,6]. They are composed of people and groups who modify themselves and their reactions in response to what they *perceive* as happening in the system, and by what happens to them. Societies have their own internal systems such as economies, knowledge, and politics and interact with other complex systems such as ecologies, climate systems, and technical systems. These complex systems overlap with each other, and cannot be easily isolated in analysis, hence the discussion of factors in this paper which some might consider relatively unimportant to the CCS process. As a result of these overlapping interactions, technological projects may increase in complexity (and difficulty of control and prediction) as other parts, and social organisations, are added to them, often leading to "tipping points" or breakdown [7]. Supportive of this position, it has been argued that experimental rigs which work at a small scale may have problems when expanded and that the bigger the carbon capture project, the more likely it is to fail [8]. This does not bode well for building a series of carbon capture projects adequate to curtail carbon pollution.

1.1. Paper Structure

The paper proceeds by briefly describing its methodology and the previous work on the history of particular carbon capture projects and their social embedding. Then it puts forward the proposition that climate change is socially generated and driven, and tied into maintaining existing patterns of power, development, and consumption. Social excess produces pollution beyond the capacity of world ecologies to process, particularly when those ecologies are being further damaged by extraction. Section 2 very briefly describes carbon capture in general, then describes carbon capture in Australia, which has a long history of encouragement and funding, but little relative success. Section 3 gives the case study history and analysis of the Gorgon project, arguing that while it is an excellent exemplar for CCS, it has missed its targets and failed to significantly reduce the emissions from the use of its products. This arises from the commercial imperatives of the operation, the lack of incentives, the complexity of the process, the presence of routine problems, geological and ecological issues, technical failures, regulatory threats, tax issues, and the project increasing emissions and consuming carbon budgets despite claims otherwise. While the Chevron Gorgon project should be straightforward, it is overwhelmed by complexity and avoidance of the problem of increased GHG emissions from its operation and products.

1.2. Methodology

The methodology involved tracking news articles on the Gorgon project and following up references in those articles to official documents, or other pieces of nonduplicating journalism, to check their accuracy where possible. I collected a total of 213 news articles and reports stretching over the period 2006–2021 together with other background material. My main interest was in the political, managerial, and economic processes involved, but it was impossible to read these documents without realisation of recurring technical problems, which might not have been expected. There is bias in my analytic procedure as I was looking for disorderly processes and problems. The normal bias is to ignore or play down disorder, blame it on unique circumstances, or condemn it. For instance, the in-house history of the project appears to downplay problems despite being a "lessons learnt" piece [9]. As I have argued previously [10], repeated or expectable disorder is a socially significant part of any process, indicating the way things are done, the systems they interact with, and the problems and processes that organisations wish to avoid.

All social and historical research on the Gorgon project is indebted to the journalist Peter Milne, of *BoilingCold*, who obtained many apparently hidden, or nonavailable, documents from Chevron or the West Australian government, through freedom of information requests. Secrecy, whether intentional or otherwise, seems an established part of the project process.

When conducting case studies through history, sociology, or anthropology it is difficult to separate "data" from "discussion". Data involves interpretation [11]. Rather than "seeing the events" with their own senses, or interpreting those events directly, the scholar is dependent upon other people's interpretations of events, and these methods of others (and the analysts own methods) can create interpretations and hence affect the way reality is perceived and acted upon [12]. "Objectivity" comes with social filters. The reports I read may be trying to justify or criticise the project. The reporters almost certainly hold existing views and purposes which influence reports; they may be writing for a specific audience, and so may the analyst. Hence, these reports have to be fitted together through discussion to see what sense they make as a pattern. The data parts become meaningful in terms of the whole narrative, and the whole narrative becomes the "results". Any interpretation can be overturned by more data and more refined processes of interpretation. Case studies also require a recognition of the potential uniqueness of the case and its context. Comparison is useful but should come after consideration of a number of case studies; otherwise, important factors can be more easily missed as the analysts are not expecting them. This paper aims at presenting a set of hypotheses and interpretations which can guide further interpretation and investigation.

1.3. Previous Work

I was unable to find many detailed histories of particular CCS projects, let alone many which investigated their social context in any depth. Most of the articles in the premier journals for sociological research into energy (*Energy Research and Social Science*) concerning CCS seem to be about public opinion, public evaluation, and communicating acceptance of carbon capture [13–15]. Likewise, an anonymous corporate case study of the ZEPP [16] project in the Netherlands seems primarily interested in how to reduce social opposition in advance of the project.

However, some previous studies show the use of historical case studies. We are fortunate to have the Trupp piece about the Gorgon project, mentioned previously [9], but it does not go into social or economic details, and it seems to avoid fairly well known problems with the project. The best technical history or case study of an individual CCS project is Cook's edited collection about the Otway Project [17]; however, it tells us more or less nothing about the economics. The Otway CCS project was primarily a research project (which implies an unusual social set-up for normal CCS), and it limited the social side of the research to consultation with the local community, which largely seems to have been oriented at persuasion rather than research. Ackerboom et al. [18] write an important paper which includes a short history of CCS in the Netherlands, rather than of individual projects, which remarks that "while CCS is technically a straightforward proposition, its deployment has historically been hindered by the lack of a sound business case and a compelling and stable socio-technical narrative". They also indicate significant governmental support for the projects, which may render those projects similar to the Gorgon project, although there is also significant social opposition (partly because the projects are near habitation) and questions over liability, which are missing in Australia. The absence of a profit motive for doing CCS also seemed important to them, as will be argued here. A previous paper by myself on the general history of CCS in Australia [10] argued that despite political and monetary support over the last 20–30 years, CCS has not made any noticeable impact on Australia's emissions and primarily functions as rhetoric to justify sales of fossil fuels and as a fantasy to defend against real climate action or emissions reduction. A case study by the National Consumer Research Centre in Finland of the Snøhvit liquid natural gas facility in Norway [19] found the site had been caught in controversies about the gas field and ongoing political uncertainty over fossil fuels. "As a consequence of its high ambition level and the controversies surrounding it, the project has experienced a sequence of delays and cost overruns". They remark that even "even local *support* cannot be totally controlled by the project managers", which is unsurprising in a complex human system but appears to indicate the idea that societies are easily manipulated into agreement with technology and are thus separate from the technological process.

This previous research gives at least some indication that it may be fruitful to pursue the social embedding of CCS projects.

1.4. The Problem: Emissions as Social Excess

Currently, some parts of *some* (not all) human societies are significantly disrupting global ecologies and climate systems [20]. They are consuming resources faster than the planet regenerates them, while simultaneously polluting and disrupting the planet's regenerative capacities, producing instability. Societies seem on the edge of a vast series of (probably rapid) chaotic changes including sea level rises, droughts, floods, wild storms, people movement, and wars. As we are dealing with interacting complex systems, uncertainties about when we will cross the line are normal [21]. Consequently, it seems safer to be cautious than not.

Carbon dioxide and methane (or "natural gas") are currently the main greenhouse gas (GHG) pollutants. CO_2 and methane are normally processed by the global ecology in a "reasonable time frame", being broken down into carbon and oxygen by metabolic processes. CO_2 has also been absorbed by the oceans, gradually increasing acidification and creating harsher conditions for some ocean life, with possibly compounding effects. GHGs are only a problem because industries are producing far more than can be processed by the global ecology within that "reasonable time frame", especially given the simultaneous destruction of ecologies through other forms of pollution or extraction (such as deforestation, fossil fuel mining, and some forms of agriculture). It has been repeatedly estimated that dominant societies, through their social organisation, industries, development, and profit drives, consume, disperse, and destroy in a year more than the planet can regenerate [22–24]. This process, known as "overshoot" or the "metabolic rift", is often seen as a hallmark of capitalist and developmentalist organisation dependent on "economic growth" [25,26].

Dominant societies seem dependent upon, and structured around, pollution and ecological destruction. The dire paradox we face is that pollution from burning fossil fuels both enables modern societies, their science, technology, business, prosperity, and military capacity, and produces climate change which could become catastrophic enough to destroy those societies. By being considered as an "externality", pollution also makes production cheaper, and profits higher for powerful social groups. The increase in CO_2 emissions over the last 70 years of "development" is marked. While there are differences in estimates, the Oxford University *Our World In Data* website, estimates that, without factoring in land use changes, humans released "only" 6 billion tonnes of CO_2 during the year 1950. This increased to 22 billion tonnes during 1990 and reached over 36 billion tonnes in 2019 [27]. The IEA tells us that emissions declined in 2020, due to COVID-19 [28], but 2021 is "set to be the second largest annual increase in history" [29].

A recent study in Nature's *Communications Earth and Environment* journal estimates that "the [carbon] budget for a 67% chance of remaining below the [1.5 °C] target is [a total of] 230 $GtCO_2$ from the year 2020 onwards" [30] (p. 3). Commenting on the article, the authors add "This is equivalent to between six and 11 years of global emissions, if they remain at current rates and do not start declining" [31]. The chance of a decline with current action is minimal. The updated UN NDC Synthesis Report predicts "a sizable increase, of about 16%, in global GHG emissions in 2030" while "limiting global average temperature increases to 1.5C requires a reduction of CO_2 emissions of 45% in 2030 or a 25% reduction by 2030 to limit warming to 2C" [32]. There is relatively little sign of social and political will to reduce GHG pollution as dramatically as needed, and some signs the social systems will continue to increase it.

Given the overt dangers, and the scientific advice, this reluctance to reduce emissions almost certainly arises from a social "lock-in" by powerful decision-makers and companies, making it harder to reduce fossil fuel burning than to increase it. Lowering fossil fuel usage threatens organisations which have depended upon those fuels for their success. It is unlikely in this scenario that one technological innovation which preserves current social organisation will be enough to solve the entire complex system of problem generation. We may need a change in social organisation to succeed [33].

In particular, polluting societies need to avoid misleading situations in which emissions from fossil fuels increase at the same time as renewable energy increases so that the increase in emissions is hidden by a lowering of "carbon density", "emissions intensity", or "emissions per unit of energy", or a small fraction of new emissions being caught and stored. Reducing the effects of climate change needs *actual* decreases in greenhouse gas (GHG) emissions: otherwise, harsh changes are inevitable. The idea of a "carbon budget", or amounts of GHG we can emit before likely generating uncontrollable damage, makes the situation clear.

2. Carbon Capture
2.1. Carbon Capture in General

It seems logical that if we could *capture* most of the GHG emissions from burning fossil fuels, or *extract* those emissions from the atmosphere, *store* them somewhere safely out of the atmosphere forever, or *turn them into something useful or harmless*, then some climate change pressure might be lessened. The pressures could also be reduced by *stopping* emissions, but the social ordering and lock-in discussed above can make this seem improbable, adding further strength to the importance of CCS.

The IPCC and the IEA have suggested that carbon capture and storage (CCS), in which CO_2 is stored underground; carbon capture utilisation and storage (CCUS), in which the carbon is utilised for some other project; and carbon dioxide removal (CDR) from the atmosphere with storage are essential for keeping climate change within socially survivable bounds. (I shall use the term CCS to cover all these ideas for convenience.) The IPCC 2021 report talks of "anthropogenic removals [of CO_2] *exceed[ing]* anthropogenic emissions, to lower surface temperature" [34] (p. 29) (emphasis added). The 2018 IPCC *Special Report: Global Warming of 1.5 °C* states that the "shares of nuclear and fossil fuels with carbon dioxide capture and storage (CCS) . . . increase in most 1.5 °C pathways" [35]. Fateh Birol, head of the IEA, is reported as saying the following: "Without [CCS], our energy and climate goals will become virtually impossible to reach", even if CCS's record was "one of unmet expectations" [36]. Many more expressions of the importance of CCS could easily be given. Whether it is sensible to put hope in long-term unmet expectations is another matter.

In 2021, the IEA reinforced the consequences of a limited carbon budget: "Net zero means huge declines in the use of coal, oil and gas . . . . Beyond projects already committed as of 2021, there are no new oil and gas fields approved for development in our pathway, and no new coal mines or mine extensions are required" [37].

That is, there should be no new sources of emissions at all. In this view, CCS with increased emissions is not useful. A study in Nature [38] also insists that to maintain a *50% chance* of remaining under 1.5 °C, nearly 60% of oil and methane, and 90% of coal, must remain unextracted, or, presumably, their emissions must be completely stored.

While it is theoretically possible for CCS to solve the emissions problem, this does not mean it is capable of solving the problem, solving it quickly or cheaply enough without significant risk, or is being used to solve the problems. There are no working examples of CCS operating at the scale needed. The IEA said in 2021: "Only one commercial power plant equipped with CCUS remains in operation today. Based on projects currently in early and advanced deployment, the potential capture capacity of all CCUS deployment in power is projected to reach ~60 $MtCO_2$ in 2030—well short of the 430 $MtCO_2$ per year in the Net Zero Emissions by 2050 Scenario" [39].

The Carbon Capture and Storage Institute is more optimistic and estimates that the capacity of CCS projects in development (not completed) grew to 111 million tonnes per annum in 2021, a tiny proportion of 36 billion tonnes of emissions per year. Much of that CO_2 is being used for enhanced oil recovery, which further increases emissions [40]. At the same, time members of the Institute write "the number of projects is far lower than what is needed to make a significant impact on climate change", although they suggest "organisational competency" is increasing [41] (pp. 4, 6). A suggestion from 2013 [42] that not enough CCS is happening to be useful is still relevant.

Even when successful, the amount of emissions stored from a project can be trivial compared with the emissions released by the companies involved. For example: "Any progress Shell demonstrates in removing carbon from the atmosphere using CCS (1 m tonnes per annum at Quest and up to 4 m tonnes at Gorgon) should be seen in light of Shell's total emissions of 656 million tonnes per annum (80 Mt scope 1 and 2; 576 Mt scope 3)" [43].

It is generally assumed that technologies become cheaper and easier to use over time, but this is not always the case [44]. CCS is an established technology, with little rapid improvement likely. It has been used since at least 1972 "when several natural-gas processing plants in the Val Verde area of Texas began employing carbon capture to supply CO_2 for enhanced oil recovery" [43]. The first international conference on carbon dioxide removal was held in the Netherlands in 1992 [45]. The Sleipner project, in Norway, began in 1996. The IPCC first reported on CCS in 2005 [46]. By 2012, the EU had committed USD 10 billion in taxpayer support [47] (p. 249). Given this history, it should be relatively easy to discover whether CCS is useful, a fantasy with regular failure, or even a mode of locking in GHG pollution.

CCS is probably also affected by the reluctance of governments to get involved in problem solving, and the neoliberal belief that development should be left to subsidised private enterprise. This turns CCS into a commercial activity with no obvious commercial co-benefits, such as profit, unless it involves activities such as extracting more oil, which expands emissions. Lack of profit and a potential increase in liability costs inhibit commercial action, although this could possibly be rectified by financial incentives, robust measures of GHG removal, or cheap pipelines to storage fields [44,48,49]. Later, this paper shall discuss problems of profit (especially as CCS adds to costs and energy use), taxation, liability costs, regulatory ambiguities, carbon accounting, and the politics of trade, in relation to CCS construction. These points resemble the four primary barriers to successful CCS put forward by Davies et al. [50]: (1) cost and cost recovery, (2) lack of financial incentive or profit, (3) long-term liability risks, and (4) lack of coherent regulations.

2.2. CCS in Australia

Australia is a major coal and gas exporter. It is currently second in the world to Qatar in gas exports and second to Indonesia in coal exports. In a media release after COP 26, Angus Taylor, Minister for Energy and Emissions Reduction, said: "Australia's economy is almost unique amongst developed countries, with an economy specialised in the production of energy- and emissions-intensive commodities. We are the world's fourth largest energy exporter, after Saudi Arabia, Russia and the United States" [51].

He previously made government backing for methane very clear: "The Government backs the gas industry, backs Australians who use gas and it backs the 850,000 Australians who rely on gas for a job. Gas is a critical enabler of Australia's economy" [52].

Eight hundred fifty thousand seems to be the number of Australians who work in "all sectors of manufacturing and not all those sectors use gas as a feedstock". There seem to be close to 8000 employees directly dependent on gas. The indirect number is harder to calculate [53].

Taylor also remarked that the emissions aims for 2030, which were not clarified in response to requests by COP26, were "fixed". Subsequently, more new large gas fields have been announced, and the Government has issued the *2021 National Gas Infrastructure Plan*, which states: "Unlocking new sources of [gas] supply will be a key focus for industry and governments out to the 2040s" [54] (p. 10).

Australia also has the highest per capita GHG emissions in the OECD [55]. Consequently, Australia has a major incentive to support CCS, so fossil fuel sales can continue to expand. Some people estimate that taxpayers have contributed over AUD 1 billion to CCS out of the AUD 3.5 billion promised [56]. Australian Governments may be classified as maintaining what Arranz [47] calls an "enthusiastic framing" of CCS, seeing it as a way to solve problems of instability in transition as the population embrace rooftop solar—"Australia now leads the world in solar per capita with 810 W/person, ahead of Germany with 650 W/person" [57] (p. 5)—and (perhaps more importantly) to maintain economic competitiveness and development. This enthusiastic focus encourages "blind spots" to the difficulties, such as CCS in Australia not reducing emissions significantly. The coal industry was previously largely uninterested in CCS, as an attempt to save coal exports. Most projects initiated have been small-scale and subsequently abandoned [4]. The largest has been the Chevron Gorgon gas fields, the subject of this paper.

In 2020, the Australian government proposed new ways of funding CCS. This included changing the scope of its AUD 2.5 billion Climate Solutions Fund, the investment guidelines for the Clean Energy Finance Corporation (CEFC) and the Australian Renewable Energy Agency (ARENA), to become "technology neutral". "Technology neutral", as used by the Coalition government, tends to mean pro-fossil fuels. The Labor opposition queried the Government's attempts to allow the CEFC to fund CCS by saying that "to pretend that a bank [the CEFC] that requires a commercial rate of return can lend to a technology that has not been commercially deployed anywhere in the world is just a fantasy" [58]. Grant King, head of the review making these recommendations, was the former head of Origin Energy (user of gas and coal) and board member of the Australian Petroleum Production & Exploration Association (APPEA), a body which has campaigned strongly against the curtailment of fossil fuels, describing itself as "the effective voice of Australia's upstream oil and gas industry on the issues that matter" [59].

Unsurprisingly, APPEA has recommended more new gas fields and CCS. Its Chief Executive Andrew McConville said "Carbon Capture and Storage (CCS) is already well established as a safe, large scale, permanent abatement solution . . . . Accelerating the roll-out of CCS projects could assist in reducing emissions from the energy, industrial and power generation sectors" [60]. "Australia needs low-cost carbon abatement to maintain its position as a leading energy exporter and ensure international competitiveness in a cleaner energy future" [61].

Again, the aim of maintaining methane exports is clear.

In November 2021, the Prime Minister announced AUD20 billion to fund "new technologies, whether it's hydrogen, carbon capture and storage, low cost soil carbon management measurement, the green steel and aluminium" [62]. However, some of the funding may not arrive, as members of the Government who opposed climate action of any type have said they will vote against legislation enabling it [63]. Nevertheless, the Australian Government and the major opposition party have both demonstrated consistent support for CCS as part of their support for maintaining fossil fuel exports. While LNG exports may reduce emissions if gas use reduces coal burning, it is not certain if such reductions in coal use are happening in importing countries, and gas-burning continues to consume the limited carbon budget as CCS is nowhere near storing or using all emissions from this burning. More new gas fields and coal mines have been announced recently in keeping with the *Gas Infrastructure Plan*. The Prime Minister announced to the Business Council of Australia that when he heard about the new AUD 16.5 billion Scarborough gas development, he "did a bit of a jig out of the Chamber. I just could not be more thrilled about that. That is such a shot in the arm for our economy and it is going to power us into the future" [64].

The Australian government has heavily promoted CCS and can be said to have glossed over, or even delighted in, increased emissions from the new gas fields they are encouraging.

3. The Gorgon Project

As stated earlier, the project does not exist in isolation from social practices and corporate organisation, and it needs to be considered through the way it is embedded in its context of other complex problem-generating systems. Complexity is routine for any project this size. This analysis will proceed via various headings, all of which should be thought of as interconnected.

3.1. Why It Is a Good Exemplar

The Chevron Gorgon project in West Australia could be considered an excellent exemplar for CCS. In 2019, Chevron said: "The Gorgon CO_2 injection project is believed to be one of the largest greenhouse gas mitigation projects undertaken by industry, which will reduce greenhouse gas emissions from the Gorgon project by around 40 per cent" [65].

That appears to translate to 80% of the CO_2 in the methane, before export.

Chevron has an economic incentive as the Gorgon gas field has too much CO_2 in the methane (14%) [9,66]. The CO_2 needs to be removed for transport, as it freezes when the gas is liquefied. Normally the gas would be released into the atmosphere. There are natural storage basins nearby, so transport is short and simple, while the storage areas contain saline water so leakage should be low. It is clearly politically welcomed, not only in keeping with the Government's promotion of gas, but also receiving AUD 60 million in government subsidy, as well as significant royalty and tax benefits, all of which increase profitability. West Australian EPA objections were bypassed [67,68], even though the Barrow Island site is a Class A nature reserve (the highest classification). Its distance from major population centres may have helped reduce protest. Western Australia is seismically stable. Chevron has conducted research at the site, possibly from 1998 with a Greenhouse Challenge Cooperative Agreement between the Gorgon Joint Venture Participants and the Australian Greenhouse Office [69]. Drilling had been carried out in the area since the 1960s, so the area is well known [70]. Chevron's partners in the project, ExxonMobil (25%) and Shell (25%), are among the most experienced fossil fuel companies in the world. The project, therefore, has much in its favour to demonstrate the possibilities, or failings, of CCS.

3.2. Rates of Construction and Use

Technological problems are normal in complicated and complex systems (see [10] for a social analysis of software problems). Technology requires social organisation, capacity, and evaluation to implement. The rate of CCS construction will be influenced by the interactions between various social and technical organisations, such as commercial exaggeration of ease, conflict between groups, technical failure, and systemic practices of ignoring problems in favour of profit.

3.2.1. The Plan

Chevron still anticipates the Gorgon project will have the lowest greenhouse gas emissions intensity of any LNG drilling project in Australia [71]. The Minister announced the plan was to store "between 3.4 and 4 million tonnes of greenhouse gas emissions each year" [72]. The project uses a solvent (activated methyl di-ethanol amine) to remove CO_2, H_2S, and other impurities, along with a mercury removal unit. Dry low NOx (DLN) burners reduce NO_x emissions [73,74]. The CO_2 is then transported close to 7 km and injected into a sandstone saline aquifer, more than 2000 m underground, where it is expected to dissolve [9], presumably making the saline acidic and possibly having some ecological or geological effect. As pressure in the aquifer increases with CO_2 injection, this is balanced by pumping water out about 4 km away. This causes some problems, as will be discussed later. This water is then pumped into a different layer of rock above the CO_2 [75]. I assume the water is checked to find out if CO_2 is present, efforts are made to prevent CO_2 escape, and tests are conducted to check for the solvent. Monitoring wells are drilled to discover the movement of the CO_2 in the aquifer [9]. These wells themselves could disturb the confinement, if not properly sealed.

3.2.2. Slow Progress

Progress on the whole project has been slow. As already stated, Chevron's research on the area possibly began in 1998. The project was first formally proposed in 2006.

It began in September 2009, and in 2012 Chevron announced it would begin storing a total of 120 million tonnes of CO_2 at a rate of 3 tons a year in 2014/15 [76]. By mid-2016, according to Chevron's annual report to the Federal Government, the CO_2 pipeline was not completely connected, and injection was delayed for a year [75,77]. In March 2016, two years late, Gorgon produced its first shipment of LNG after a budget blowout of USD 18 billion, suggesting significant problems [75]. Export was shut down due to problems with the propane cooling system [78] (see below Section 3.2.4), some of which were said to be organisational. "The procedures for operating the propane cooler required the operator to know the pressure at the inlet of the propane compressor, but no such indication existed. Other issues Chevron identified, included workers starting up the plant having an 'unclear line of management oversight' and 'inadequate technical resources to back up operations' [79].

Later it appears the vessels were imperfect to begin with.

Exports were supposed to reach 15.6 million tonnes a year by about mid-2017 [80]. Commercial gas output before the CCS was working was said to be averaging 449,000 barrels of oil equivalent per day [81]. Plans were also announced to further expand gas production in 2018–2019, although it was unclear if there were plans to expand CO_2 storage [82,83]. Some of these exports came from the nearby Jansz-Io field which has lower CO_2 content, and through releasing excess CO_2 into the atmosphere. Income was prioritised over CCS.

The first storage injection occurred on 6 August 2019, at least four years late. By the end of June 2020, 2.5 million tonnes of GHG had been stored. By September 2020, they were claiming 3 million tonnes of stored CO_2 [84,85]. Problems remained with storage due to pressure issues. A Chevron report from 2020 states that "investigations into the loss of injectivity at the pressure management water injection wells was ongoing" ([84] see next section). The CCS part of the project was expected to cost USD 2 billion [76], but by 2021 "the capital budget had increased to $3.092 billion" [84].

3.2.3. Failure to Achieve Targets

The slow progress, perhaps because of prioritising gas sales, resulted in the Gorgon project not achieving its storage targets.

In 2009, before Chevron, Shell, and ExxonMobil committed to the project, they were required to "implement all practicable means to inject underground all reservoir carbon dioxide removed during gas processing operations on Barrow Island and ensure that calculated on a 5 year rolling average, at least 80 percent of reservoir carbon dioxide removed during gas processing operations on Barrow Island and that would be otherwise vented to the atmosphere is injected" [86]. If this target was not met, then Chevron would have to offset the emissions. What was later decided to count as the initial period finished in late 2021.

Some of this disruption to targets resulted from the equipment extracting water from the injection sites failing when they clogged with sand, "despite prior studies to selectively perforate the four water production wells to avoid weak zones that might be prone to sand production" [84]. Chevron promised to "install equipment to extract the 'significant volume of sand' from the water before it is reinjected underground" [87]. Quite where the sand was to be stored, given the delicate nature of the ecology, is not clear, and it is not clear why Chevron failed to detect sand in the water in exploratory investigations. During 2020, CO_2 injection, under a series of permissions from WA's Department of Mining, Industry Regulation and Safety, averaged 70% of maximum capacity despite the water wells not functioning properly. Presumably, Chevron did not succeed in fixing the problem, and in December 2020 this failure led to regulators restricting carbon injection to a maximum of two-thirds of its supposed capacity from 1 January 2021, to avoid high pressure in the reservoirs and potential cracking and leakage (see Section 3.3). Over a year, this would mean an additional 2.64 million tonnes of pollution [88].

In a project report from 2021, Chevron said: "It is yet to be confirmed whether sand production is likely to be a long-term issue. If sand production is found to be [a] persistent issue it is possible changes to the surface facilities may be required" [84] (p. 6). A spokesperson said: "While CO_2 injection safely continues, daily injection rates have been amended which has resulted in additional CO_2 venting in the short-term" [89].

According to Professor Newman of Curtin University, the difficulties were a surprise: "The whole reason for being allowed to go on an A-Class reserve (Barrow Island) was because the sediments were perfect for this sequestration" [65].

Helpfully, the WA government determined that emissions made before an operating licence was awarded did not count, reducing Chevron's liability.

Faced with these failures to meet targets, a Chevron spokesperson said the carbon capture project was complex and bigger than anything undertaken anywhere in the world [90]. With the sand and pressure problem, they said: "Like any pioneering endeavour, it has presented some challenges and we continue to work closely with the regulator to optimise the system, with a focus on long-term, safe and reliable operation over its 40-plus year life" [91]. Innovation may not only cause delays but also act as an excuse.

This led to some political protest; for example, the Conservation Council of WA said Chevron should close the plant until it could demonstrate its CCS system was working, as it was violating its licence conditions [92].

3.2.4. Routine Problems

The project involved massive interconnected and complex infrastructure, which might be expected to generate problems with CCS targets, deadlines, and costs, especially if gas production was prioritised. Work was hampered by breakdowns on the site of processes unrelated to CCS [79,88].

One problem involved a design issue with the compressors which allowed water and CO_2 to mix, forming carbolic acid which could then corrode the equipment. In 2017, checks "found leaking valves, valves that could corrode and excess water in the pipeline from the LNG plant to the injection wells that could cause the pipeline to corrode" [93]. This produced a significant delay, officially announced at the end of 2018, and CO_2 was again vented directly to the air. Team leader for the Gorgon CCS project, Mark Trupp (coauthor of [9]), said: "Carbon dioxide is a corrosive substance. We have had some issues managing the water content of the carbon dioxide that has required modifications to our facilities. That is what has been delaying us" [94]. Chevron might have been expected to realise the presence of CO_2 and water vapour in LNG to be a recurring rather than unpredictable problem, although perhaps not if venting was routine.

As well, the project was faced with other technological mishaps which slowed production and added to complications. Cracks in propane vessels, needed to cool gas for export, were revealed through worker complaints to the media (possibly because the company had appeared to ignore safety issues) and led to a Department of Mines, Industry Regulation and Safety WorkSafe investigation [95]. "Cracks up to 1 metre long and 30 millimetres deep were found in between eight and 11 kettle heat exchangers on Train 2 of the plant" [96]. This meant that Gorgon's three LNG trains were shut down for repair for some months [88]. The precise causes appear to have remained secret, although they seem to have involved faulty welding during manufacture. It is not clear whether these cracks are related to the earlier propane cooling problems discussed above. Rumours asserted the kettles would need to be replaced [97]. Other information suggests the kettles did not comply with Australian standards. Eventually, these problems led to increased inspections [98] and shut down some parts of the plant. The problems resurfaced, and in late January 2021, Chevron warned that continuing repairs would lower output [99]. In March 2021, Chevron announced it would use the June quarter to close the third LNG production unit to check for more defective welding [100]. In May 2021, Chief Financial Officer Pierre Breber said at least one train had been out of action since mid-2020 [101].

3.3. Geology
3.3.1. Sand in Aquifers

This has been largely discussed above. However, the Dupuy Formation in which the emissions are being stored is described by the MIT Gorgon fact sheet [66] (latest revision 2016, before the problems occurred) as "a massive turbidite sand deposit" which might have been expected to be a problem in advance. Chevron claimed in 2021 that "an upgrade to the filtration system for the sand was now complete" [102]. Young [103] states that the regulators "approved Chevron to purge sand from its production wells into sandbags", so it can be hoped the sand does not contain heavy metals or other poisons which could leak out into the nature reserve. Where the filled bags are to be stored is unclear.

3.3.2. Seismic Events

There are some concerns that CCS might provoke seismic events which could break the storage and undo the effort completely [104,105]. Others analysts (still remarking on long-term uncertainty of CO_2 behaviour underground over thousands of years) seem less troubled [106]. Local geology appears to be relatively stable, which reduces the chance of leaks through crack creation. Geoscience Australia's search engine records 23 quakes above 5.0 in, or offshore, WA in the last 21 years [107]. The WA Department of Mines, Industry Regulation and Safety required Chevron to install detection equipment so that if microseismic activity seemed high, then Chevron could slow CO_2 injection. Chevron themselves said: "Seismic activity is part of the system design and was considered as part of the regulatory approvals for the system" [88]. Chevron was reported as announcing that more than 800 "micro-seismic" events "had been detected at the site, with the frequency of the events increasing with injection" [108]. This was possibly connected to the pressure issues described above, resulting from sand clogging. While the microseismic events individually could seem little threat to storage stability, it is hard to know what the cumulative effects might be and whether some kind of leakage monitoring system is required or how effective that monitoring system would be. This is a social/political decision.

3.4. Corporate Economics
3.4.1. Problems of Profit and Politics

Chevron and its partners, like all corporations, operate under social imperatives to return high profits and lower costs. They also operate with social privileges when profits are increased by tax concessions or subsidies or pollution controls are waived or ignored.

In this case, normal cost blowouts seem significant. There was a massive rise in the expected cost of the whole gasfield project from USD 19 billion in 2006, USD 37 billion in 2009, to USD 54 billion in 2015 when it was said to be over 90% complete [67,109]. For some of the production time, the price of oil and gas crashed, and plans for expansion were put on hold. In 2015, Chevron reported a 90% collapse in profits [110]. This led to massive asset sales [111], which continued into 2019 [112]. It is unclear what the final project cost will be, what the total losses were from the asset sales, or what effects market vagaries had on CCS development, but evidence suggests that maximising profit from gas sales took priority.

In 2021, Chevron and its partners invested another USD 6 billion in the project, making it "the country's largest single resources investment" [113]. This investment had nothing to do with improving CCS but involved the interconnected Jansz-Io field, the gas of which is processed at the same plant and which has less CO_2 in its methane, perhaps avoiding the capture problems or necessities. This project involves building a 27,000-tonne "floating field control station", a "subsea compression infrastructure", and a 135 kilometre underwater power cable to carry energy from the Barrow Island LNG plant. It is not clear if that energy is generated by burning gas, adding to site emissions [114,115].

Another problem is that even with CCS at the production site, burning gas by purchasers produces GHG emissions, as does burning gas to power the CCS process. I could find no information on the storage of emissions from powering the CCS. This situation could become vulnerable to governmental regulation and policy if climate change is taken seriously. There are repeated rumours that the EU will use tariffs to protect its industries from foreigners who are not reducing emissions, which could easily affect Australia's exports. The Blueprint Institute, aligned with those in the Federal Coalition who are concerned about climate, said: "It's just another reminder that we have to take climate action seriously. The choice is clear: reduce emissions to defend our exports and seize new opportunities, or cling to stubborn climate policies at the cost of our economic competitiveness" [116].

Fossil fuels also risk becoming an investment hazard, a stranded asset. In mid-2021: "Santos chief executive Kevin Gallagher led a wave of oil and gas industry leaders warning that achieving net zero emissions will be critical for the natural gas industry to avoid coal's fate of being blacklisted by equity investors and lenders" [117]. Unsurprisingly, Gallagher advocated carbon capture and storage and hydrogen manufactured from methane (with CO_2 as a by-product of manufacture) as ways the industry could reach carbon neutrality—which is only possible if *all* emissions from *all* emission stages are captured. He argued that Australia had the potential to become a "carbon storage superpower" and that Australia needed large-scale projects "to make development of our oil and gas resources viable for investors, financiers and customers so that the wealth of these resources can be unlocked for the nation" [118]. CCS seems to be part of a rhetoric to bypass increasing GHG emissions.

At the same time, the chief executive of Clough, Peter Bennett, expressed worry that "financial backers were deserting the gas industry based on an 'almost hysterical' principle that all fossil fuels were bad", despite industry claims gas was important to help reach net zero emissions [117]. Peter Coleman, former head of Woodside Petroleum, adds that investor concerns about climate change, and the risk of stranded assets, mean the era of massive new LNG projects is over. "It's difficult for me to see a Gorgon happening again, what's fundamentally changed now is the capital discipline in the industry that wasn't there before and obviously the focus on climate change". Coleman also suggested that geological conditions in Australia made CCS unsuitable for wide-scale use here [119].

The uncertain politics of climate action affect future investment. Why invest in more, or better, CCS if it cannot save existing investment? A contradictory problem may arise from the so-called "green paradox" [120], in which fears of resources becoming constrained by legislation lead companies to sell as much as possible before the value runs out. This can produce lock-in for customers which may then undermine pressure for climate action and emissions reduction. If CCS is primarily a disguise for increasing overall emissions, then it makes circumstances worse.

3.4.2. Problems of Privilege: Tax

Australia has a tax regime friendly to fossil fuel miners. In 2019, tax *credits* for oil and gas companies taking Australian fossil fuels rose to AUD 324 billion—that is AUD 324 billion in tax the companies owe but do not have to pay [121,122]. Chevron's partner Shell forecasts it will never pay Resources Rent Tax for gas extracted from the Gorgon and other gas and oil projects in Australia. Juan Carlos Boué, counsel at international law firm Curtis, said: "Shell is saying nothing that the Government and everybody in the know has not been aware of for some time now" [123]. Tax and tax avoidance are also part of the background of CCS and seem extremely favourable for its success.

One way of making CCS financially viable is to have a carbon price or tax. Carbon pricing systems in the EU and UK by July 2021 reached record levels near GBP 45 a tonne. However, according to the London *Financial Times*, some think carbon prices will need to double to make CCS viable and persuade companies to pay for carbon sequestration [124]. Gorgon was being prepared during a period when a promised carbon price gave the project extra viability; however, the carbon price was removed along with emissions targets by the incoming Coalition government in 2013. This probably added to the financial stress of the project. Given that the campaign against carbon pricing has been considered significant, by almost all political commentators, in producing the Coalition's victory, it is doubtful whether any Australian government will introduce a transparent system of pricing in the near future. Indeed the current government's slogan is "technology not taxes", the technology being largely expected, or imaginary, innovations.

There is some suggestion that plant profits were increased through tax deals, minimisation, avoidance, and transfer pricing through internal company loans. In 2018 to 2019, Chevron Australia paid Chevron US, AUD 6.3 billion capital repayments at higher than market interest rates, plus another AUD 4.2 billion of dividends, for a total of AUD 10.5 billion from Australia, all of which was tax-free [125]. Chevron also campaigned to be allowed to sell carbon credits based on the carbon they stored, which, given that they were storing their own emissions for permission to mine, appears to make the storage count twice [126]. Presumably this is allowed as part of business practice, but it adds to the possibilities of disruption, should a government change policies. It might also indicate where human energy and imagination are being expended.

Financial viability could also have been threatened by a 2015 Senate inquiry into corporate tax avoidance which began to consider closing tax loopholes allowing Chevron, ExxonMobil, and Shell to claim tax-free profits from the Gorgon gas project, through loaning to their Australian branches at higher than normal interest rates [127]. Chevron Australia had a debt-to-equity ratio of 76.2% largely in loans to its parent, which was almost 10 times the debt level of its global parent. The US parent paid a mere USD 248 in tax in the US in 2014–2015 according to Chevron itself [128,129]. Chevron and its backers had campaigned for tax concessions at the beginning of the project, despite apparent exemption from royalties for the gas. Allegations later arose that Chevron had paid larger amounts to Australian political parties than it had paid in tax [130]. The tax case was resolved with Chevron being convicted of transfer pricing in 2017 [131], but nothing appears to have changed by 2021, with APPEA arguing nothing should change [123].

There is also a question of whether the project was being subsidised by tax avoidance. Tax avoidance is not illegal, but it is an unstable way of guaranteeing an efficient business case for a project, while undermining revenue expected by the host country.

3.4.3. Liability Costs Transferred to Taxpayers

As previously stated, Barrow Island is a Class A nature reserve. While leaks could be ecologically disastrous, the project owners are only responsible for leaks occurring during the project's lifetime and for 15 years afterwards—a small window of responsibility for storage which is meant to be eternal. Before the project began, "the federal and WA governments . . . agreed to accept responsibility for any long-term liabilities". This means the taxpayer is further subsidising the ongoing cost of CCS [76], and the company has less incentive to store the carbon safely, as it will not have responsibility for leaks. While it is not that unusual for taxpayers in capitalist society to take on the cost burdens for private projects, this could stir dissent about public subsidising of private profit, and the deleterious effects of commercialising carbon storage.

3.4.4. Regulatory and Legal costs and instabilities

Chevron also engaged in a legal dispute over cost blowouts with the builders of its wharf. The dispute took nearly four years to resolve [132,133]. This seems to be part of a worldwide pattern of companies either underquoting (or underestimating) construction costs or delaying payment of debts, a normality which adds to costs, complexities, disruptions, and delays.

Legal issues also eventuated because of Chevron not meeting storage targets [93], after selling gas for 3 years without CCS [134]. This provoked mild conflict with the WA government. Effective penalties had been diminished in May 2018, when the Environment Minister asked the WA Environmental Protection Authority to decide when the beginning of the five-year period for the 80% of CO_2 in the methane storage requirement commenced. In September 2019, the EPA stated injection should not be assessed from when production began, but from when the LNG trains received their operating licence. This was after 14 July 2016 for one train and mid-2018 for the other two. This gave Gorgon a free

25.7 million tonnes of emissions [75]. Chevron apparently wanted the limits to only count after July 2018, two years after the first shipment of gas, giving them even more profit and freeloading.

In late 2020, the Conservation Council of WA used the Appeals Convener to challenge Chevron's operating licence. They made three complaints:

- Lack of public disclosure about the CCS facility's operations and emissions from the Gorgon project (Secrecy see Section 3.4.5);
- Lack of limits on the amount of pollutants the project could emit;
- The 20-year length of the operating licence before review.

The first two points were rejected, but the Minister lowered the length of the operating licence to 10 years [92]. This does put some pressure on the project, but if it is remotely successful it should meet its targets in 10 years from now. The Conservation Council of WA's (CCWA) director, Piers Verstegen, was not surprised Chevron had failed its targets. In July 2021, he called for "the Environment Minister and the state government to enforce those conditions and require Chevron to meet its promises . . . . [CCS] shouldn't be relied upon to justify the increased expansion of the oil and gas industry" [135].

The CCWA requested that general operations and production at the Gorgon project be suspended or scaled back because of these failures, or a limit on CO_2 emissions be enforced, with transparent disclosure of volumes stored [135]. The Minister rejected the request.

Chevron admitted it fell short of targets by 5.23 MT and committed to buy carbon credits and invest AUD 40 million in unspecified "low carbon energy projects" in the state [136]. There are different estimates of the penalty, but taking a contemporary spot price for Australian Carbon Credit Units and using them as offsets, the cost could be between AUD 100 and 200 million. This is relatively small, as Chevron's share of such a bill would count for a few days of its 2020 annual Australian revenue of USD 5.9 billion (USD 7.9 billion). Additionally, Chevron can buy cheaper offsets to gain less penalty [75,137].

The sandy water problems meant that the Department of Mines, Industry Regulation and Safety (DMRS) had issued multiple extensions for Chevron to keep operating, as deadlines for repair were broken [75]. A Chevron report stated: "Field injection rates were curtailed [from an achieved maximum of 147 kg/s] from the 18 December 2020 to meet the CO_2 injection rate restriction of 42 kg/s whilst the pressure management system was offline and being remediated" [138] (p. 2).

Without the water being removed, there was a risk that the increasing pressure required to pump the CO_2 underground would fracture the rock around the injection wells, permanently damaging the system's performance [75].

Sympathetic government means that the potential penalties for failure are low compared to profit, with little incentive to prioritise storage over profit.

3.4.5. Business Hype, Marketing, and Secrecy

(Dis)information is part of market action [10,139]. Business, much like the State, attempts to spin the best result for its action, hide embarrassing events, attack its competitors, build political support, defuse political hostility, and promise to render all competing products obsolete. The problems of hidden data should already be apparent, and the less anyone knows what is going on, the more will probably be hidden.

Typically, Chevron claims: "To advance a lower-carbon future, we are focused on cost efficiently lowering our carbon intensity, increasing renewables and offsets in support of our business, and investing in low-carbon technologies that enable commercial solutions" [115].

However, lowering carbon intensity does not *have* to lower emissions, as pointed out earlier. Chevron's claims about CCS also seem misleading. They state that their targets are equivalent to "taking more than 1 million passenger vehicles off the road each year" [140] (although the Federal Minister Matt Canavan said it was the "equivalent of removing 680,000 cars from the roads each year" [72]), but the project does not remove previously *existing* pollution. While it is better that *some* CO_2 extracted in production be stored, when taken *as a whole*, even without counting emissions from subsequent burning, the project increases global emissions. A report for the Global CCS Institute and the WA government claimed that Gorgon and another CCS project would store "more than eight million tonnes of CO_2 annually, approximately 11 percent of the State's annual current emissions", again implying emissions reduction, when there would be a net increase in WA's emissions through the projects [141] (p. 9). Similarly, it has been suggested that gas is better than coal (by Chevron for example), but gas is only reducing emissions if coal emissions are phased out faster than gas emissions are phased in.

Chevron has compounded this misdirection by joining other fossil fuel companies in 2018 to campaign to keep their GHG emissions secret, on the grounds that releasing data could help overseas competitors [142]. Chevron specifically remarked that reporting was expensive and "costs must be kept as low as practicable" [143]. If this secrecy is successful, it undoes claims of verifiable storage. Chevron was still declining to provide any data on its storage rates in Feb 2020, while emphasising its supposed future rates of storage [134], although it did announce it had stored its millionth ton soon after [144]. Similarly, when the Department of Mines, Industry Regulation and Safety limited the amount of CO_2 that can be injected because of the pressure problems, "neither Chevron nor the WA government would disclose the cap or the amount by which emissions [had] increased" [92]. Furthermore, much of the information about the project seemed to be hidden and required journalists to make freedom of information applications (see Section 1.2). There is also hardly any mention of carbon capture in Chevron's 2020 Annual Report [145] or 2020 Corporate Sustainability Report [146]. There does not seem to be any easy comparison of what they store compared to their complete three-scope emissions.

Even failure can be promoted as success. The company's Australian boss Mark Hatfield said the company was "deploying technology, innovation and skills to deliver cleaner energy and reduce our carbon footprint. The road hasn't always been smooth, but the challenges we've faced and overcome make it easier for those who aspire to reduce their emissions through CCS" [147]. Misleading information also comes from politicians and industry support groups. Angus Taylor, the federal energy and emissions reduction minister, in 2020 cited Gorgon as an "already working" example of CCS [148], while an APPEA press release said Chevron showed the industry was "continuing to walk the walk when it comes to reducing emissions" and "Chevron's announcement is on top of all the work our industry is already doing to combat climate change" [149]. These claims distract from the project's failure to produce net emissions reduction.

It is notable that Chevron, Shell, and Exxon are frequently implicated for promoting doubt about climate science to justify continuing sales of fossil fuels [150–152]. This suggests that their treatment of CCS may be a continuing part of that strategy.

3.4.6. Net Zero

In 2019, the WA Environmental Protection Authority argued that large gas projects had to be zero emissions, or buy offsets; otherwise, Australia would not fulfil its Paris commitments. WA's emissions had increased by 27% from 2000 to 2016 [153]. The recommendation was denounced by fossil fuel companies, including Chevron who threatened to end projects, and by the WA and Federal Governments. WA Premier Mark McGowan indicated the government would ignore EPA advice, just as it did in approving the Gorgon project in 2006 [154,155]. This approach could show that companies do not think zero emissions is feasible, or worth moving towards through CCS, and that Australian governments support this position. Piers Verstegen, director of the Conservation Council of Western Australia, remarked that the actions of fossil fuel companies showed their "only plan is to bully governments into letting them get away with doing nothing" [156].

3.5. Effectiveness

Chevron expects CCS to reduce its *production* emissions by about 40%, storing 80% of CO_2 in the extracted methane. They expect to store 100 million tonnes of CO_2 over the life of the project while, given the 40% figure, presumably releasing another 150 million tonnes in production during that period. This project does not lower baseline emissions production in Australia. Furthermore, storage in WA does not lower GHG generated by burning the gas elsewhere, so the proportion of CO_2 stored, compared with that released in use, is likely to be insignificant. Significant reductions would need CCS wherever the gas was burned or released. Mark Ogge of the Australia Institute, which is not pro-CCS, argued in 2021 that the Gorgon project would capture just 1.7% of its total emissions (Scope 1, 2, and 3) over 5 years [157], while a report to the US Congress states: "While Chevron claims that its carbon-capture projects will reduce its greenhouse gas emissions by roughly 5 million tonnes per year, this would account for only a minuscule fraction of the company's emissions, which in 2019 amounted to 697 million tonnes of carbon dioxide equivalent" [150].

According to the same source, "Chevron did not report *any* lobbying on the Paris Agreement, despite spending $54 million on lobbying since 2015" and despite support for the Agreement being a supposed key corporate goal [150].

Perhaps more significant than Gorgon's failure to reach its targets, Clean Energy Regulator (CER) data "shows the facility produced over 9 million tonnes of CO_2-equivalent emissions" for 2017–2018, making it "Australia's highest CO_2 emitting gas facility" (Kilvert 2019). Physicist and climate scientist Bill Hare told the ABC that the "volume of pollution coming out of the Chevron project far outweighs the savings of carbon pollution from rooftop solar". The ABC reports "Chevron declined to comment on the comparison" and that a former adviser to Margaret Thatcher and regional president for BP Australasia said: "These sorts of issues can set very, very dangerous precedents. They should be required to purchase [carbon] offsets equivalent to the same volume they were expected to inject over the first five-year period" [65]. This would not reduce their emissions, just price them.

Chevron was later classified by the CER as the country's sixth-biggest polluter [158], capturing only one-third of the GHGs that the approvals required while venting millions of tonnes a year more [159]. An estimate in November 2020 states "Gorgon emitted almost 34 million tonnes of greenhouse gases in the five years to June 2020 from the reservoir CO_2 that was vented instead of buried, as well as gas combusted to power the plant and excess gas burnt in a flare" [87,160]. However, as already seen, in Australia, these increased emissions were largely not a problem for the company, as it already had the right to emit 25 million tonnes from 2017 to 2020 before having to buy offsets, diminishing its incentive to fix the pollution problem [161,162]. If we are serious about reducing greenhouse gas emissions, then increased methane production, release, and burning will not solve the problem, especially when new fields are coming online.

4. Conclusions

The limitations of the research are obvious. Data could be expanded by uncovering more records (perhaps needing more freedom of information requests, or archival exploration in Chevron and Government offices), through interviews with administrators, and workers, and possibly through day-to-day fieldwork, although the problems of "studying up", describing the intricacies of real corporate processes, and gaining permissions are well documented and increased by normal business secrecy. Research could also be usefully expanded into studying the dynamics of CCS within Chevron's role as a major fossil fuel energy producer and its competition and cooperation with other fossil fuel companies (including its partners on this project). There are numerous technical details that I could not uncover.

However, this study has demonstrated that CCS is not simply just a technological problem. Technology is embedded in social and ecological relations, particularly in corporate and developmental organisations. It cannot be separated from these relations. If analysts do not consider the social background, then they will miss important dynamics of the CCS projects and expectable blocks in their effectiveness. Therefore it seems useful to look for disorders in the narratives of success promoted by those engaged in CCS and in the way that social processes and technical processes are intertwined. The processes turn out to be far more complex than a narrow focus on the need for CCS, or the technical possibilities of CCS, would suggest. Problems compound. Problems with sand, cracked equipment, acid formation in pipes, on-site energy consumption and emissions, the cumulative effect of seismic events, workers, wharves, regulators (even friendly ones), tax avoidance, secrecy, propaganda, liability evasion, and export-oriented governments, along with a primary business focus on profit, have the capacity to disrupt a project's significance in generating real emissions reduction. This is the case, even if regulatory issues were not significant for the project because of governmental enthusiasm, and low levels of protest. Due to standardised corporate secrecy, there were almost certainly more problems with the project than I have described here. However, given the expertise of the companies involved, and the length of their presence at the site, these kinds of problems cannot be considered to be secondary, or easily resolved by requests for increased competence.

Decisions about CCS are social, political, and economic decisions about profit. They involve an emphasis on mining and sales rather than storage, corporate reactions to losing prior capital investments in fossil fuels, and attempts to persist in burning gas and "unintentionally" increasing emissions in so doing. Focus on profit can lead to undue simplification. This may explain why in 2017, the departing Australian director said that Chevron clearly had not done enough background work: "We have to verify every single aspect of these projects in advance, because we're on the hook for them, regardless of the kind of contract that we sign" [163]. This understanding may be impossible given the complexities, but it does indicate some recognition of a lack of awareness about potential problems.

Social context means that while technology can work in theory, there is no reason to assume it will be used properly, no matter how essential it seems. Technology can be used as a mode of rhetoric or fantasy to reinforce, or hide, social relations and destructive inclinations. CCS seems to be being used in this way. Rather than reducing total emissions, or coming under necessary carbon budgets, it seems to be used to contend that increasing emissions can be ignored or to distract from those increasing emissions. I see nothing in the evidence which suggests that Australian governments are going to use CCS to enforce, or encourage, lowering of total emissions or to promote a universal and high carbon price which would seem to be needed to provide an economic rationale for CCS.

These fundamental problems can be seen in the Gorgon project, which should otherwise be an example of easy success. The relevant governments provide support to increase gas exports, are largely relaxed about tax avoidance and broken regulations, and accept long-term taxpayer-funded liabilities. The geological/ecological situation seemed straightforward, with storage that was nearby, but potential problems were not recognised during exploration. The project was unambitious, in only attempting to store excess CO_2 in the methane which needed to be removed for transport. It did not store gas burnt or released at the site, nor gas burned at the customer's site. Nevertheless, Chevron faced significant difficulties, made slow progress, was troubled by routine problems and cost blowouts, released considerable emissions, and failed to produce anything like net zero. There is no indication that the Gorgon project, even if it is fully successful, will reduce the emissions from the fossil fuels it excavates and sells, and given the problems it faced, it seems unlikely that storing a significant amount of emissions produced by burning would be possible. Given that this project is recent and using the best knowledge available, it seems to suggest that it is unlikely that enough large-scale CCS projects will be built for emissions reduction purposes, in the current social order.

It seems that the social drives promoting "free" GHG pollution, and promoting profit, disrupt CCS or make it unlikely to be a significant contributor to reaching zero net emissions or to solving the problems of climate change—in fact, possibly quite the opposite. The social organisation of CCS is perhaps fatal to its success, irrespective of technical difficulties.

Funding: The research in this paper was funded by an Australian Research Council 'Future Fellowship' (FT160100301).

Institutional Review Board Statement: Ethical review and approval were waived for this study due to it using documents which were already in the public domain.

Informed Consent Statement: All research was done through public domain data.

Data Availability Statement: Data sources are indicated in the reference list.

Conflicts of Interest: The author declares no conflict of interest. The views expressed in the article may not be the views of the ARC.

References

1. Feenberg, A. *Transforming Technology: A Critical Theory Revised*; OUP: Oxford, UK, 2002.
2. Hackert, E.J.; Amsterdamaska, O.; Lynch, M.; Wajcman, J. *The Handbook of Science and Technology Studies*, 3rd ed.; MIT Press: Cambridge, MA, USA, 2008.
3. Hughes, T.P. *Human-Built World: How to Think about Technology and Culture*; Chicago UP: Chicago, IL, USA, 2004.
4. Marshall, J.P. Disordering fantasies of coal and technology: Carbon capture and storage in Australia. *Energy Policy* **2016**, *99*, 288–298. [CrossRef]
5. Jackson, M.C. *Critical Systems Thinking and the Management of Complexity*; Wiley: Hoboken, NJ, USA, 2019.
6. Williams, A. *Political Hegemony and Social Complexity*; Palgrave Macmillan: London, UK, 2020.
7. Tainter, J. Energy, complexity, and sustainability: A historical Perspective. *Environ. Innov. Soc. Transit.* **2011**, *1*, 89–95. [CrossRef]
8. Wang, N.; Akimoto, K.; Nemet, G.F. What went wrong? Learning from three decades of carbon capture, utilization and sequestration (CCUS) pilot and demonstration projects. *Energy Policy* **2021**, *158*, 112546.
9. Trupp, M.; Ryan, S.; Barranco, I.; Leon, D.; Scoby-Smith, L. Developing the World's Largest CO_2 Injection System—A History of the Gorgon Carbon Dioxide Injection System. In Proceedings of the 15th International Conference on Greenhouse Gas Control Technologies, GHGT-15, Abu Dhabi, United Arab Emirates, 15–18 March 2021. Available online: https://papers.ssrn.com/sol3/papers.cfm?abstract_id=3815492 (accessed on 2 December 2021).
10. Marshall, J.P.; Goodman, J.; Zowghi, D.; Da Rimini, F. *Disorder and the Disinformation Society: The Social Dynamics of Information, Networks and Software*; Routledge: New York, NY, USA, 2015.
11. Geertz, C. *The Interpretation of Cultures*; Basic Books: New York, NY, USA, 1975.
12. Law, J. *After Method: Mess in Social Science Research*; Routledge: Oxford, UK, 2004.
13. Broecks, K.; Jack, C.; Ter Mors, E.; Boomsma, C.; Shackley, S. How do people perceive carbon capture and storage for industrial processes? Examining factors underlying public opinion in the Netherlands and the United Kingdom. *Energy Res. Soc. Sci.* **2021**, *81*, 102236.
14. Linzenich, A.; Arning, K.; Offermann-van Heek, J.; Ziefle, M. Uncovering attitudes towards carbon capture storage and utilization technologies in Germany: Insights into affective-cognitive evaluations of benefits and risks. *Energy Res. Soc. Sci.* **2019**, *48*, 205–218. [CrossRef]
15. Otto, D.; Gross, M. Stuck on coal and persuasion? A critical review of carbon capture and storage communication. *Energy Res. Soc. Sci.* **2021**, *82*, 102306.
16. Anon. ZEPP Introducing CO_2 Capture and Storage in The Netherlands. Available online: https://www.esteem-tool.eu/fileadmin/esteem-tool/docs/ZEPP.pdf (accessed on 2 December 2021).
17. Cook, P.J. *Geologically Storing Carbon: Learning from the Otway Project Experience*; CSIRO Publishing: Melbourne, Australia, 2014.
18. Akerboom, S.; Waldmann, S.; Mukherjee, A.; Agaton, C.; Sanders, M.; Kramer, G.J. Different This Time? The Prospects of CCS in the Netherlands in the 2020s. *Front. Energy Res.* **2021**, *9*, 644796. Available online: https://www.frontiersin.org/articles/10.3389/fenrg.2021.644796/full (accessed on 2 December 2021). [CrossRef]
19. Heiskanen, E. *Case 24: Snøhvit CO_2 Capture & Storage Project*; National Consumer Research Centre: Helsinki, Finland, 2008. Available online: https://www.esteem-tool.eu/fileadmin/esteem-tool/docs/CASE_24_def.pdf (accessed on 2 December 2021).
20. Kenner, D. *Carbon Inequality*; Routledge: Oxford, UK, 2019.
21. Incropera, F.P. *Climate Change: A Wicked Problem Complexity and Uncertainty at the Intersection of Science, Economics, Politics, and Human Behavior*; Cambridge UP: New York, NY, USA, 2016.
22. Catton, W.R. *Overshoot: The Ecological Basis of Revolutionary Change*; University of Illinois Press: Champagne, IL, USA, 1982.
23. Cock, D. *Global Overshoot: Contemplating the World's Converging Problems*; Springer: New York, NY, USA, 2013.
24. Earth Overshoot Day. Available online: https://www.overshootday.org/ (accessed on 4 December 2021).
25. Clack, B.; York, R. Carbon Metabolism: Global Capitalism, Climate Change, and the Biospheric Rift. *Theory Soc.* **2005**, *34*, 391–428.
26. Foster, J.B.; Clark, B. *The Robbery of Nature: Capitalism and the Ecological Rift*; Monthly Review Press: New York, NY, USA, 2021.
27. Ritchie, H.; Rosa, M. CO_2 and Greenhouse Gas Emissions. 2020 Update. Available online: https://ourworldindata.org/co2-and-other-greenhouse-gas-emissions (accessed on 2 December 2021).
28. IEA. Global Energy Review: CO_2 Emissions in 2020. *IEA Website*, 2 March 2021. Available online: https://www.iea.org/articles/global-energy-review-co2-emissions-in-2020 (accessed on 2 December 2021).

29. IEA. Despite Some Increases in Clean Energy Investment, World is in Midst of 'Uneven and Unsustainable Economic Recovery'—With Emissions Set for 2nd Largest Rebound in History. *IEA Website*, 28 October 2021. Available online: https://www.iea.org/news/despite-some-increases-in-clean-energy-investment-world-is-in-midst-of-uneven-and-unsustainable-economic-recovery-with-emissions-set-for-2nd-largest-rebound-in-history (accessed on 2 December 2021).
30. Matthews, H.D.; Tokarska, K.B.; Rogelj, J.; Smith, C.J.; MacDougall, A.H.; Haustein, K.; Mengis, N.; Sippel, S.; Forster, P.M.; Knutti, R. An Integrated Approach to Quantifying Uncertainties in the Remaining Carbon Budget. *Commun. Earth Environ.* **2021**, *2*, 7. Available online: https://www.nature.com/articles/s43247-020-00064-9 (accessed on 2 December 2021). [CrossRef]
31. Tokarsaka, K.B.; Matthews, H.D. Guest Post: Refining the Remaining 1.5C 'Carbon Budget'. *Carbon Brief*, 19 January 2021. Available online: https://www.carbonbrief.org/guest-post-refining-the-remaining-1-5c-carbon-budget (accessed on 2 December 2021).
32. UN. Updated NDC Synthesis Report: Worrying Trends Confirmed. *UN Press Release*, 21 October 2021. Available online: https://unfccc.int/news/updated-ndc-synthesis-report-worrying-trends-confirmed (accessed on 2 December 2021).
33. Baer, H. *Climate Change and Capitalism in Australia: An Eco-Socialist Vision for the Future*; Routledge: Oxford, UK, 2022.
34. IPCC. Climate Change 2021. The Physical Science Basis: Summary for Policymakers. 2021. Available online: https://www.ipcc.ch/report/ar6/wg1/#SPM (accessed on 2 December 2021).
35. IPCC Special Report: Global Warming of 1.5 °C: Summary for Policy Makers. 2018. Available online: https://www.ipcc.ch/sr15/chapter/spm/ (accessed on 2 December 2021).
36. Reuters. Global Climate Goals 'Virtually Impossible' Without Carbon Capture: IEA. *Reuters*, 24 September 2020. Available online: https://www.reuters.com/article/us-iea-carboncapture-idUSKCN26F0IB (accessed on 2 December 2021).
37. IEA. Net Zero by 2050: A Roadmap for the Global Energy Sector. *IEA Website*, May 2021. Available online: https://www.iea.org/reports/net-zero-by-2050 (accessed on 2 December 2021).
38. Welsby, D.; Price, J.; Pye, S.; Ekins, P. Unextractable fossil fuels in a 1.5 °C world. *Nature* **2021**, *597*, 230–234. [CrossRef]
39. IEA. CCUS in Power: Tracking Report—November 2021. IEA Website. Available online: https://www.iea.org/reports/ccus-in-power (accessed on 2 December 2021).
40. Global CCS Institute. Global Status of CCS 2021. Available online: https://www.globalccsinstitute.com/resources/global-status-report/ (accessed on 2 December 2021).
41. Loria, P.; Bright, M.B.H. Lessons captured from 50 years of CCS projects. *Electr. J.* **2021**, *34*, 106998. [CrossRef]
42. Nykvist, B. Ten times more difficult: Quantifying the carbon capture and storage challenge. *Energy Policy* **2013**, *55*, 683–689. [CrossRef]
43. Butler, C. Carbon Capture and Storage Is About Reputation, Not Economics: Supermajors Saving Face More than Reducing Emissions. Institute for Energy Economics and Financial Analysis. July 2020. Available online: https://ieefa.org/wp-content/uploads/2020/07/CCS-Is-About-Reputation-Not-Economics_July-2020.pdf (accessed on 2 December 2021).
44. Rai, V.; Victor, D.G.; Thurber, M.C. Carbon capture and storage at scale: Lessons from the growth of analogous energy technologies. *Energy Policy* **2010**, *38*, 4089–4098. [CrossRef]
45. Alders, J.G.M. Opening speech on the occasion of the First International Conference on Carbon Dioxide Removal. *Energy Convers. Manag.* **1992**, *33*, 283–286. [CrossRef]
46. IPCC. Carbon Dioxide Capture and Storage. 2005. Available online: https://www.ipcc.ch/site/assets/uploads/2018/03/srccs_wholereport-1.pdf (accessed on 2 December 2021).
47. Arranz, A.M. Carbon capture and storage: Frames and blind spots. *Energy Policy* **2015**, *82*, 249–259. [CrossRef]
48. Honegger, M.; Reiner, D. The political economy of negative emissions technologies: Consequences for international policy design. *Clim. Policy* **2018**, *18*, 306–321. [CrossRef]
49. Mack, J.; Endemann, B. Making carbon dioxide sequestration feasible: Toward federal regulation of CO_2 sequestration pipelines. *Energy Policy* **2010**, *38*, 735–743. [CrossRef]
50. Davies, L.L.; Uchitel, K.; Ruple, J. Understanding barriers to commercial-scale carbon capture and sequestration in the United States: An empirical assessment. *Energy Policy* **2013**, *59*, 745–761. [CrossRef]
51. Taylor, A.; Payne, M. Australia Welcomes Positive Outcomes at COP26. *Ministerial Media Release*, 14 November 2021. Available online: https://www.minister.industry.gov.au/ministers/taylor/media-releases/australia-welcomes-positive-outcomes-cop26 (accessed on 2 December 2021).
52. Taylor, A. Australia's Energy Future. *Ministerial Speech*, 29 October 2020. Available online: https://www.minister.industry.gov.au/ministers/taylor/speeches/australias-energy-future (accessed on 2 December 2021).
53. Thomas, S. Former Trump Adviser Andrew Liveris Admits 'Incorrect' Jobs Claim from Natural Gas on Q+A. *RMIT ABC Fact Check*, 27 May 2021. Available online: https://www.abc.net.au/news/2021-05-26/andrew-liveris-incorrect-claim-feedstock-natural-gas-fact-check/100160172 (accessed on 2 December 2021).
54. Commonwealth of Australia. The 2021 National Gas Infrastructure Plan. Available online: https://www.energy.gov.au/sites/default/files/2021%20National%20Gas%20Infrastructure%20Plan.pdf (accessed on 2 December 2021).
55. Statistica. Per Capita Greenhouse Gas Emissions in 2019, by Select Country (in Metric Tons Per Capita). *Statistica Website*, 8 October 2021. Available online: https://www.statista.com/statistics/478783/leading-countries-based-on-per-capita-greenhouse-gas-emissions/ (accessed on 2 December 2021).

56. Browne, B.; Swann, T. Money for Nothing. Australia Institute. May 2017. Available online: https://australiainstitute.org.au/wp-content/uploads/2020/12/P357-Money-for-nothing_0.pdf (accessed on 2 December 2021).
57. IEA; PVPS; TCP. National Survey Report of PV Power Applications in AUSTRALIA 2020. Australian PV Institute. Available online: https://apvi.org.au/wp-content/uploads/2021/10/PViA-Report-27.9.21-AUS.pdf (accessed on 2 December 2021).
58. The Hon. Mark Butler MP, Radio Interview. *ABC Adelaide*; 31 May 2017. Available online: https://parlinfo.aph.gov.au/parlInfo/download/media/pressrel/5310515/upload_binary/5310515.pdf (accessed on 2 December 2021).
59. APPEA. About: The Voice of Oil and Gas. Available online: https://www.appea.com.au/ (accessed on 2 December 2021).
60. APPEA. APPEA Welcomes Labor's CO_2 Storage Commitment. 24 June 2020. Available online: https://www.appea.com.au/all_news/appea-welcomes-labors-co2-storage-commitment/ (accessed on 2 December 2021).
61. APPEA Media Release—Good Move to Unlock Carbon Capture and Storage. 29 June 2021. Available online: https://www.appea.com.au/all_news/media-release-good-move-to-unlock-carbon-capture-and-storage/ (accessed on 2 December 2021).
62. Morrison, S. Press Conference Port of Newcastle Carrington, NSW. 8 November 2021. Available online: https://www.pm.gov.au/media/press-conference-port-newcastle-carrington-nsw (accessed on 2 December 2021).
63. Mazengarb, M. Taylor's Latest $500m Carbon Capture Fund May Already be Headed for Scrapheap. *RenewEconomy*, 12 November 2021. Available online: https://reneweconomy.com.au/taylors-latest-500m-carbon-capture-fund-may-already-be-headed-for-scrapheap/ (accessed on 2 December 2021).
64. Morrison, S. Virtual Address, Business Council of Australia Annual General Meeting. 24 November 2021. Available online: https://www.pm.gov.au/media/virtual-address-business-council-australia-annual-general-meeting (accessed on 2 December 2021).
65. Kilvert, N. Green Groups Accuse Chevron of 'Deliberate Mismanagement' of Its Own Carbon Storage Project. *ABC News, Science*, 17 July 2019. Available online: https://www.abc.net.au/news/science/2019-07-17/chevron-gorgon-gas-sequestration-mismanagement/11309076 (accessed on 2 December 2021).
66. MIT Gorgon Fact Sheet: Carbon Dioxide Capture and Storage Project. 2016. Available online: https://sequestration.mit.edu/tools/projects/gorgon.html (accessed on 2 December 2021).
67. AAP. WA Approves Controversial $15b Gorgon Gas Project. *Sydney Morning Herald*, 13 December 2006. Available online: https://www.smh.com.au/business/wa-approves-controversial-15b-gorgon-gas-project-20061213-gdp1br.html (accessed on 2 December 2021).
68. Lawson, R. Chevron Considers EPA Gorgon Conditions. *Business News*, 4 April 2009. Available online: https://www.businessnews.com.au/article/Chevron-considers-EPA-Gorgon-conditions (accessed on 2 December 2021).
69. Chevron. Gorgon Gas Development and Jansz Feed Gas Pipeline. Chevron Australia Pty Ltd. Document No: G1-NT-PLNX00000. 12 May 2015. Available online: https://australia.chevron.com/-/media/australia/our-businesses/documents/gorgon-emp-greenhouse-gas-abatement-program.pdf (accessed on 2 December 2021).
70. Department of Resources, Energy and Tourism. Barrow Sub-Basin Exploration Overview. Exploration History. 2008. Available online: https://web.archive.org/web/20090911081639/http://www.ret.gov.au/resources/Documents/acreage_releases/2008/site/page203.htm (accessed on 2 December 2021).
71. Chevron. Explore the Largest Single-Resource Development in Australia's History. Chevron Website. Available online: https://www.chevron.com/projects/gorgon (accessed on 2 December 2021).
72. Canavan, M. World Leading CO_2 Injection Project Starts Operations. *Ministerial Media Release*, 8 August 2019. Available online: https://parlinfo.aph.gov.au/parlInfo/download/media/pressrel/7348320/upload_binary/7348320.pdf (accessed on 2 December 2021).
73. Chevron. Gorgon Project Application for a Licence to Operate LNG Trains 1 to 3 and their Associated Facilities. *Chevron Australia*, 20 October 2017. Available online: https://pdfslide.net/documents/gorgon-project-department-of-environment-regulation-gorgon-project-application.html (accessed on 2 December 2021).
74. Chevron. Gorgon Gas Development and Jansz Feed Gas Pipeline Best Practice Pollution Control Design Report. G1-NT-REPX0001730. 15 January 2020. Available online: https://australia.chevron.com/-/media/australia/our-businesses/documents/gorgon-emp-best-practice-pollution-control-design-report.pdf (accessed on 2 December 2021).
75. Milne, P. Time's up on Gorgon's Five Years of Carbon STORAGE failure. *BoilingCold*, 16 July 2021. Available online: https://www.boilingcold.com.au/times-up-on-gorgons-five-years-of-carbon-storage-failure/ (accessed on 2 December 2021).
76. Kemp, J. World's Largest Carbon Capture Begins Even as Abbott Tax Repeal Looms. *Sydney Morning Herald*, 11 September 2013. Available online: https://www.smh.com.au/business/the-economy/worlds-largest-carbon-capture-begins-even-as-abbott-tax-repeal-looms-20130911-2tj0c.html (accessed on 2 December 2021).
77. Chevron. Gorgon Project. Gorgon Project Carbon Dioxide Injection Project: Low Emissions Technology Demonstration Fund. Annual Report for the Period 1 July 2015–30 June 2016. Available online: https://s3.documentcloud.org/documents/20509164/gorgon-project-carbon-dioxide-injection-project-low-emissions-technology-demonstration-fund-annual-report-1-july-2015-30-june-2016.pdf (accessed on 2 December 2021).
78. Macdonald-Smith, A. Gorgon Shut Down after One Shipment. *The Australian Financial Review*, 7 April 2016; p. 21.
79. Milne, P. The Day Disaster Struck Gorgon. *The West Australian*, 10 April 2017. Available online: https://thewest.com.au/business/energy/the-day-disaster-struck-gorgon-ng-b88423525z (accessed on 2 December 2021).
80. Macdonald-Smith, A.; Sprague, J. First Shipment of Gorgon gas prepares to sail. *The Australian Financial Review*, 9 March 2016; p. 28.

81. MacDonald-Smith, A. Chevron turns around performance at Gorgon venture. *The Australian Financial Review*, 5 February 2018; p. 15.
82. MacDonald-Smith, A. Chevron lifts Gorgon gas investment. *The Australian Financial Review*, 16 April 2018; p. 1.
83. MacDonald-Smith, A. Chevron ponders expansion in WA to meet LNG demand. *The Australian Financial Review*, 30 April 2018; p. 17.
84. Chevron Gorgon Project Carbon Dioxide Injection Project Low Emissions Technology Demonstration Fund Annual Report 1 July 2019–30 June 2020. Available online: https://s3.documentcloud.org/documents/20440488/foi-2-gorgon-project-2020-letdf-annual-report-rev-1-ar.pdf (accessed on 2 December 2021).
85. Macdonald-Smith, A. Gorgon CCS: 3 m Tonnes of CO_2 and Counting. *The Australian Financial Review*, 25 September 2020. Available online: https://www.afr.com/companies/energy/gorgon-ccs-3m-tonnes-of-co2-and-counting-20200925-p55z9k (accessed on 2 December 2021).
86. Vogel, P. Gorgon Gas Development Revised and Expanded Proposal: Barrow island Nature Reserve. West Australian Environmental Protection Authority. 29 April 2011. Available online: https://www.epa.wa.gov.au/sites/default/files/Ministerial_Statement/00800.pdf (accessed on 2 December 2021).
87. Milne, P. Chevron's Gorgon Emissions to Rise after Sand Clogs $3.1B CO_2 injection system. *BoilingCold*, 7 January 2021. Available online: https://www.boilingcold.com.au/chevrons-gorgon-co2-emissions-to-rise-sand-clogs/ (accessed on 2 December 2021).
88. Milne, P. Gorgon Emissions to Soar until Chevron Fixes Restricted CO_2 Injection. *BoilingCold*, 10 February 2021. Available online: https://www.boilingcold.com.au/regulator-limits-chevrons-troubled-gorgon-co2-injection-to-one-third-capacity/ (accessed on 2 December 2021).
89. MacDonald-Smith, A. Emissions bill poised to rise at Gorgon LNG. *The Australian Financial Review*, 15 January 2021; p. 15.
90. Milne, P. Gorgon CO_2 Injection Stopped by Leaks and Corrosion. *BoilingCold*, 19 December 2017. Available online: https://www.boilingcold.com.au/gorgon-co2-injection-stopped-by-leaks-and-corrosion/ (accessed on 2 December 2021).
91. Young, E. WA's Gorgon Project Fails to Deliver on Pollution Deal, Adding Millions of Tonnes of Carbon a Year. *WA Today*, 16 February 2021. Available online: https://www.watoday.com.au/national/millions-of-tonnes-of-carbon-added-to-pollution-as-gorgon-project-fails-capture-deal-20210215-p572na.html (accessed on 2 December 2021).
92. Cox, L. Western Australia LNG Plant Faces Calls to Shut down until Faulty Carbon Capture System is Fixed. *The Guardian*, 15 January 2021. Available online: https://www.theguardian.com/environment/2021/jan/15/western-australia-lng-plant-faces-calls-to-shut-down-until-faulty-carbon-capture-system-is-fixed (accessed on 2 December 2021).
93. Milne, P. Carbon Hiccup for Chevron with 5 Million-Tonne Greenhouse Gas Problem at Gorgon LNG Plant. *The West Australian*, 19 December 2017. Available online: https://thewest.com.au/business/oil-gas/carbon-hiccup-for-chevron-with-5-million-tonne-greenhouse-gas-problem-at-gorgon-lng-plant-ng-b88694565z (accessed on 2 December 2021).
94. Ker, P. Gorgon set to capture all carbon in 2020: Cleaner LNG. *The Australian Financial Review*, 20 November 2019; p. 11.
95. Milne, P. Chevron Cops 33 Orders from Regulators to Make Gorgon Safe. *Boiling Cold*, 7 August 2020. Available online: https://www.boilingcold.com.au/safety-cop-orders-chevron-to-fix-gorgon/ (accessed on 2 December 2021).
96. Hastie, H. Gorgon Project: Calls for Chevron to Shut Down Pilbara Gas Plant after Cracks Found in Critical Components. *Sydney Morning Herald*, 22 July 2020. Available online: https://www.smh.com.au/business/companies/workers-fear-for-their-safety-calls-for-gorgon-shutdown-after-cracks-found-in-critical-plant-components-20200722-p55eel.html (accessed on 2 December 2021).
97. Thompson, B. Union fears Gorgon kettle cracks are beyond repair: LNG. *The Australian Financial Review*, 25 July 2020; p. 23.
98. MacDonald-Smith, A. Gorgon avoids total shutdown: LNG. *The Australian Financial Review*, 22 August 2020; p. 23.
99. MacDonald-Smith, A. Repairs to hit Chevron's LNG output. *The Australian Financial Review*, 1 February 2021; p. 18.
100. MacDonald-Smith, A. Third Gorgon train to shut. *The Australian Financial Review*, 11 March 2021; p. 17.
101. Thompson, B. Chevron Sees Light at End of $70 Billion Tunnel at Gorgon. *The Australian Financial Review*, 2 May 2021. Available online: https://www.afr.com/companies/energy/chevron-sees-light-at-end-of-70-billion-tunnel-at-gorgon-20210502-p57o3n (accessed on 2 December 2021).
102. Milne, P. Chevron's Gorgon CO_2 Injection Fix Needs More Time, so More Emissions. *BoilingCold*, 2 July 2021. Available online: https://www.boilingcold.com.au/chevrons-gorgon-co2-injection-fix-needs-more-time-so-more-emissions/ (accessed on 2 December 2021).
103. Young, E. More Carbon to be Vented in Further Embarrassment for Chevron's Gorgon. *WA Today*, 13 January 2021. Available online: https://www.watoday.com.au/national/western-australia/carbon-capture-woes-mean-more-emissions-to-be-vented-in-further-embarrassment-for-chevron-s-gorgon-20210112-p56tk5.html (accessed on 2 December 2021).
104. Verdon, J.P. Significance for secure CO_2 storage of earthquakes induced by fluid injection. *Environ. Res. Lett.* **2014**, *9*, 064022. [CrossRef]
105. Zoback, M.D.; Gorelick, S.M. Earthquake triggering and large-scale geologic storage of carbon dioxide. *Proc. Natl. Acad. Sci. USA* **2012**, *109*, 10164–10168. [CrossRef]
106. Alcalde, J.; Flude, S.; Wilkinson, M.; Johnson, G.; Edlmann, K.; Bond, C.E.; Scott, V.; Gilfillan, S.M.V.; Ogaya, X.; Haszeldine, R.S. Estimating geological CO_2 storage security to deliver on climate mitigation. *Nat. Commun.* **2018**, *9*, 2201. [CrossRef] [PubMed]
107. Geosciences Australia. Earthquakes@ GA. Available online: https://earthquakes.ga.gov.au/ (accessed on 12 November 2021).

108. Mazengarb, M. Sand Clogs up Australia's Only Operating Carbon Capture Project. *RenewEconomy*, 13 January 2021. Available online: https://reneweconomy.com.au/sand-clogs-up-australias-only-operating-carbon-capture-project/ (accessed on 2 December 2021).
109. MacDonald-Smith, A. Chevron plans first Gorgon LNG shipment: Resources. *The Australian Financial Review*, 12 March 2015; p. 25.
110. Stevens, M. Gorgon highlights reform need. *The Australian Financial Review*, 21 July 2015; p. 34.
111. MacDonald Smith, A. Chevron close to Gorgon LNG restart. *The Australian Financial Review*, 2 May 2016; p. 15.
112. MacDonald-Smith, A. Chevron NW Shelf exit plays into Woodside's hand: LNG processing. *The Australian Financial Review*, 20 June 2020; p. 23.
113. MacDonald-Smith, A. Gorgon set for another $6b boost: LNG. *The Australian Financial Review*, 3 July 2021; p. 20.
114. Ianucci, E. Chevron Takes A$6bn Plunge at Gorgon. *Mining Weekly*, 2 July 2021. Available online: https://www.miningweekly.com/print-version/chevron-takes-a6bn-plunge-at-gorgon-2021-07-02 (accessed on 2 December 2021).
115. Chevron. Jansz-Io Compression Project to Proceed. Available online: https://australia.chevron.com/news/2021/jansz-io-compression-project-to-proceed (accessed on 2 December 2021).
116. Beal, E.; Heeney, L. EU Carbon Border Tax is a Warning to Australia: Cut Emissions or Lose Exports. *Sydney Morning Herald*, 15 July 2021. Available online: https://www.smh.com.au/environment/climate-change/eu-carbon-border-tax-is-a-warning-to-australia-cut-emissions-or-lose-exports-20210715-p58a2y.html (accessed on 2 December 2021).
117. Macdonald-Smith, A.; Thompson, B. Net zero critical for gas to avoid fate of coal: Santos chief. *The Australian Financial Review*, 16 June 2021; p. 1.
118. Gallagher, K. Australia can become a carbon storage superpower. *The Australian Financial Review*, 16 June 2021; p. 47.
119. Hewett, J. Coleman calls time on big new LNG projects: Exclusive. *The Australian Financial Review*, 23 April 2021; p. 1.
120. Sinn, H.-W. *The Green Paradox: A Supply-Side Approach to Global Warming*; MIT Press: Cambridge, MA, USA, 2012.
121. Khadem, N. Tax credits for oil and gas giants rise to $324 billion. *ABC News*, 1 April 2019. Available online: https://www.abc.net.au/news/2019-04-01/tax-credits-for-oil-and-gas-giants-rise-to-324-billion/10959236 (accessed on 2 December 2021).
122. Kraal, D. In the Midst of an LNG Export Boom, Why are We Getting so Little for Our Gas? *The Conversation*, 17 February 2020. Available online: https://theconversation.com/in-the-midst-of-an-lng-export-boom-why-are-we-getting-so-little-for-our-gas-131461 (accessed on 2 December 2021).
123. Milne, P. Shell Predicts Free Gas Forever for Gorgon and Prelude LNG. *BoilingCold*, 30 March 2021. Available online: https://www.boilingcold.com.au/shell-predicts-free-gas-forever-for-gorgon-and-prelude-lng/ (accessed on 2 December 2021).
124. Smyth, J.; Sheppard, D. Monster Problem: Australia's $54-Billion Gorgon LNG Project is a Test Case for Carbon Capture. *Financial Times*, 26 July 2021. Available online: https://www.ft.com/content/428e60ee-56cc-4e75-88d5-2b880a9b854a (accessed on 2 December 2021).
125. Myriam, R.; Chenoweth, N. Chevron's thankless $9.5 billion. *The Australian Financial Review*, 26 August 2019; p. 40.
126. Ker, P. Energy sector seeks carbon storage credits: Exclusive. *The Australian Financial Review*, 18 October 2019; p. 8.
127. Chenoweth, N. Move to close $3b Gorgon loophole. *The Australian Financial Review*, 18 August 2015; p. 7.
128. Chenoweth, N. Chevron paid $248 tax on $1.7b profit. *The Australian Financial Review*, 10 November 2015; p. 3.
129. Chenoweth, N. Chevron says Gorgon's $1.7b profit not taxable. *The Australian Financial Review*, 19 November 2015; p. 6.
130. Butler, B.; Evershed, N. Some of Australia's Biggest Companies Paid More in Political Donations than Tax in 2018–2019. *The Guardian*, 11 December 2020. Available online: https://www.theguardian.com/australia-news/2020/dec/11/australian-arm-of-fossil-fuel-giant-chevron-paid-no-tax-on-income-of-900m-in-2019 (accessed on 2 December 2021).
131. Mather, J. Chevron predicts up to $140b super profits bill: Tax. *The Australian Financial Review*, 29 April 2017; p. 6.
132. Wiggins, J. CIMIC's $1.1b Chevron jetty dispute to drag on until 2019. *The Australian Financial Review*, 19 July 2017; p. 15.
133. Wiggins, J. CIMIC's $1b Write-off Shows its Accounts are Unreliable: Analyst. *The Australian Financial Review*, 20 October 2020. Available online: https://www.afr.com/companies/infrastructure/cimic-loses-1b-dispute-with-chevron-20201020-p566pz (accessed on 2 December 2021).
134. Young, E. No Idea What Enforcement or When: Flying Blind on Chevron's Gorgon. *WA Today*, 4 February 2020. Available online: https://www.watoday.com.au/business/companies/no-idea-what-enforcement-or-when-flying-blind-on-chevron-s-gorgon-20200203-p53xej.html (accessed on 2 December 2021).
135. Gorman, V.; Searson, A. Environmental Group Says Chevron Failure will Undermine Public Confidence. *ABC News*, 20 July 2021. Available online: https://www.abc.net.au/news/2021-07-20/environmental-group-says-chevron-failure-undermines-confidence/100306730 (accessed on 2 December 2021).
136. Chevron. Chevron Announces $40 Million Western Australian Lower Carbon Investment. Media Release. Available online: https://australia.chevron.com/news/2021/chevron-announces-aud40-million-western-australian-lower-carbon-investment (accessed on 2 December 2021).
137. Readfearn, G. Australia's Only Working Carbon Capture and Storage Project Fails to Meet Target. *The Guardian*, 12 November 2021. Available online: https://www.theguardian.com/australia-news/2021/nov/12/australias-only-working-carbon-capture-and-storage-project-fails-to-meet-target (accessed on 2 December 2021).

138. Chevron. Gorgon Project Carbon Dioxide Injection Project Barrow Island Act 2003 Section 13 Approval Annual Operational Report (1 January 2020–31 December 2020). Available online: https://s3.documentcloud.org/documents/20793746/gorgon-project-carbon-dioxide-injection-annual-operational-report-to-wa-state-government-1-january-2020-31-december-2020.pdf (accessed on 2 December 2021).
139. Fong, B. Analysing the Behavioural Finance Impact of 'Fake News' Phenomena on Financial Markets: A Representative Agent Model and Empirical Validation. *Financ. Innov.* **2021**, *7*, 53. Available online: https://jfin-swufe.springeropen.com/articles/10.1186/s40854-021-00271-z (accessed on 2 December 2021). [CrossRef]
140. Chevron. The Human Energy Company: Gorgon Carbon Capture and Storage Fact Sheet. Chevron Australia. 2021. Available online: https://australia.chevron.com/-/media/australia/publications/documents/gorgon-co2-injection-project.pdf (accessed on 2 December 2021).
141. Smith, F.; Van Gent, D.; Sewell, M. Western Australia Greenhouse Gas Capture and Storage: A Tale of Two Projects. Australian Government Department of Resources, Energy and Tourism, and Government of Western Australia, Department of Mines and Petroleum. Available online: https://www.globalccsinstitute.com/archive/hub/publications/39961/ccsinwareport-opt.pdf (accessed on 2 December 2021).
142. Morton, A. 'Nothing to hide?' Oil and Gas Lobby Pushes to Limit Data on Its Emissions. *The Guardian*, 16 November 2018. Available online: https://www.theguardian.com/environment/2018/nov/16/nothing-to-hide-oil-and-gas-lobby-pushes-to-limit-data-on-its-emissions (accessed on 2 December 2021).
143. Chevron. Chevron Submission to the Climate Change Authority Review of the National Greenhouse and Energy Reporting Legislation. Available online: https://www.climatechangeauthority.gov.au/sites/default/files/2020-06/2018%20NGER%20Review/Submissions/Chevron.pdf (accessed on 2 December 2021).
144. MacDonald-Smith, A. Gorgon Captures its Millionth Tonne of Carbon: Climate Mitigation. *The Australian Financial Review*, 15 February 2020; p. 27.
145. Chevron. The Human Energy Company: 2020 Annual Report. Available online: https://www.chevron.com/-/media/chevron/annual-report/2020/documents/2020-Annual-Report.pdf (accessed on 2 December 2021).
146. Chevron. 2020 Corporate Sustainability Report. Available online: https://www.chevron.com/-/media/shared-media/documents/chevron-sustainability-report-2020.pdf (accessed on 2 December 2021).
147. Chevron. Media Statement Chevron Australia CO_2 Injection Milestone. *PERTH, Western Australia*, 19 July 2021. Available online: https://australia.chevron.com/news/2021/co2-injection-milestone (accessed on 2 December 2021).
148. Kilvert, N. Angus Taylor says Australia has the world's largest carbon capture and storage project. Here's what he's not saying. *ABC News*, 19 September 2020. Available online: https://www.abc.net.au/news/science/2020-09-19/angus-taylor-carbon-capture-storage-gorgon-chevron/12676732 (accessed on 2 December 2021).
149. APPEA. Gas Industry Investment and Technology Helping Reduce Emissions. *Media Release*, 19 July 2021. Available online: https://www.appea.com.au/wp-content/uploads/2021/07/Media-release-Gas-Industry-investment-and-technology-helping-reduce-emissions.pdf (accessed on 2 December 2021).
150. Congressional Committee on Oversight and Reform. Memorandum: Analysis of the Fossil Fuel Industry's Legislative Lobbying and Capital Expenditures Related to Climate Change. 28 October 2021. Available online: https://oversight.house.gov/sites/democrats.oversight.house.gov/files/Analysis%20of%20the%20Fossil%20Fuel%20Industrys%20Legislative%20Lobbying%20and%20Capital%20Expenditures%20Related%20to%20Climate%20Change%20-%20Staff%20Memo%20%2810.28.21%29.pdf (accessed on 2 December 2021).
151. Supran, G.; Oreskes, N. Rhetoric and Frame Analysis of ExxonMobil's Climate Change Communications. *One Earth* **2021**, *4*, 696–719. Available online: https://www.cell.com/one-earth/fulltext/S2590-3322(21)00233-5 (accessed on 2 December 2021). [CrossRef]
152. Bush, M.J. Chapter: Denial and Deception. In *Climate Change and Renewable Energy*; Springer International Publishing: Berlin, Germany, 2019; pp. 373–420.
153. Cox, L. Western Australia Environment Watchdog Plans Tougher Curbs on Emissions. *The Guardian*, 7 March 2019. Available online: https://www.theguardian.com/australia-news/2019/mar/07/western-australia-environment-watchdog-plans-tougher-curbs-on-emissions (accessed on 2 December 2021).
154. MacDonald-Smith, A. Chevron's warning on carbon and costs. *The Australian Financial Review*, 28 May 2019; p. 17.
155. MacDonald-Smith, A.; Thompson, B. Woodside decries carbon 'red tape': Resources. *The Australian Financial Review*, 9 March 2019; p. 23.
156. Morton, A. WA's Rejection of Carbon-Neutral Guidelines Leaves LNG Emissions Booming: Protect Ningaloo. *The Guardian*, 20 March 2019. Available online: https://www.theguardian.com/environment/2019/mar/20/lng-redux (accessed on 2 December 2021).
157. Ogge, M. Regulatory Carbon Capture Submission on the Proposed Methodology Determination for Carbon Capture and Storage. The Australia Institute. 2021. Available online: https://australiainstitute.org.au/wp-content/uploads/2021/10/P1110-Australia-Institute-submission-to-ERF-method-consultation-WEB.pdf (accessed on 2 December 2021).
158. Clean Energy Regulator. Australia's 10 Highest Greenhouse Gas Emitters 2018-19. *CER Website*, 28 February 2020.
159. Available online: http://www.cleanenergyregulator.gov.au/NGER/Pages/Published%20information/Data%20highlights/2018-19%20factsheets/Australia%27s-10-highest-greenhouse-gas-emitters-2018-19.aspx (accessed on 2 December 2021).

160. Milne, P. Chevron Faces Little Grief from Gorgon LNG's 7 Million Tonnes Emissions Miss. *BoilingCold*, 26 November 2020. Available online: https://www.boilingcold.com.au/chevron-faces-little-grief-from-gorgon-lng-emissions-miss/ (accessed on 2 December 2021).
161. Morton, A. 'A Shocking Failure': Chevron Criticised for Missing Carbon Capture Target at WA Gas Project. *The Guardian*, 19 July 2021. Available online: https://www.theguardian.com/environment/2021/jul/20/a-shocking-failure-chevron-criticised-for-missing-carbon-capture-target-at-wa-gas-project (accessed on 2 December 2021).
162. Morton, A. WA's Rejection of Carbon-Neutral Guidelines Leaves LNG Emissions Booming: Protect Ningaloo. *The Guardian*, 20 March 2019. Available online: https://www.theguardian.com/environment/2018/nov/14/half-of-australias-emissions-increase-linked-to-was-gorgon-lng-plant (accessed on 2 December 2021).
163. Milne, P. Do Your Homework: Chevron's Outgoing Chief John Watson Learns From Gorgon Mistakes. *BoilingCold*, 30 October 2017. Available online: https://thewest.com.au/business/oil-gas/do-your-homework-chevrons-outgoing-chief-john-watson-learns-from-gorgon-mistakes-ng-b88643593z (accessed on 2 December 2021).

Article

Planning a Notable CCS Pilot-Scale Project: A Case Study in France, Paris Basin—Ile-de-France

Fernanda M. L. Veloso *, Isaline Gravaud, Frédéric A. Mathurin and Sabrine Ben Rhouma

French Geological Survey (Bureau de Recherche Géologique et Minière—BRGM), 3 Avenue Claude Guillemin, CEDEX 2, 45060 Orléans, France; i.gravaud@brgm.fr (I.G.); f.mathurin@brgm.fr (F.A.M.); s.benrhouma@brgm.fr (S.B.R.)
* Correspondence: f.veloso@brgm.fr

Abstract: Few commercial-scale carbon capture and storage (CCS) projects are currently operating in the world, with almost all in the USA and China. Despite a high number of CCS pilot-scale projects achieved in Europe, only two commercial-scale projects are operating today. The goal of this study is to present a case study in France to select a promising location to deploy a notable CCS pilot-scale project based on a multicriteria regional-scale approach. The methodology applied in this case study describes and assesses different aspects involved in CCS technology at the regional scale, and then an evaluation of economic key performance indicators (KPI) of CCS is carried out. The assessment at the regional scale gives an overview of where CCS could be applied, when CCS could be deployed and how to launch CCS considering the needs and concerns of stakeholders in the region. Technical aspects were mapped, such as the location of irreducible CO_2 sources and long-lasting emissions and the location of storage resources and existing potential transport infrastructures. We identified the waste-to-energy and chemical sectors as the main CO_2 sources in the region. An economic analysis of a hypothetical scenario of CCS deployment was elaborated considering three of the higher emitters in the region. A CCS scenario in the Paris Basin region with a deployment between 2027 and 2050 indicates a low CO_2 cost per ton avoided between 43 EUR/t and 70 EUR/t for a cumulated total of 25 Mt and 16 Mt, respectively, of CO_2 captured and stored for 26 years, including 7.7 Mt of CO_2 from biomass (potential negative emissions). Storage maturity and availability of the resource are the most uncertain parameters of the scenario, although they are the key elements to push investment in capture facilities and transport. Geological storage pilot projects are mandatory to prove storage resource and should be located in strategic locations close to potential CO_2 sources in case of confirmation of proven resources. Well-perceived pilot-scale projects are the first step to start engaging in deciding and investing in commercial-scale CCS projects.

Keywords: CCS pilot-scale; CO_2 reduction; Ile-de-France; regional scale; waste to energy; decarbonizing industry; Paris Basin; key performance indicators; economic evaluation

Citation: Veloso, F.M.L.; Gravaud, I.; Mathurin, F.A.; Ben Rhouma, S. Planning a Notable CCS Pilot-Scale Project: A Case Study in France, Paris Basin—Ile-de-France. *Clean Technol.* 2022, 4, 458–476. https://doi.org/10.3390/cleantechnol4020028

Academic Editor: Diganta B. Das

Received: 8 March 2022
Accepted: 12 May 2022
Published: 18 May 2022

Publisher's Note: MDPI stays neutral with regard to jurisdictional claims in published maps and institutional affiliations.

Copyright: © 2022 by the authors. Licensee MDPI, Basel, Switzerland. This article is an open access article distributed under the terms and conditions of the Creative Commons Attribution (CC BY) license (https://creativecommons.org/licenses/by/4.0/).

1. Introduction

The development of carbon capture and storage (CCS) has been slow in the last decade in Europe. Only two CCS projects are currently operating within the European Economic Area, mainly off the Norwegian coast. The main reasons include a low CO_2 price on the EU Emissions Trading System (EU ETS). Well below 10 EUR/tCO_2 prior to 2017, the CO_2 ETS price has increased since 2018, reaching the highest ETS price of 95 EUR/tCO_2 on 13 February 2022 [1]. Negative perceptions of CCS projects in several nations also contributed to delaying CCS deployment [2,3]. Projects were set on hold or even cancelled due to reasons such as financing gaps, resistance of the local populations, or lack of political support [4,5]. CO_2 sources would partly influence social acceptance of CCS technology [6]. Adapting the identity of a project to local factors such as the presence of industry, transport network, or benefit from the exploitation of underground resources should play a key role in public opinion about these projects [7].

Today, few commercial-scale CCS projects are operating in the world, with almost all in the USA and China. Commercial-scale projects are those capturing, transporting and storing at least 500,000 tons of CO_2 per year. Enhanced hydrocarbon recovery (EHR) is the dominant type of project, injecting more than 500 kt of CO_2 per year [8,9]. The Global CCS Institute [8] proposes a new CCS facility classification to differentiate large-scale CCS projects and pilot-demonstration-scale projects. The proposed classification considers smaller capture facilities, which can be commercially viable. In that respect, CCS hubs are regarded as opportunities to create economies of scale that lower costs of transport and storage to multiple smaller CO_2 sources. Thus, CCS facilities must support a commercial return while operating and meeting the national regulatory requirement. Pilot and demonstration facilities capture CO_2 for testing, enhancing or demonstrating CCS technology or processes without the obligation to store CO_2 permanently.

Looking at current CCS operating and in-development projects in Europe using the proposed classification of the Global CCS Institute (Figure 1A), all CCS large-scale facilities operating and in development are located around the North Sea. Other countries in Southern and Eastern Europe completed or are operating pilot-scale projects, with approximatively half of them without CO_2 storage (Figure 1B).

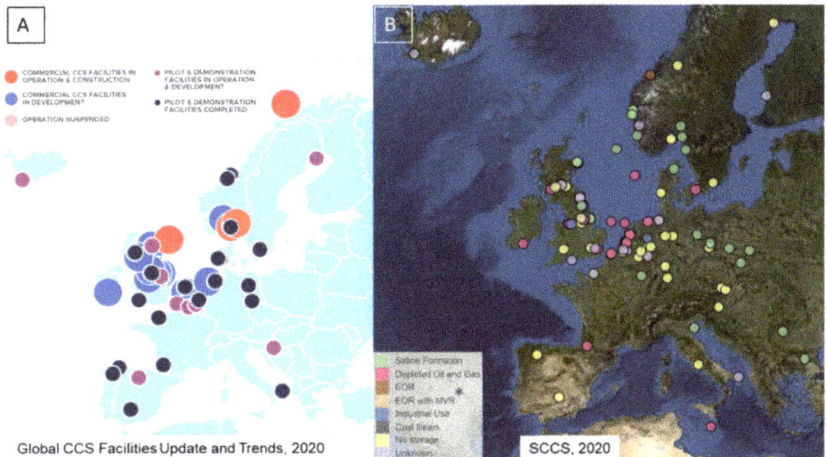

Figure 1. European map of CCS facilities completed, in development and ongoing. In (**A**), the overview of Global CCS facilities in 2020 classified by size of the project: commercial (large) scale or pilot scale. The status of these projects is represented by the color bubble. Map (**B**) indicates the storage status of all CCS projects regardless of their size. Yellow circles are projects without geological storage of CO_2. (* EOR = enhanced oil recovery; MVR = monitoring, verification and reporting).

Countries with a policy to create a business case for investment in CCS projects, such as Norway, UK, the Netherlands and the USA, are leading ongoing and in-development CCS commercial-scale projects, but other technical aspects pushed these regions up to leading in the field CCS deployment technology. These countries have a good knowledge of their storage resources from oil and gas history and government support. An atlas of CO_2 storage resources, such as the CO_2 Storage Evaluation Database (CO_2 Stored) in the UK or the Norwegian CO_2 Storage Atlas [10,11], are accurate public information based on seismic coverage, data wells and published research. The knowledge and maturity of storage resources seems to be a crucial element for the development of commercial-scale CCS facilities.

The goal of this study is to present a case study in France to select a promising location to deploy a notable CCS pilot-scale project based on technical, economic and societal aspects. This study explains and justifies the choice to locate a pilot-scale CCS project focusing on the storage element of the CCS chain and the optimization of transport for regional

CO$_2$ sources. Before investing in a capture facility, the emitter plant needs guarantees on where the CO$_2$ would be stored and how it would be transported. The technology of CO$_2$ capture has improved in the last 20 years, with costs depending on the gas stream and CO$_2$ concentration. In a wide range of industry sectors (refinery, cement, iron and steel), the cost of capture is between USD 40 and USD 120 per ton of CO$_2$ [12].

The maturity and confidence of storage resources in Europe are low, except around the North Sea [13], which seems to be the driver of CCS operational projects. The significant lead time for the development and permitting of CO$_2$ geological storage sites is in the order of 7–10 years, which implies a selection of potential sites to be developed well in advance of when they are predicted to be needed. Today, CCS pilot-scale projects would play a key role in enabling CCS commercial-scale projects around Europe. Through pilot projects, storage capacity could be proven, ensuring availability of storage resources to the trajectory of investment in capture facilities.

2. Materials and Methods

A notable CCS pilot-scale project should demonstrate the technical feasibility of the technology to engage regional and national stakeholders in further developments. Today, in Europe, the maturity and confidence of storage resources seem to be the major challenge to elaborate plans for the deployment of CCS outside the North Sea. Indeed, policy also plays a key role in accelerating the technology, as well as the societal engagement to deploy it. The methodology applied in this case study in France describes and assesses different aspects involved in CCS technology at the regional scale and carries out an evaluation of economic key performance indicators (KPIs) of CCS. The assessment at the regional scale gives an overview on where CCS could be applied, when CCS could be deployed and how to launch CCS considering the needs and concerns of stakeholders in the region.

The proposed methodology is based on the mapping of technical aspects at the regional scale to define the most promising industrial clusters and hubs which would benefit from CCS technology. After this first screening of emission sources, transport infrastructures and storage site options, a second step of economic evaluation assessed the economical key performance indicators (KPIs) of deploying CCUS at the regional scale. The societal perception of some regional stakeholders is also considered as part of the mapping aspects (Figure 2) in this early exercise of planning CCS.

Figure 2. Schematic methodology chart illustrating the workflow to select potential areas to deploy a notable pilot-scale CCS project. * KPI: key performance indicators.

2.1. Mapping CCUS Aspects

Technical and societal aspects involved in the CCS technology were mapped in the frame of STRATEGY CCUS project (H2020, grant agreement: No 837754) for the Paris Basin region, mainly inside the Ile-de-France Department. The data gathered in STRATEGY CCUS aimed at providing the technical basis on capture, transport, and storage conditions for assessing the viability of defining and implementing CCUS clusters and hubs. Storage capacity maturity and its confidence level was assessed for two preliminary candidates. The mapping of spatial conditions for network development considers the geographic distribution of the source, sinks and transport opportunities.

The technical potential for implementing CCUS in the Paris Basin region was assessed on the basic premise that industrial CCUS clusters provide synergies, either in the capture facilities, at the transport networks or at injection and storage facilities, that result in decreasing costs for implementing the technology [14].

The mapping of emissions sources determines what CO_2 may be captured to develop an understanding of the CO_2 as part of an industrial emissions reduction program using CCUS. The starting point was the definition of current CO_2 emission quantities in the area, the locations of emitters and related details. Distinctions between the fossil fuel combustion emissions, biomass emissions and process emissions were conducted whenever there was sufficient information. Once this inventory of the CO_2 emissions was established, it was necessary to consider what portion of that would be appropriate to address using CCUS [15].

The mapping of a CO_2 transport infrastructure is the identification and planning of a CO_2 transport network within the cluster to send the CO_2 from each capture facility to a consolidation point, a hub. The transport network can be composed of a pipeline system, but for very small-capacity sources, the collection network can be composed of a modular system including road truck, rail tank-car, shipping or barge transport on inland waterways. Captured CO_2 is collected in the cluster, then conditioning facilities (e.g., compressing, liquefaction, etc.) prepare the CO_2 for transportation by truck, pipeline or ship to the injection and storage site where further reconditioning of the stream may be necessary. The available geological storage capacity and its distribution with respect to the sources result in scenarios to assess the optimal transport network development.

The storage capacities reported here were calculated using a volumetric approach for the Dogger Fm. and reservoir simulation approach for the Trias Fm. [16]. Capacity estimated by volumetric approach is dependent on standard parameters (bulk volume, porosity, net-to-gross and CO_2 density) and a modifying term, the storage efficiency factor (SEF). Storage efficiency values also reflect general geologic characteristics and boundary conditions. For example, carbonates and open systems have a higher efficiency than clastic reservoirs and closed systems. Capacity estimates were ranked using a quantitative resource pyramid approach (Table 1). Based on four tiers, the classification captures the maturity level of existing data and the understanding of the potential storage capacity. Each tier introduces gradual knowledge of the reservoir—i.e., influencing the accuracy of the storage estimate —starting from regional approximations to the evaluation of specific targeted sites. The requirements for each tier reflect this maturation. The described tiers are compatible with existing schemes, allowing outcomes to be transferred to equivalent classifications if required [13].

The CO_2 utilization opportunities are here regarded as those uses with a clear mitigation impact, either with a greenhouse gas contribution or that clearly enable other low-carbon actions, leaving out those technologies that have a negligible impact [13].

The diffusion of a technology is also a social challenge. A dedicated work package within the STRATEGY CCUS project focuses on—all kinds of—actors involved in CCUS applications. Mapping societal aspects of CCUS technology in the Paris Basin region provides a first statement of the actor structure in the innovation system for CCUS [6] at the national and regional levels. Stakeholders are defined as individuals (e.g., employee, customer and citizen) who can be concerned by the development of a CCUS project, either with respect to demands or responsibilities towards it. A mapping of stakeholders' perceptions, attitudes and interests led to defining the scope of relevant issues and specific needs to be considered locally. Following the identification of relevant actors, semistructured interviews were conducted. These interviews and broadly based discussions around CCUS, involving both representatives of the stakeholder group from the Paris Basin region and some stakeholders at the national level [17].

Table 1. Tier classification/definition and suitability criteria defining the maturity of geological CO_2 storage resource capacities.

Tiers	Classification	Suitability Criteria
Tier 1	Regional assessment; equivalent to prospective (theoretical)	Generic SEFs (storage efficiency factor). Formation and storage unit estimate. First approximation. Low data burden and global storage efficiency values where boundary conditions are poorly constrained or uncertain.
Tier 2	Discovery assessment; equivalent to low contingent (effective)	Tailored SEFs. Daughter unit estimates. Second approximation. Moderate data burden and lithology-specific regional storage efficiency factors. Distinction between deep saline aquifers, depleted hydrocarbon fields and coal beds. Boundary conditions are established.
Tier 3	Prospect assessment; equivalent to pending/on hold (practical)	Detailed data prospective candidates. Third approximation with a more taxing data burden, including subattributes of the main factors used to estimate capacity and lithology-specific local SEFs. Each candidate prospect requires either existing or targeted data acquisition sufficient to build a simple geomodel for first-pass simulation and well location consideration.
Tier 4	Site assessment; equivalent to justified/approved/on injection (matched), project.	Targeted storage sites. The final approximation prior to operation. This has the highest data burden and requires a detailed geomodel for reservoir simulation studies. Outcomes from the simulations test the accuracy of the storage efficiency factors and provide scenarios for maximizing capacity based on well planning and scheduling.

2.2. Economic Key Performance Indicators (KPIs)

The CCUS scenario deployment at the Paris Basin described the business case of CCUS technology until 2050 based on technoeconomic modelling and hypothesis. The regional CCUS scenarios are based on both the performances of local industries in operation and for which CCUS is a relevant mitigation alternative, as well as the regional storage capacities known to date. For each of the regional scenarios evaluated, the cost difference between investing in CCUS or paying the carbon penalties to remain in compliance with the EU ETS is calculated to estimate the CCUS costs in terms of CO_2 avoided for each of the scenarios deployed [18].

A scenario evaluation tool was developed in the STRATEGY CCUS project to evaluate future CCUS value chains [19], where CO_2 is captured from point emissions and transported to utilization industries or for permanent storage. The tool uses the data gathered from the mapping aspects at the regional level and the key technological and economical parameters for implementing the CCUS technology related to:

1. Energy consumption.
2. Net present costs for the capture, transport and storage.
3. Amount of CO_2 emissions avoided and negative emissions.
4. Revenue created by the down-stream utilization industries.

Scenario analysis examines the results of how future events are laid out in time. Despite the inherent uncertainty in the predictions, regional evaluations provide a first glance at possible future decision paths. A better planning of the project development enables proactive actions to be taken (e.g., with regards to total energy consumption) and allows decision makers to avoid foreseeable risks.

3. Results

The Paris Basin—Ile-de-France (IDF), as studied in the STRATEGY CCUS project (EU H2020 project, grant agreement: 837754) and showed in Figure 3, is located in the center-northern part of France around the French capital—Paris—and it covers the administrative region of Ile-de-France and the Loiret department (storage option). It is the most populated region of France with more than 12 million inhabitants (20% of the French population). The Paris Basin IDF is still largely rural: nearly 11 million people live in the Paris agglomeration, which represents 24% of the Ile-de-France surface area, the rest of the region is made up of agricultural land, forest and natural spaces. The Ile-de-France department is an economically active region, producing nearly 30% of the French gross domestic product (GDP).

Figure 3. Geographic location of the Ile-de-France department and Paris Basin region as studied in STRATEGY CCUS project. Copyright: Google images @2022. Geographic National Institute of France.

Demography and land occupation is the first concern of the Paris Basin region, with CO_2 emissions mainly related to waste from energy plants, heat (power) plants and the chemical industry, which corresponded to 54%, 23% and 12%, respectively, of the total CO_2 emission of this region in 2019 (Appendix A).

3.1. Mapping Results
3.1.1. Emissions Sources

Emissions of CO_2 in the Paris Basin amounted to 5.5 Mt in 2019 [20]. This places the region well behind the French port regions (Dunkirk, Le Havre and Marseille-Fos), despite its high population rate. The emissions pattern is also very different, as the 5.5 Mt of CO_2 is split into 39 emitters, with almost 40% of these facilities emitting less than 50 kt of CO_2 in 2019 (Figure 4 and Appendix A). Only about 10% of facilities emitted more than 300 kt of CO_2 in 2019 and around 30% of facilities emitted between 100–300 kt of CO_2 (Figure 5A).

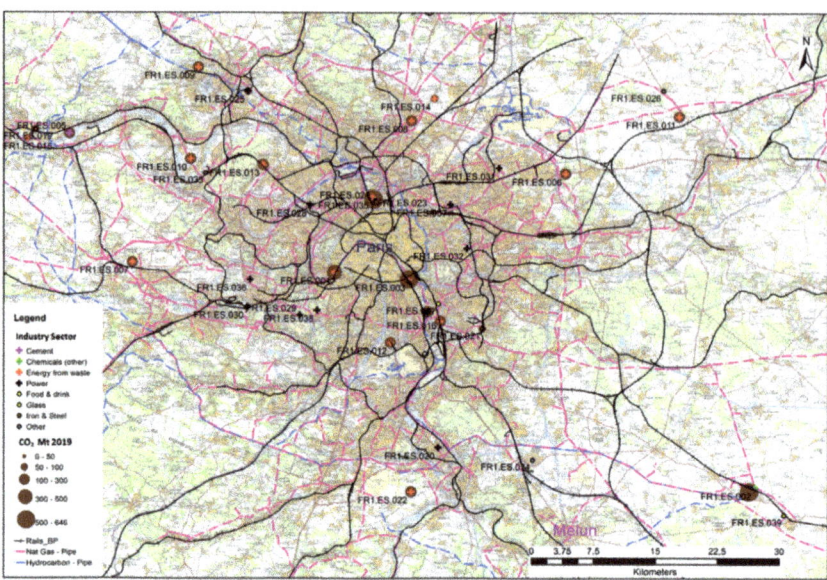

Figure 4. Geographical location of emission sources of the Paris Basin region. Color of symbol indicates the industrial sector and the size of symbols (in brown) the CO_2 emission in Mt in 2019 (data from [20]). The railways and existing natural gas and hydrocarbon pipelines are also indicated in the map (pipeline data from: GRTgaz and Data Gouv.).

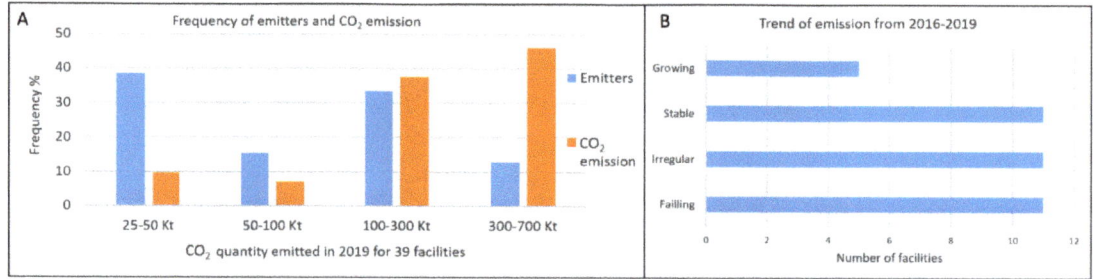

Figure 5. (**A**) CO_2 emitter facilities and CO_2 emissions frequency classed by CO_2 quantity ranges for 39 industrial facilities in the Paris Basin region. (**B**) Emission trend of these 39 facilities between 2016 and 2019. Data from [20].

In terms of the emission trends of CO_2 between 2016 and 2019 (Figure 5B), sources in the area show a decreasing trend in the emissions in recent years for eleven facilities, the other eleven facilities showed an irregular tendency and eleven others had a stable trend; only five facilities have increased their emissions (waste to energy and power). Twenty-eight facilities representing almost 80% of region's emissions (Figure 6A,B) are energy-from-waste and power (heat) facilities, which is consistent with the high-population pattern of the region. Another large part of the emissions come from one chemical facility (12%). CO_2 emissions from non-fossil-fuel combustion are an important proportion of the total emissions in the region, being estimated at up to 2.1 Mt/y, with 38% of the total emissions related to biomass combustion possibly raising the case for bioenergy with carbon capture and storage (BECCS). This alternative may be particularly interesting for the two large energy-from-waste plants south-west of Paris, FR1.ES.003 and FR1.ES.004, where CO_2 emissions from biomass are estimated, respectively, at 0.34 Mt/y and 0.28 Mt/y.

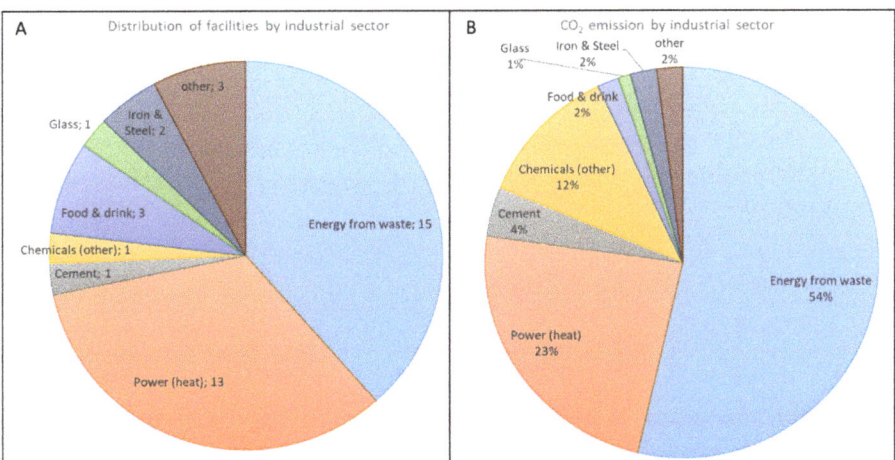

Figure 6. (**A**) Number of facilities by industrial sector and (**B**) the respective percentage of these sectors in the global amount of CO_2 emission of the region in 2019.

3.1.2. Storage Options

The Paris Basin in France is the largest onshore French sedimentary basin. First volumetric estimations of CO_2 storage capacity in the Paris Basin ranged from 800 Mt up to 27 Gt of CO_2. Two sedimentary formations, the Dogger Fm. of the Middle Jurassic and the Keuper Fm. of the Triassic, have known and good reservoir levels in the Paris Basin region [21].

The France Nord project (2013) [16] carried out detailed modeling of Keuper Fm., including the Donnemarie, Chaunoy and Boissy sedimentary members, which are mainly composed of silici-clastic sediments. Capacity estimates resulted in an assessment of the effective storage capacity, appropriate to the Tier 2 definition (Table 1). The resulting estimates relied on (i) refined geological and dynamic models in the investigated injection areas, (ii) scenarios for the commissioning of CO_2 injectors and (iii) preassessment of the long-term behavior and fate of the CO_2. The overall main objective was to reach 200 Mt of injected CO_2 in the reservoir over 40 years. Two areas of the Paris Basin were evaluated for storage in the Keuper Fm., one in the North of Paris—Keuper Nord—and another in the South of Paris—Keuper Sud (Figure 7). The effective storage capacity for Keuper Sud and Keuper Nord were, respectively, estimated to be up to 140 Mt and 81 Mt of CO_2 through dynamic modelling, after a 40 year period of injection in the—optimized combination of—injector wells. Water production was considered among the optimization scenarios but was finally dismissed for the estimates of the effective storage capacity, mostly due to the limited knowledge of the hydraulic connectivity in the deep sandstone formations.

The Dogger reservoir has been an important oil-reservoir target since the 1950s. Since the 1970s, the Dogger Fm. has progressively become the main geothermal aquifer exploited in the Paris region, with up to forty geothermal plants currently in operation. As a deep and productive aquifer (1500–2000 m depth), the hot groundwater (55 °C to 85 °C) is locally extracted from the Dogger Fm. to supply heat for up to 210,000 housing units. However, the performance of some wells has been affected by corrosion processes and the deposition of scale (i.e., secondary mineral precipitates). Moreover, the geothermal exploitation of the Dogger Fm. over decades has led to a gradual development of "cold bubbles" in the aquifer around and nearby the re-injection wells, progressively reducing the heat productivity over time.

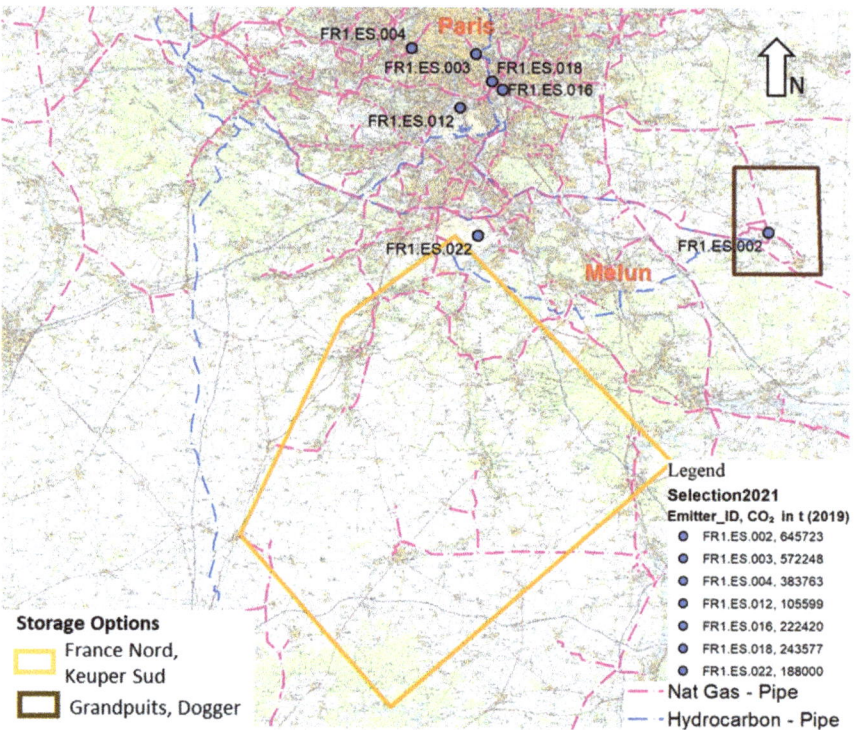

Figure 7. Storage options in both potential geological formations of the Paris Basin: the Keuper Fm. studied in the France Nord project and the Dogger Fm. at Grandpuits around the biggest CO_2 emitter of the region. Other CO_2 emission sources are also indicated by blue dots. These other sources are waste-to-energy facilities.

Apart from oil and gas and geothermal energy, the Dogger Fm. was previously studied with respect to the CO_2 storage capacity in a different research project. Within the France Nord project, the carbonates (limestones) of the Dogger Fm. displayed a limited thickness (<30 m) and a likely cemented primary porosity in the investigated areas [16]. In the GESTCO [22] and Geocapacity [23] projects, the theoretical capacity (Tier 1) of carbonate rocks was estimated to be up to 4320 Mt for a storage efficiency factor (SEF) of 6% and up to 1440 Mt for a SEF of 2%. As a result, the storage efficiencies were calculated to 6% and 2% (conservative approach) for each respective estimate. The density of CO_2 used for calculation was 400 kg/m^3, which corresponds to an approximate depth of 1400 m and a temperature of 70 °C. The significant discrepancy of the storage capacity between the identified structural traps, and the broad aquifer taken as a whole, illustrates the required necessity for large suitable geological structures in front of the CO_2 productions.

In order to study an alternative option for storage, and taking into account the very good potential of the Paris Basin in providing storage resources, a screening of the Grandpuits area (Figure 7) close to the emission source already capturing CO_2 (Emitter ID: FR1.ES.002 in Figure 4) explored possibilities to optimize and reduce CO_2 transport. This screening concerned technical geological aspects and a gap analysis of available data [24].

The Keuper Fm. in the Grandpuits area is deeper and is being exploited currently for oil-field production in the boundaries of the selected area (Figure 7). Oil fields are likely compartmentalized by sedimentary heterogeneity linked to the fluvial system or by faults. Seven old wellbores in the area reached the Keuper Fm. with few cores available. Keuper reservoirs are more than 2500 m deep in the Grandpuits area (Figure 7). The Dogger Fm. is also known as a good reservoir in this area. The top of the (Bathonian) Dogger reservoir around the emitter FR1.ES.002 is around 1700–1800 m deep. The geothermal potential linked to the high permeability and porosity of the Bathonian (Middle Dogger) is well known around Paris and Melun, which are located at 100 km and 20 km from the Grandpuits area, respectively. Nine old wellbores are available in the area, and many cores were drilled close to the investigated area.

The CO_2 storage capacity of the Dogger Fm. in the Grandpuits area using an analytical formula was estimated using the Equation (1).

$$M_{CO_2} = 1 \times 10^{-9} \times [(A \times 1 \times 10^6) \times h \times Phi] \times \rho_{CO_2} \times SEF$$
With
$$[(A \times 1 \times 10^6) \times h \times Phi] = \text{Reservoir Pore volume} \tag{1}$$

where:
M_{CO_2} is the CO_2 storage capacity of a prospect field as a mass (Mega ton). **A** is the total area of prospect reservoir (km^2).
h is the gross reservoir thickness (m).
Phi is the average porosity (decimal).
ρ_{CO_2} is the CO_2 density at reservoir storage conditions (kg/m^3).
SEF is the storage efficiency factor (decimal).

The total area (A), the gross reservoir thickness (h) and the average porosity (Phi) of the prospect reservoir for the Dogger Fm. in this area was obtained from the volume calculation of the reservoir pore volume using a porosity value of 10%. The geological model of the Dogger Fm. elaborated in the ANR project SHPCO2 (2010) [25] at the regional scale was used to calculate reservoir pore volume. The resolution of the model is low, therefore, a SEF of 2% was used as the efficiency factor. The capacity estimate as Tier 2 using the regional-scale geomodel of the Dogger is 165 Mt of CO_2 for a reservoir pore volume of 1.61×10^{10} rm^3.

3.1.3. Spatial Condition for Cluster and Network

The proximity of the French capital, Paris, makes the area well served by natural gas and hydrocarbon pipelines, rails and important road axes (Figure 4). Despite the good possibility of a transport network, two aspects should be considered: pipeline availability and railway connection and availability. The CO_2 sources are spread across the whole promising region (Figure 4); however, only at Grandpuits, with the chemical plant (FR1.ES.002) and at the south-western part of Paris, with the two largest energy-from-waste plants (FR1.ES.003 and FR1.ES.004), does there seem to exist the locus for the onset of an industrial CCS cluster based on large emitters aggregating other minor sources to build a common network at the south of Paris (Figure 8).

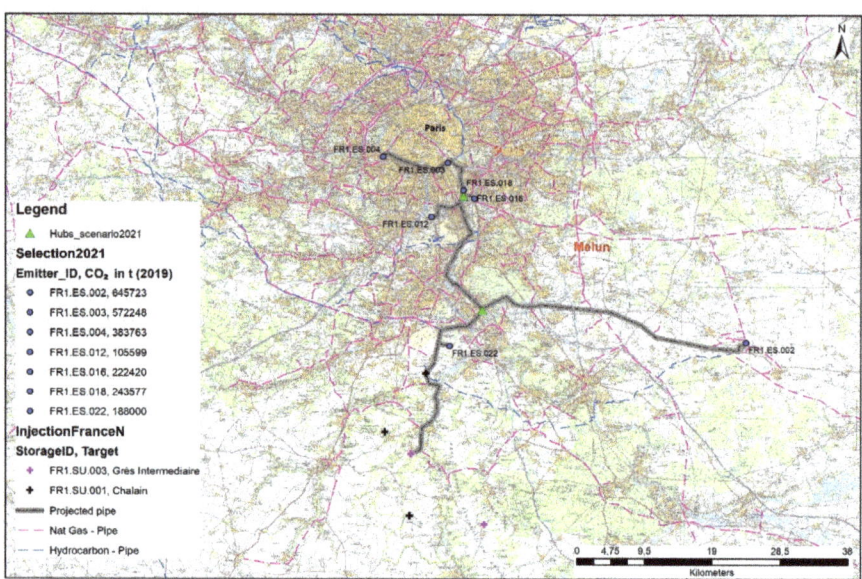

Figure 8. Spatial conditions for clusters and the transport network of CO_2. New pipelines following existing ones are considered "projected pipelines". Storage locations are represented by simulation injection points carried out in the France Nord project (FR1.SU.003 and FR1.SU.001) for the Keuper Fm. and for the Dogger Fm.; storage location is on the emission point FR1. ES.002.

3.1.4. National Low-Carbon Strategy and Emission Profile

The French National Low-Carbon Strategy (SNBC) serves as France's policymaking road map in terms of climate change mitigation [26]. The SNBC roadmap considered around 80 Mt CO_2 as inevitable or irreducible emissions by 2050. The carbon neutrality for 2050 therefore involves carbon being permanently stored to compensate for these emissions. Land-sector sink (forest and agricultural land) and CO_2 capture and storage (industrial processes) are permanent storage options with an estimate of around 15% for CCS in the schema.

The industrial sector accounted for ~15% of French GHG (green house gas) emissions in 2018. Around 84% of the sector's emissions operate under the European Union Emissions Trading Scheme (EU ETS). Industrial emissions correspond mainly to the combustion of fossil fuel or biomass required to produce energy and to the industrial process itself (i.e., chemical industries). The roadmap aims to reduce emissions of the industrial sector by 2050. Taking the emission levels of the year 2015 as the basis of reference for comparison, a gradual reduction in emissions of 35% and 81% are targeted by 2030 and 2050, respectively. According to the current state of knowledge, irreducible emissions in 2050 are related to nonenergy sectors. Apart from agriculture, the mineral production, primary metallurgy, certain chemical processes and fluorinated gases represent the main targeted emitters. The energy consumption is assumed to become entirely decarbonized. The waste-to-energy sector contributed ~3% of CO_2 emissions in 2018. The SNBC roadmap accounts to reduce the sector's emissions by 37% and 66% by 2030 and 2050, respectively, taking year 2015 as the basis of reference for comparison.

CCUS technologies could contribute to avoiding 15 $MtCO_2$ per year by 2050, including around 10 $MtCO_2$ of negative emissions with energy production installations using biomass. Such technology is referred as bio-energy with carbon capture and storage (BECCS). In 2009, the adaptation of the European CCS Directive established a legislative framework to facilitate the development of the CCUS technology.

3.1.5. Mapping Societal Aspects

The mapping of societal aspects aims to study the attitude towards CCUS development and its level of acceptance of selected members of the stakeholder group. Semistructured interviews collect (i) opinions about sources of concern, (ii) perceived benefits and risks (Table 2) and (iii) conditions for acceptance and perceived barriers, each with respect to the regional development of CCUS [17]. Preferences and expectations for energy futures among stakeholders were also raised and gathered during the interviews [17].

Table 2. List of cited benefits and risks established from interviews in the Paris Basin region. At the top, the most mentioned arguments for both categories, benefits and risks are listed.

Benefits	Risks
Environmental benefits (climate change mitigation, carbon neutrality in the industries in the region and pollution reduction in the region)	Economic viability (increase in cost and decrease in competitiveness for industries)
Economic development in the region (new industries, employment, investments and allowing power plants to keep working)	Environmental risks (risk of underground storage)
Other (financial benefits for companies, beneficial for company image and promotion of a circular economy)	Social impacts (public opposition)

Twelve interviews were carried out in the Paris Basin region with regional and national stakeholders from: industry (three people); politics and policies (four people); research and education (three people); and support organization (two people). The profile of stakeholders identified for the interviews were based on the analysis of actor structures in the innovation system for CCUS [6].

Three key ideas arose from interviews in the Paris Basin:

1. The majority of interviewees considered CCUS technologies as a potential option to fight against climate change.
2. Interviewees often underline that CCUS is only one option among other solutions to reduce carbon dioxide emissions.
3. CCU is particularly well-perceived by interviewees and appears to them to offer higher potential than CCS, regardless of the current limited volumes concerned by CO_2 valorization.

3.2. Economic KPIs

The economic simulation of the region's scenario gives the main economic key performance indicators (KPIs) of CCS business cases for the period from now to the Horizon 2050. The volume of CO_2 avoided and/or removed at the regional scale and the costs associated illustrate the technoeconomic potential of the CCS technology. The regional scenarios evaluate cost differences between investing in CCUS or paying carbon penalties related to compliance with the EU ETS, giving an estimate of the breakeven price of CO_2 for each of the studied scenarios. The scenarios are elaborated for the Horizon 2050 considering the construction time for the infrastructures as capture systems, drilling wellbores for injection and monitoring and conditioning stations for transport (compressor, pumping station, etc.).

The scenario is based on the three largest carbon emitters in the south of Paris, since the storage site is located in the southern part of the region. None of the CO_2 utilization technologies were identified in the region. The fertilizer plant in Grandpuits (emitter FR1.ES.002) emitted 646 ktCO_2 in 2019. It is located in the south-eastern part of the Paris Basin region in an agricultural area, in the vicinity of the closed Grandpuits refinery. The main part of the emissions of the plant come from the SMR unit on site, which produces H_2 for an ammonia synthesis process. As methane reforming produces a H_2/CO_2 mix, the plant already has a carbon capture installation to remove CO_2 and produce pure H_2. Actually, a part of the captured CO_2 is sold to industrial gases companies, but the largest part is released into the atmosphere. Consequently, approximately 360 $ktCO_2$ would be already available for storage.

The installation in Ivry (emitter FR1.ES.003) is the biggest waste incineration plant of the Paris area. In 2019, 661,593 tons of waste were treated with the production of 20,393 MWh of electricity and 1,124,190 MWh of vapor injected into the Parisian heating network (CPCU). The corresponding carbon emissions amounted to 572 $ktCO_2$. However, the plant will be replaced by 2023–2024 by a new installation currently under construction on the same site. Anticipating the waste reduction objectives, this new plant will have half the capacity of the current one (a valorization of 350,000 tons of waste per year). The carbon emissions of this new facility should broadly amount to 300 kt/y from 2024.

The waste valorization plant in Issy-les-Moulineaux (emitter FR1.ES.004) is the most recent incineration plant in the Paris area, as it started up in 2007. It has a capacity of 510 000 tons of waste per year. In 2019, the plant incinerated 469,097 tons of waste, emitted 384 $ktCO_2$, produced 705,379 MWh of steam for the CPCU urban heating network and sold 34,016 MWh of electricity. The area around this emitter has high demography density. There is no physical place to install a current CO_2 capture system for this facility.

Features and carbon emissions of these three sites are gathered in the table below (Table 3). A total of 25.2 Mt of CO_2 could be captured from 2027 to 2050 with these three emitters, including 7.7 Mt of CO_2 from biomass.

Table 3. Industries considered in the scenario with their features and carbon emissions detailed after capture.

Industries	Sector	Location	Capture Start Year	Annual CO_2 Emissions Considered—$MtCO_2/y$	CO_2 Capture Rate (%)	Annual CO_2 Captured (Mt/y)	Total CO_2 Captured (Mt/y)	Part of CO_2 Captured from Biomass (Mt/y)
E#01 (FR1.ES.002)	Chemistry	Grandpuits	2027	0.65	n/a	0.36	9.7	0.0
E#02 (FR1.ES.003)	Energy from waste	Ivry-sur-Seine	2030	0.30	0.90	0.27	7.4	3.7
E#03 (FR1.ES.004)	Energy from waste	Issy-les-Moulineaux	2032	0.38	0.85	0.33	8.1	4.0

Costs related to the scenario were calculated for each stage of the chain: capture, transport and storage for the three installations. Global CCUS CAPEX and OPEX for each installation are summarized in Table 4. The excess of energy consumption for capturing CO_2 is given in TJ.

Table 4. Summary of CAPEX, OPEX and Energy consumption of CCUS for the three selected emitters of the Paris Basin region.

Industries with Capture Medium Term	CAPEX (M EUR)	Fixed OPEX (M EUR)	Variable OPEX (M EUR)	Total Costs (M EUR)	Excess of Energy Consumption for Capture (TJ)
E#01 (FR1.ES.002)	4.1	2.9	1.3	8.3	n/a
E#02 (FR1.ES.003)	76.4	360.1	0.3	436.8	24,413.0
E#03 (FR1.ES.004)	84.9	362.2	0.4	447.5	28,255.0

Table 5 shows the analysis of EU ETS allowance for regional expenses of the scenario with CCUS and without CCUS. The energy costs for the capture technology are taken into account in terms of TWh/year using current costs of electricity and its evolution for 2050. The regional expense in ETS allowance without CCUS is EUR 2 270 M EUR for the scenario from 2027 to 2050, whereas costs of CCUS (including remain ETS costs) are of EUR 1131 MEUR for the period. The CCUS costs represents around half of the ETS costs of allowances for the scenario without CCUS.

Table 5. Analysis of EU ETS allowance in the scenario and energy consumption.

EU ETS Parameters (EUR/tCO$_2$)	Price of Allowances in 2025	70.1
	Price of Allowances in 2045	212.4
Whole regional expense without CCUS (M EUR)	ETS costs without CCUS	2270.0
Whole region expense with CCUS (M EUR)	ETS costs with CCUS and remaining emissions	89.8
	Costs of CCUS	1041.2
	TOTAL costs with CCUS	1131.0

The CCUS value chain of the scenario is calculated in terms of EUR/t of CO$_2$ avoided (Table 6), taking into account the EU ETS analysis of Table 5. The breakeven CO$_2$ price of the scenario is 43 EUR/t of CO$_2$ to have a positive economic impact of CCUS in the period between 2027 and 2050. The breakeven of CO$_2$ price of the CCUS value chain without the emitter FR1.ES.002, which is already capturing CO$_2$, gives a price of around 70 EUR/t of CO$_2$ avoided for 16 Mt of CO$_2$ captured and stored.

Table 6. Analysis of CCUS system in terms of EUR/tCO$_2$ avoided using the EU ETS parameters of Table 5.

CCS Value Chain (EUR/tCO$_2$ Avoided)		−42
CAPEX (EUR/tCO$_2$ avoided)	Total per block	−8.3
	Cost of Capture	−2.9
	Cost of Transport	−1.1
	Cost of Storage	−4.3
OPEX (EUR/tCO$_2$ avoided)	OPEX per block	−33.4
	Cost of Capture	−24.7
	Cost of Transport	−0.6
	Cost of Storage	−8.1
Transport cost (EUR/tCO$_2$ transported)		−1.1
Utilization (income from CO$_2$ sales) (M EUR)		0
EU ETS credit savings in the region (M EUR)		2180

Waste-to-Energy Challenge

Although the waste-to-energy (WtE) plants are not currently included in the EU ETS in France, these facilities are important emission sources in the region, as well sources of heat energy to houses and buildings in the vicinity. Furthermore, these facilities have great potential for providing negative emissions, as part of the CO$_2$ emission comes from burning biomass. The emission trend of WtE facilities around high demographic zones is uncertain and could likely increase by 2030 and 2040. Most European WtE plants emit from 100 to 500 ktCO$_2$ yearly, for a production of heat and power equivalent to about 90 and 39 TWh, respectively. The WtE plants are mainly located in urban areas or in proximity, usually being the biggest CO$_2$ emission sources in these areas. According to carbon limits [27], emissions from the incineration of waste are irreducible once the waste streams have been created, and CCS is the abatement technology applicable. European statistics on incinerated waste showed an increase of 30% from 2006 to 2016. The European waste-to-energy association (CEWEP) analyzed the EU recycling targets for 2035 [28] and estimated a residual nonrecycled waste stream of 142 Mt/year of waste in 2035. This amount of waste at the European scale corresponds to an increase of about 40 Mt of current incineration capacities.

Regarding the perspective of CO$_2$ emission from waste-to-energy facilities in France, the number of incinerators decreased since 2004, passing from 131 facilities to 121 facilities in 2018, whereas the quantity of waste showed a slight increase of ~1.2 Mt, with 14.7 Mt of waste being burned in 2018 [29]. Demography in the Ile-de-France department increased by 0.4% between 2013 and 2018, passing from 11,959,807 habitants to 12,213,447. The WtE plants are currently working at 94% of their legal capacities. Landfilling options in France counted for 18 Mt of nondangerous waste in 2018 [29]. The reduction in waste quantity sent every year to WtE plants seems to be the major challenge, as WtE plants are an alternative to landfilling options which become unsustainable and uneconomic while the living standard and waste production grows [30].

CO_2 capture technology for waste-to-energy plants uses similar technology as those used for coal-fired power stations. Some examples in the Netherlands and Norway showed the feasibility of capture systems for WtE plants [30]. The WtE plant in Twence, the Netherlands, converts 1 million tons of waste to energy every year [31]. The Ministry of Economic Affairs and Climate Policy is providing a subsidy of 14.3 million for the capture system. In Norway, the WtE plant Klemetsrud is seeking to capture 400,000 Mt/year of CO_2, corresponding to 90% of the plant's emissions by 2025. The Klemetsrud plant has a capacity to process around 350 Kt of waste and emits 385 $ktCO_2$ per year [32]. Both of these projects demonstrate the applicability and feasibility of current CO_2 capture technologies to WtE plants with similar capacities of waste processing and CO_2 emissions as the main WtE plants of the Paris Basin region, the emitter FR.ES.003 (730 kt of waste in 2017), FR.ES.005 (650 kt of waste in 2017) and FR.ES.006 (510 kt of waste in 2017). A technoeconomic analysis of the CCS implementation for the WtE plant in Klemetsrud estimated a P50 cost of 153 EUR/t of CO_2 avoided for the capture part of CCS chain, 208 EUR/t of CO_2 avoided including different parts of the chain CCS (steam consumption, energy, conditioning, transport and storage) and 186 EUR/t of CO_2 including CCS with EOR.

In France, the WtE plants pay several taxes related to polluting activities. The inclusion of WtE facilities in EU ETS is economically unfeasible today in France without a review of the current and future taxes applied to WtE plants as a public service. The TGAP (general tax for polluting activities) is an important tax concerning the tons of incoming of nondangerous waste received for storage and incineration processing. In 2016, 86.4% of incoming waste was household and similar waste. It is important to notice an increase of incoming waste refused from the waste treatment and disposal centers. The TGAP is paid by ton-of-waste received and its amount is a function of three factors: to have an ISO 50001 certificate on energy management systems; an NOx content in the emissions of less than 80 mg/Nm^3; an energy utilization higher than 0.65 of the energy outturn. This tax is increasing quickly, from 3 EUR/t to 11 EUR/t in the past two years (2021 and 2022). In 2025, the three categories defining the amount of the tax will be replaced by a fixed amount of 15 EUR/t of waste for any facility.

The CCS for French WtE plants could drastically reduce the CO_2 emissions around high demographic areas and provide negative emissions. Although, without financial compensation or government support, the inclusion of WtE facilities in the EU ETS means adding another tax to citizens related to polluting activities. The main difference between TGAP and EU ETS is the environmental benefit of installing BECCS to avoid CO_2 emissions.

4. Discussion and Conclusions

The geographic location of sources is the first concern in the elaboration of long-term CCS scenarios. Three important emission sources are located at the south–southwest of the Paris metropolis with a low demographic area in between (Figure 4). The high demographic area around the Paris metropolis emitters would imply installing CO_2 capture systems using current technologies in a limited geographic area. Studied storage possibilities are located in the south of the Ile-de-France Department.

The key performance indicators of a CCS scenario in the Paris Basin region for a deployment between 2027 and 2050 indicates a low CO_2 cost per ton/avoided between 43 EUR/t and 70 EUR/t, for a cumulated total of 25 Mt and 16 Mt, respectively, of CO_2 captured and stored for 26 years, including 7.7 Mt of CO_2 from biomass (potential negative emissions). The low CO_2 price for the scenario would be seen as an opportunity to apply CCS in the regional scale to reach regional objectives and the ambition of CO_2 reduction for

Horizon 2050. CCS should be seen as a regional option for decarbonizing industries and not as an individual facility option.

Despite the clear statement of the SNBC (French National Low Carbon Strategy) about the benefit of CCS for irreducible emissions from industries and the benefit of deploying BECCS (negative emissions), waste-to-energy (WtE) plants are not included in the EU ETS system in France. At the perspective of reducing CO_2 emission in the Paris Basin region, the deployment of CCS and its environmental benefit for WtE installations should be considered by the French authority. Today, without the support of the government as in the Netherlands and Norway, the WtE installations are unable to consider CCS as a solution for decarbonizing the territories around big cities such as Paris, despite the low cost of about 70 EUR/t of CO_2 avoided at the regional scale.

The biggest emitter of the Paris Basin region, the fertilizers plant (FR1.ES.002), is already capturing CO_2 from its industrial process and venting it to the atmosphere. This configuration places this emitter as the candidate to launch CCS technology in the region, as CO_2 is available. The capture system represents half of the total costs of CAPEX and OPEX for CCS in this region. The geological storage capacities of Dogger Fm. around this emitter are an effective capacity (Tier 2) estimate of 165 Mt of CO_2 and seem to be enough to store its emissions of 9.7 Mt cumulated for almost 30 years. The area around the emitter is mostly rural, with the land being used for wheat crops (Figure 7). The oil and gas industry has been present for decades. Three licenses of hydrocarbon exploitation in the Keuper Fm. are being operated around the Grandpuits area, with one licensing in the Dogger Fm. Although these hydrocarbon fields are currently operating, they should stop their research and exploitation by 1 January 2040 [33]. These hydrocarbon fields being depleted would provide additional storage resources for the Horizon 2050.

In terms of infrastructures, this area is well-served by hydrocarbon pipelines, which have been exploited by the oil and gas industry since 1950. The development of a pilot-scale CCS in this area would become a notable CCS project with a perspective for large-scale development. The CCS pilot-scale project would demonstrate to local and national stakeholders the feasibility and environmental impact of the technology in terms of reducing emissions and associated risks. The reusing of oil and gas infrastructures and the high potential of geological storage for both resources, deep saline aquifers and depleted hydrocarbon reservoirs, make this location promising for further CCS development aiming to decarbonize industries around the Paris metropolis.

Although more research is needed concerning the social aspects of CCS technology and how it is perceived by national and regional stakeholders, a first overview of CCUS perception showed a positive attitude towards the technology, which was recognized as one of the tools to reduce CO_2 emissions.

Author Contributions: Conceptualization, F.M.L.V. and I.G.; methodology and data gathering, F.M.L.V., I.G., S.B.R. and F.A.M.; writing, F.M.L.V. and I.G.; review and editing, F.M.L.V., I.G. and F.A.M. All authors have read and agreed to the published version of the manuscript.

Funding: This research was funded by the European Union's Horizon 2020 program. STRATEGY CCUS project. Grant Agreement: No 837754. https://www.strategyccus.eu/ (accessed on 1 March 2022).

Institutional Review Board Statement: Not applicable.

Informed Consent Statement: Informed consent was obtained from all subjects involved in the study.

Data Availability Statement: Pipeline data in France are available in https://www.data.gouv.fr/fr/datasets/carte-du-reseau-de-transport-de-gaz-sur-la-france-metropolitaine-1/ (accessed on 1 March 2022).

Acknowledgments: The authors acknowledge discussion of the French team under actions of the EU H2020 project PilotSTRATEGY (grant Agreement: 101022664), which carried out the gap analysis of geological data in the area of Grandpuits, close to the largest emitter of the Paris Basin region.

Conflicts of Interest: The authors declare no conflict of interest.

Appendix A

Table A1. Quantity reported and the emission trend between 2016 and 2019.

Emitter ID	Facility Name	Industry Sector	CO_2 from Biomass Combustion (Ton)	CO_2 Reported (Ton)	Year Reported	Emission Trend (2016–2019)
FR1.ES.002	Borealis Grandpuits	Chemicals (other)		645,723	2019	Irregular
FR1.ES.003	IVRY PARIS XIII	Energy from waste	330,683	572,248	2019	Stable
FR1.ES.004	CPCU chaufferies de ST-OUEN I et ST-OUEN II	Power	150,949	522,182	2019	Stable
FR1.ES.005	DALKIA WASTENERGY	Energy from waste	231,791	416,366	2019	
FR1.ES.006	TSI	Energy from waste	217,779	383,763	2019	Growing
FR1.ES.007	SNC Cogé VITRY	Power		243,577	2019	Stable
FR1.ES.008	Ciments Calcia usine de Gargenville	Cement	100,275	224,897	2019	Falling
FR1.ES.009	VALO'MARNE	Energy from waste	123,700	222,420	2019	Stable
FR1.ES.010	SEMARIV-CITD	Energy from waste	107,000	188,000	2018	Falling
FR1.ES.011	CPCU ST-OUEN III	Power		163,579	2019	Stable
FR1.ES.012	SIAAP Site Seine Aval	Energy from waste	143,847	144,299	2019	Falling
FR1.ES.013	SAREN	Energy from waste	81,893	143,672	2019	Growing
FR1.ES.014	Routière de l'Est Parisien (ISDND de Claye Souilly)	Energy from waste	140,933	140,933	2019	Falling
FR1.ES.015	AUROR'ENVIRONNEMENT	Energy from waste	78,501	137,944	2019	Falling
FR1.ES.016	CVD Thiverval-Grignon	Energy from waste	76,000	133,000	2018	Irregular
FR1.ES.017	AZALYS	Energy from waste	67,860	119,053	2019	Falling
FR1.ES.018	SOMOVAL	Energy from waste	60,085	106,088	2019	Stable
FR1.ES.019	GENERIS—Site de Rungis	Energy from waste	60,033	105,599	2019	Falling
FR1.ES.020	BOUQUEVAL ENERGIE	Energy from waste	86,736	86,736	2019	Falling
FR1.ES.021	SARP Industries	Energy from waste		72,764	2019	Irregular
FR1.ES.022	SAM MONTEREAU	Iron & Steel		68,948	2019	Irregular
FR1.ES.023	SGD Usine de SUCY EN BRIE	Glass		56,851	2019	Stable
FR1.ES.024	CYEL	Power	32,042	54,489	2019	Irregular
FR1.ES.025	ALPA	Iron & Steel		50,398	2019	Growing
FR1.ES.026	KNAUF Plâtres	Other		48,995	2019	Stable
FR1.ES.027	BIO SPRINGER	Food & drink		45,223	2019	Stable
FR1.ES.028	GRAND PARIS SUD ENERGIE POSITIVE	Power		44,095	2019	Growing
FR1.ES.029	ENERTHERM Noël Pons	Power		40,437	2019	Irregular
FR1.ES.030	VELIDIS Chaufferie Vélizy V3	Power		39,226	2019	Falling
FR1.ES.031	VERSEO	Power		37,512	2019	Irregular
FR1.ES.032	chaufferie zup de fontenay	Power		35,777	2019	Falling
FR1.ES.033	SAFRAN AIRCRAFT ENGINES	Other		35,666	2019	Irregular
FR1.ES.034	Chaufferie de Parly 2	Power		32,344	2019	Stable
FR1.ES.035	PEUGEOT CITROËN POISSY SNC	Other		31,713	2019	Falling
FR1.ES.036	SEMECO (et IDEX ENERGIES)	Power		30,916	2019	Stable
FR1.ES.037	chaufferie zup de sevran	Power	16,938	30,738	2019	Irregular
FR1.ES.038	LESAFFRE FRERES	Food and drink		27,850	2019	Irregular
FR1.ES.039	ENGIE Chaufferie de Meudon	Power		26,585	2019	Growing
FR1.ES.040	OUVRE FILS Sucrerie et Distillerie	Food and drink		23,812	2019	Irregular

References

1. Carbon European Trade System Viewer. Available online: https://ember-climate.org/data/carbon-price-viewer/ (accessed on 1 March 2022).
2. Oltra, C.; Upham, P.; Riesch, H.; Boso, A.; Brunsting, S.; Dütschke, E.; Lis, A. Public Responses to CO_2 Storage Sites: Lessons from Five European Cases. *Energy Environ.* **2012**, *23*, 227–248. [CrossRef]
3. Anderson, C.; Schirmer, J.; Abjorensen, N. Exploring CCS community acceptance and public participation from a hu-man and social capital perspective. *Mitig. Adapt. Strateg. Glob. Chang.* **2012**, *17*, 687–706. [CrossRef]
4. Wallquist, L.; Orange Seigo, S.; Visschers, V.; Siegrist, M. Public acceptance of CCS system elements: A conjoint measurement. *Int. J. Greenh. Gas Control.* **2012**, *6*, 77–83. [CrossRef]
5. Hammond, J.; Shackley, S. Towards a public communication and engagement strategy for carbon dioxide capture and storage projects in Scotland. In *Scottish Centre for Carbon Capture Working Paper*; British Geological Survey: Edinburgh, UK, 2010.
6. Dütschke, E.; Wesche, J.; Oltra, C.; Prades, A.; Álvarez, F.C.; Carneiro, J.F.; Gravaud, I.; Vulin, D. Stakeholder Mapping Report WP3-Deliverable 3.1, Report, 2019. EU H2020 STRATEGY CCUS Project 837754. Available online: https://cordis.europa.eu/project/id/837754/results/fr (accessed on 15 March 2022).
7. Rothkirch, V.J.; Ejderyan, O. Anticipating the social fit of CCS projects by looking at place factors. *Int. J. Greenh. Gas Control* **2021**, *110*, 103399. [CrossRef]
8. Page, B.; Turan, G.; Zapantis, A.; Burrows, J.; Zhang, T. The Global Status of CCS. In *Technical Report*; Global CCS Institute: Melbourne, Australia, 2020. Available online: https://www.globalccsinstitute.com/resources/global-status-report/ (accessed on 1 January 2022).
9. Scottish Carbon Center (SCCS) Map of European Map of CCS Facilities. Available online: https://www.sccs.org.uk/expertise/global-ccs-map (accessed on 1 March 2022).
10. Bentham, M.; Mallows, T.; Lowndes, J.; Green, A. CO_2 STORage Evaluation Database (CO_2 Stored). The UK's online storage atlas. *Energy Procedia* **2014**, *63*, 5103–5113. [CrossRef]
11. Halland, E.K.; Johansen, W.T.; Riis, F. Geographical data as shape and rasterfiles in zip-archive, Norwegian North Sea. In *CO_2 Storage Atlas North Sea. Norwegian Petroleum*; Directorate: Stavanger, Norway, 2014; p. 72. [CrossRef]
12. Kearns, D.; Liu, H.; Consoli, C. *Technology Readiness and Costs of CCS*; Global CCS Institute: Brussels, Belgium, 2021. Available online: https://www.globalccsinstitute.com/wp-content/uploads/2021/04/CCS-Tech-and-Costs.pdf (accessed on 1 March 2022).
13. Cavanagh, A.J.; Wilkinson, M.; Haszeldine, R.S. Bridging the Gap, Storage Resource Assessment Methodologies. In *Methodologies for Cluster Development and Best Practices for Data Collection in the Promising Regions*; EU H2020 STRATEGY CCUS Project 837754, Report; Brownsort, P.A., Cavanagh, A.J., Wilkinson, M., Haszeldine, R.S., Eds.; 2020; p. 67. Available online: https://cordis.europa.eu/project/id/837754/results/fr (accessed on 15 March 2022).
14. Brownsort, P.A. Part 1, Industrial CCUS Clusters and CO_2 transport systems: Methodologies for characterisation and definition. In *Methodologies for Cluster Development and Best Practices for Data Collection in the Promising Regions*; EU H2020 STRATEGY CCUS Project 837754, Report; Brownsort, P.A., Cavanagh, A.J., Wilkinson, M., Haszeldine, R.S., Eds.; 2020; p. 68. Available online: https://cordis.europa.eu/project/id/837754/results/fr (accessed on 15 March 2022).
15. Carneiro, J.F.; Mesquita, P. *Key Data for Characterising Sources, Transport Options, Storage and Uses in Promising Regions*; EU H2020 STRATEGY CCUS Project 837754, Report; 2020; p. 146. Available online: https://cordis.europa.eu/project/id/837754/results/fr (accessed on 15 March 2022).
16. France Nord Project. Final Report Confidential. 2013. Available online: https://www.ademe.fr/france-nord (accessed on 1 December 2020).
17. Oltra, C.; Preuß, S.; Gérman, S.; Gravaud, I.; Vulin, D. *Stakeholders' Views on CCUS Developments in the Studied Regions*; EU H2020 STRATEGY CCUS Project 837754. Report; 2020; p. 96. Available online: https://cordis.europa.eu/project/id/837754/results/fr (accessed on 15 March 2022).
18. Coussy, P. *Deliverable D5.2: Description of CCUS Business Cases in Eight Southern European Regions*; EU H2020 STRATEGY CCUS. Project 837754. Report; 2021; p. 133. Available online: https://cordis.europa.eu/project/id/837754/results/fr (accessed on 15 March 2022)(after European Commission Approval).
19. Berenblyum, R. *Deliverable D5.1 Elaboration and Implementation of Data Collected of the Business Case for Each Region*; EU-H2020 STRATEGY CCUS Project 837754. Deliverable Report; 2021; p. 75, (confidential).
20. IREP—French Register of Polluting Emissions. 2021. Available online: https://www.georisques.gouv.fr/risques/registre-des-emissions-polluantes (accessed on 1 December 2021).
21. Veloso, F.M.L. *Maturity Level and Confidence of Storage Capacities Estimates in the Promising Regions*; EU H2020 STRATEGY CCUS Project 837754, Deliverable Report; 2021; p. 125. Available online: https://cordis.europa.eu/project/id/837754/results/fr (accessed on 15 March 2022).
22. Robelin, C.; Matray, J.M. Geology of Paris Basin reservoirs. In *Feasibility of CO_2 Storage in Geothermal Reservoirs Example of the Paris Basin—France. GESTCO Report*; Bonijoly, D., Ed.; 2003; pp. 31–60. Available online: https://cordis.europa.eu/project/id/ENK6-CT-1999-00010/fr (accessed on 15 March 2022).
23. Geocapacity European Project (SES6-518318–6th Framework Programme): D16 WP2 Report Storage Capacity. 2009. Available online: http://www.geology.cz/geocapacity/publications (accessed on 1 December 2020).

24. Veloso, F.M.L.; Estublier, A.; Bonte, D.; Mathurin, F.; Frey, J.; Maury, J.; Poumadère, M.; Stephant, S. *Target Area of Paris Basin Region—FR: Data Inventory, Seismic Target and GAP Analysis*; EU H2020 PilotSTRATEGY project 101022664, Report; 2021; p. 29. Available online: https://cordis.europa.eu/project/id/101022664 (accessed on 15 March 2022) (after Europe Commission approval).
25. ANR-SHPCO29. French National Agency—Simulation Haute Performance du Stockage Géologique de CO_2. 2012. Available online: https://anr.fr/Colloques/Energies2012/presentations/SHPCO2.pdf (accessed on 1 March 2022).
26. French National Low Carbon Strategy. 2021. Available online: https://www.ecologie.gouv.fr/sites/default/files/en_SNBC-2_complete.pdf (accessed on 1 January 2022).
27. Carbon Limits. The Role of Carbon Capture and Storage in a Carbon Neutral Europe. Assessment of the Norwegian Full-Scale Carbon Capture and Storage Project's Benefits. 2020. Available online: https://www.regjeringen.no/contentassets/971e2b1859054d0d87df9593acb660b8/the-role-of-ccs-in-a-carbon-neutral-europe.pdf (accessed on 1 March 2022).
28. Directive (EU) 2018/851 on Waste Amending EU Directive 2008/98/EC. Available online: https://eur-lex.europa.eu/legal-content/en/TXT/PDF/?uri=CELEX:32018L0851&from=EN (accessed on 1 January 2022).
29. ADEME, Le Traitement des Déchets Ménagers et Assimilés—ITOM. Available online: https://librairie.ademe.fr/dechets-economie-circulaire/4336-le-traitement-des-dechets-menagers-et-assimiles-itom.html (accessed on 1 March 2022).
30. Kearns, D.T. *Waste-to-Energy with CCS: A pathway to Carbon-Negative Power Generation*; Perspective Global CCS Institute. 2019. Available online: https://www.globalccsinstitute.com/wp-content/uploads/2019/10/Waste-to-energy-with-CCS_A-pathway-to-carbon-negative-power-generation_Oct2019-4.pdf (accessed on 1 March 2022).
31. Aker Carbon Capture. Aker Carbon Capture Ready to Start CCUS Project at Twence's Waste-to-Energy Plant in the Netherlands. Press Release. 2021. Available online: https://akercarboncapture.com/?cision_id=9DDF1859C78B5320 (accessed on 1 January 2022).
32. KlimaOslo. Sustainable Waste Management for a Carbon Neutral Europe. Press Release. 2021. Available online: https://www.klimaoslo.no/2021/02/26/the-klemetsrud-carbon-capture-project/ (accessed on 1 March 2022).
33. French Decree. 2017. Available online: https://www.legifrance.gouv.fr/jorf/id/JORFTEXT000036339396 (accessed on 1 March 2022).

Article

Understanding the Anomalous Corrosion Behaviour of 17% Chromium Martensitic Stainless Steel in Laboratory CCS-Environment—A Descriptive Approach

Anja Pfennig [1,*] and Axel Kranzmann [2]

[1] Department of Engineering and Life Sciences, University of Applied Science—HTW Berlin, 12459 Berlin, Germany
[2] BAM Federal Institute of Materials Research and Testing, 12205 Berlin, Germany; axel.kranzmann@bam.de
* Correspondence: anja.pfennig@htw-berlin.de; Tel.: +49-5019-4231

Abstract: To mitigate carbon dioxide emissions CO_2 is compressed and sequestrated into deep geological layers (Carbon Capture and Storage CCS). The corrosion of injection pipe steels is induced when the metal is in contact with CO_2 and at the same time the geological saline formation water. Stainless steels X35CrMo17 and X5CrNiCuNb16-4 with approximately 17% Cr show potential as injection pipes to engineer the Northern German Basin geological onshore CCS-site. Static laboratory experiments (T = 60 °C, p = 100 bar, 700–8000 h exposure time, aquifer water, CO_2-flow rate of 9 L/h) were conducted to evaluate corrosion kinetics. The anomalous surface corrosion phenomena were found to be independent of heat treatment prior to exposure. The corrosion process is described as a function of the atmosphere and diffusion process of ionic species to explain the precipitation mechanism and better estimate the reliability of these particular steels in a downhole CCS environment.

Keywords: corrosion; steel; high alloyed steel; corrosion mechanism; CCS; carbon capture and storage

1. Introduction

The sequestration of carbon (carbon capture and storage (CCS [1,2]) comprises the sequestration, transport and injection of emission gasses into a deep geological layer. This technique is well acknowledged to mitigate climate change. Safe deep onshore or offshore geological layers—mainly saline aquifers (brine)—offer storage sites for emission gases that arose mostly from combustion processes of cement production or power plants [1–4]. Due to the highly corrosive environment, especially at the phase boundaries of metal, CO_2 and saline aquifer water injection pipe steels are highly exposed to CO_2 corrosion [3–9] directly dependent on multiple criteria [5,6,10–31]:

- Temperature (60 °C is a severe damaging temperature region);
- CO_2 partial pressure;
- alloy composition;
- heat treatment of steels (austenitizing temperature and durance as well as annealing
- element distribution in the corrosive media);
- purity of alloy and aquifer media;
- conditions of flow;
- pressure during injection and;
- protecting corrosion scales.

Alloy composition [22] and heat treatment [23–30] are the main determining factors influencing corrosive phenomena. Surface corrosion is mainly reduced by high nickel and chromium percentages [26,27]. Local corrosion of martensitic steels is reduced through the presence of retained austenite [26], the higher temperature during austenitizing [28–30] and annealing [22,23,28]. Surface corrosion recedes as a function of increasing austenitizing time [16] but is neglectable regarding local corrosion [32–34], compared to the ferritic or

ferritic-bainitic microstructure martensitic steels containing carbon and manganese, which show low corrosion resistance because grain boundaries are highly reactive in NaCl containing H_2S [31]. Different authors describe an immediate dependence of the corrosion behaviour on the surface condition after machining processes [35–40]. In general, the corrosion resistance increases with receding vertical height on the surface for carbon steel [35,36], austenitic stainless steel and ferritic stainless steel when the roughness exceeds 0.5 µm [38] and after shot peeing [39]. The initial surface roughness, however, has less effect than the relative humidity. In terms of protection and inhibition of internal pipeline corrosion, it is more beneficial to decrease the humidity than the initial surface roughness [37].

The potential of stainless steel X35CrMo17 (1.4122) is discussed and compared to the results of earlier studies with different high alloyed steels [16,17,41–43]. It is a heat treatable chromium steel that is highly resistant to a high number of organic and inorganic acids because of the high percentage of molybdenum. X35CrMo17 shows fairly good resistance to salt water. Moreover, its resistance to crevice corrosion up to 500 °C (working temperature) is improved.

The hardened martensitic stainless steel precipitation contains about 3% copper X5CrNiCuNb16-4 (1.4542, AISI 630) and is characterized by small copper precipitates that are distributed within the matrix which ensure the mechanism of precipitation hardening [44]. Small niobium and copper carbides are embedded in the martensitic bcc-structured microstructure [41]. This increases the alloys' strength and permits excellent mechanical properties and, at the same time, good resistance against corrosive attack [16]. However, martensitic 1.4542 is prone to stress corrosion cracking (SCC) and the martensitic microstructure is less corrosion resistant than the solution-treated microstructure that (as a drawback) shows reduced strength [45–52]; (Note that the resistance against corrosive attack is higher although the strength is low in the solution-treated state) [51,52]).

Surface corrosion rates at 60 °C are generally independent of heat treatment prior to exposure at ambient pressure and neglectable at 100 bar. Corrosion rates below 0.005 mm/year are reported after long exposure to CCS environment (8000 h) [16]. The corrosion behaviour is rather attributed to chromium content and atmosphere than heat treatment.

Low corrosion rates in the liquid (CO_2-saturated aquifer water) and even lower in the supercritical phase (water-saturated CO_2) are linked to passivation and possibly insufficient electrolytes [48,49]. In the supercritical phase, cathodic reactions result in a higher H_2CO_3 concentration (after a solution of CO_2 in water) and therefore in a higher acidic and more reactive surrounding as in the CO_2 saturated liquid phase [7,26]. As a function of time corrosion, rates increase at 60 °C and 100 bar in the supercritical phase and remain stable in the liquid phase (0.003 mm/year after 4000 h) [16]. Sufficient surface corrosion resistance at ambient pressure is related to the microstructure of hardened or hardened and tempered alloys [16]. Surface corrosion resistance at 100 bar under supercritical CO_2 conditions is mentioned for hardened and tempered alloys at 670 °C (<0.001 mm/year, martensitic microstructure). By normalizing the microstructure, good corrosion resistance in the liquid phase is offered (ca. 0.004 mm/year, ferritic-pearlitic microstructure) [16,51,53].

The authors relate depassivation after long exposure (100 h) in the supercritical phase to fast reaction kinetics and carbide precipitation in earlier studies [16]. Because depassivation is accompanied by depleting the matrix of chromium, new passivation is prohibited and the material degrades [16,51,53]. Consequently, both phenomena lead to the unusual formation of a surface corrosion layer (Figure 1).

 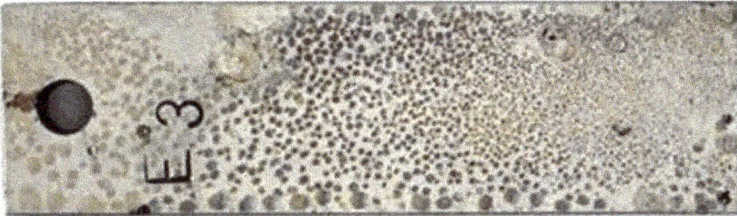

Figure 1. Left: SEM micrographs (8000 h at 60 °C/100 bar exposed to water-saturated supercritical CO_2) of the corrosion layer formed on X5CrNiCuNb16-4 with ellipsoidal peculiarity on hardened and tempered at 670 °C. Reprintetd with permission from [16]. 2021 MDPI, A. Pfennig.

When these 17% chromium steels are exposed to the carbon dioxide environment, the corrosion layer produced on both, the pits and surface are compared to each other [15,17], usually composed of siderite $FeCO_3$ [3,16,51]. $FeCO_3$ shows low solubility in water (p_{Ksp} = 10.54 at 25 °C [16,26,29,43,54]), which causes anodic iron dissolution that is initialized by the formation of transient iron hydroxide $Fe(OH)_2$ [6,16,49]. The pH elevates locally and causes reactions [15,29] to form a ferrous carbonate film internally as well as externally. This paper derives a descriptive approach to better understand this corrosion mechanism and offers a descriptive approach when this laboratory research is extended to small-scale applied research, for example, to monitor injection sights in CCS-sights. Revision times of the plant may be scheduled according to the corrosion type and scale formation with a possibly positive influence on the corrosion resistance of pipe steels in a geothermal environment.

2. Materials and Methods

To better understand corrosion behaviour in CCS, environment steel coupons were statically immersed in environments as existing during carbon capture and storage.

2.1. Steels

Static corrosion tests at ambient and high pressure (100 bar) were conducted with samples of:
1. No AISI (X35CrMo17, 1.4122) (Table 1);
2. AISI 630 (X5CrNiCuNb 16-4, 1.4542) (Table 2).

Table 1. 1.4122 (X35CrMo17): chemical composition in mass per cent.

Elements	C	Si	Mn	P	S	Cr	Mo	Ni	Co	Fe
acc standard [a]	0.33–0.45	<1.00	≤1.00	≤0.045	≤0.03	15.5–17.5	0.8–1.3	≤1.00		0.20–0.45

[a] Elements as specified according to DIN EN 10088-3 in %.

Table 2. 1.4542 (X5CrNiCuNb16-4, AISI 630), chemical composition in mass per cent.

Elements	C	Si	Mn	P	S	Cr	Mo	Ni	Cu	Nb
acc standard [a]	≤0.07	≤0.70	≤1.50	≤0.04	≤0.015	15.0–17.0	≤0.60	3.00–5.00	3.00–5.00	0.20–0.45
analysed [b]	0.03	0.42	0.68	0.018	0.002	15.75	0.11	4.54	3.00	0.242

[a] Elements as specified according to DIN EN 10088-3 in %; [b] spark emission spectrometry.

The chemical composition was reassured by spark emission spectrometry SPEKTROLAB M and by the electron probe microanalyzer JXA8900-RLn, JEOL, Tokyo, Japan (Tables 1 and 2).

2.2. Aquifer Water

The geothermal condition (in-situ) requested for synthesized laboratory geothermal aquifer water (Stuttgart Aquifer [55,56] and Northern German Basin (NGB) [56,57]). This had to be conducted strictly ordered to avoid salts and carbonates precipitating early (Table 3).

Table 3. Northern German Basin (NGB) and Stuttgart Formation electrolyte: Chemical composition.

	\multicolumn{10}{c}{According to the Northern German Basin or According to Stuttgart Formation}
	NaCl	KCl	$CaCl_2 \times 2H_2O$	$MgCl_2 \times 6H_2O$	NH_4Cl	$ZnCl_2$	$SrCl_2 \times 6H_2O$	$PbCl_2$	Na_2SO_4	pH value
g/L	98.22	5.93	207.24	4.18	0.59	0.33	4.72	0.30	0.07	5.4–6
	NaCl	KCl	$CaCl_2 \times 2H_2O$	$MgCl_2 \times 6H_2O$	$Na_2SO_4 \times 10H_2O$		KOH		$NaHCO_3$	
g/L	224.6	0.39	6.45	10.62	12.07		0.321		0.048	
	Ca^+	K^{2+}	Mg^{2+}	Na^{2+}	Cl^-		SO_4^{2-}		HCO_3^-	pH value
g/L	1.76	0.43	1.27	90.1	14.33		3.6		0.04	8.2–9

2.3. Heat Treatment and Static Corrosion Experiments

As-received and thermally treated steel coupons with 8 mm thickness, 20 mm width, 50 mm length were immersed in 1. CO_2-saturated aquifer brine and 2. Water-saturated CO_2. For each exposure time, 4 coupons were tested. Depicted coupons were heat-treated following the protocol of Table 4 [9,15–17,32,34,41,49,51,52,58–60].

Table 4. X5CrNiCuNb16-4: heat treatment.

Heat Treatment	$T_{Austenitizing}/°C$	$T_{Annealing}/°C$	Time	Cooling
			Min	Medium
HT1 normalizing HT1	850		30	oil
HT2 hardening	1040		30	oil
HT3 hardening plus tempering 1	100	655	30	oil
HT4 hardening plus tempering 2	1000	670	30	oil
HT5 hardening plus tempering 3	1000	755	30	oil

Specimens were tested in both the vapour and liquid phase, fixed through a hole of 3.9 mm. A capillary meter GDX600_man by QCAL Messtechnik GmbH, Munic surveyed the CO_2 flow (purity 99,995 vol%) into the aquifer water in ambient pressure experiments at 3 NL/h. Specimens immersed for 700 to 8000 h at 60 °C and 100 bar in a high-pressure vessel [9,15–17,32,34,41,49,51,52,58,59] and additionally in a low pressure vessel at ambient pressure [9,15,16].

The surface of the steel coupons was ground under water down to 120 µm using SiC-paper. After executing corrosion experiments, samples were dissected, leaving the corrosion scales attached to the surface. After surface analysis, they were descaled with 37% HCl to conduct kinetic analysis). Embedding samples in Epoxicure, Buehler cold resin, allowing for smooth cutting and polishing (180 to 1200 µm) with SiC paper under water. Coupons were finished with 6 µm and 1 µm diamond paste. [16]

2.4. Analysis

Light optical and electron microscopy ensured analysis of morphology and layer structure of the corrosion scales. The double optical system MicroProf®TTV by FRT GmbH,

Bergisch Gladbach, Germany uses three-dimensional images to characterize local corrosion. X-ray diffraction with CoK α-radiation and automatic slit adjustment, step 0.03° and count of 5 s in a URD-6 (Seifert-FPM) enabled phase analysis. The PDF-2 (2005) powder patterns were used to automatically identify peak positions. The most likely structures were matched with the inorganic crystal structural database ICSD and the POWDERCELL 2.4 program by the authors of [61] and the AUTOQUAN® by Seifert FPM Holding GmbH, Freiberg, Germany helped refine the fitting of raw data files. The image analysis program Analysis Docu ax-4 Aquinto Olympus Corporation, Olympus Deutschland GmbH, Hamburg, Germany a semi-automatic analyzing program, was used to predict corrosion kinetics. Therefore, the corrosion scale was measured according to the plane fraction of 3 microsections or according to a set of 30 line measurements of each 3 microsection frames, then deriving an estimated scale thickness. Material loss due to lateral spallation and/or corrosive attack was acquired via the mass change method using 4 coupons for each exposure time. The mass change of the coupons before and after exposure to the corrosive environment allowed for estimating surface corrosion rates according to DIN 50 905 part 1–4 (Equation (1)).

$$\text{corrosion rate}\left[\frac{mm}{year}\right] = \frac{8760\left[\frac{hours}{year}\right] \times 10\left[\frac{mm}{cm}\right] \times \text{weight loss}[g]}{\text{area}[cm^2] \times \text{density}\left[\frac{g}{cm^3}\right] \times \text{time}[hour]} \quad (1)$$

3. Results and Discussion

CO_2 is generally injected into saline aquifer water reservoirs in the supercritical state [9,15–17], where it reacts with brine salts and mineralizes quickly [55–57]. During technical revisions, the injection process is intermitted and the pressure in the injection pipe is reduced, which then leads to the raising of the water level into the pipe, and the brine may flow back into the borehole. The resulting three-phase boundary comprises of gaseous/supercritical CO_2, liquid aquifer water, and solid-state steel from the injection pipe and enhances severe corrosive attack [16]. In laboratory experiments, one-year exposure to an artificial aquifer environment is sufficient to obtain meaningful corrosion data to reproduce the CCS environment and describe the corrosion mechanism [9,15].

3.1. Comprehensive Demonstration of Corrosion Kinetics

Checked against other possible injection pipe steels (42CrMo4, X20/46Cr13, X5CrNiCu Nb16-4) X35CrMo17 shows very good corrosion resistance at ambient pressure in the liquid phase and 100 bar in both the supercritical and liquid phases (Figures 2 and 3).

Surface corrosion rates accrete with elongated exposure time and are higher at ambient pressure compared to rates obtained at 100 bar—most likely a consequence of excess oxygen in the open test circuit [16,42]. Moreover, higher corrosion rates at ambient pressure could be attributed to an open capillary system drawing through the corrosion layer that is closed at 100 bar [16,41]. Open capillaries that are required for scale growth enable ionic species to interdependently diffuse fast [16,41]. In general, corrosion rates for X35CrMo17 (supercritical phase: max. 0.0065 mm/year after 8000 h of exposure at 100 bar and 0.096 mm/year after 8000 h of exposure at ambient pressure) are much lower compared to other steel qualities at 100 bar and with the exception of the intermediate phase also at ambient pressure. These generally lower corrosion rates are independent of the atmosphere (water-saturated supercritical CO_2 intermediate (phase boundary) and CO_2 saturated saline aquifer water) and indicate that the CO_2 partial pressure is not sufficient to initiate the corrosive reactions described in the following chapters.

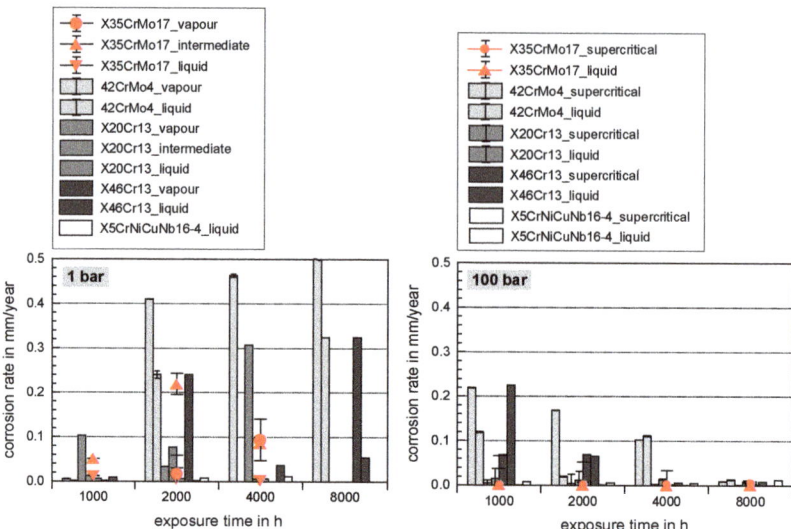

Figure 2. Comparison of corrosion rates of X35CrMo17 to X20Cr13, X46Cr13, 42CrMo4 and X5CrNiCuNb16-4 exposed to liquid and vapour/supercritical CO_2-saturated geothermal environment at ambient pressure (**left**) and 100 bar (**right**) after exposure for 8000 h to aquifer brine water at 60 °C. Results were taken from [9,15–17,42] and combined.

Note that independent of pressure (ambient pressure and at 100 bar), the corrosion rate of X35CrMo17 in water-saturated supercritical CO_2 increases with exposure time, while the corrosion rate of samples exposed to CO_2-saturated aquifer water decreases slightly, assuming that passivating corrosion layer precipitates (incubation time) (Figure 3). There are three possible reasons:

1. In general, the relative supersaturation of water-saturated CO_2 (supercritical/vapour phase) is higher compared to CO_2-saturated brine (liquid phase) because the concentration of reactive corrosion ions in the supercritical phase is higher than in the brine [16,41].
2. A possible final failure of the passivating layer exposes the newly formed metal surface to an electrolyte with high CO_2 partial pressure that then accelerates the corrosion reactions.
3. Long exposure times enhance carbide precipitation that depletes the surrounding metal matrix of chromium and prohibit surface passivation. Although independent of the pressure, the CO_3^{2-} concentration remains the same [3], the higher corrosion rates in supercritical CO_2 result in increased formation rate of Fe^{2+} ions, offering a high number of carbides precipitating on the steel's surface. These are more susceptible to decomposing reactions, but carbides also affect the scale growth mechanism [3].
4. At high pressure with lower CO_2 supersaturation in the liquid phase than in the supercritical phase, nucleation reactions are slow and stable crystal growth of siderite dominates the kinetics. A stable and dense siderite layer is formed, giving low corrosion rates in water-saturated supercritical CO_2 as shown in Figures 2 and 3.

As a consequence, the base metal decays after elongated exposure and corrosive reactions are accelerated in water-saturated supercritical CO_2.

The impact of heat treatment on the corrosion behaviour of steels was shown earlier [11,16,17,22,28,29,41,51]. The heat treatment shows a stronger influence on the corrosion behaviour at 100 bar than at ambient pressure [16]. Good corrosion resistance at 100 bar in water-saturated supercritical CO_2 (lowest surface corrosion rates: <0.001 mm/year) regarding surface corrosion in water-saturated supercritical CO_2 and CO_2-saturated saline water was attributed to martensitic microstructure, when steels are hardened and then annealed at 600–670 °C. However, it was shown that the normalized ferritic-pearlitic microstructure performs better in the CO_2-saturated aquifer (ca. 0.004 mm/year) [16,41,51].

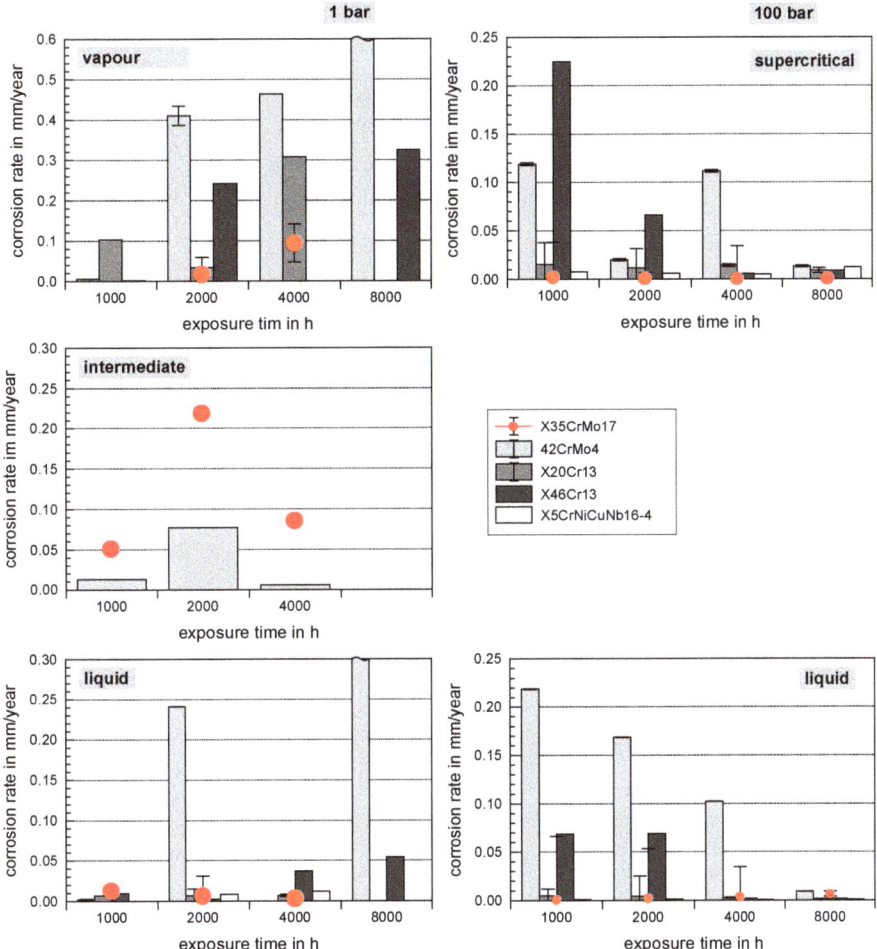

Figure 3. Arrangement of corrosion rates of X35CrMo17 to X20Cr13, X46Cr13, 42CrMo4 and X5CrNiCuNb16-4 with regard to atmosphere: the liquid, intermediate, vapor/supercritical phase at ambient pressure (**left**) and 100 bar (**right**) after 8000 h of exposure to aquifer brine water at 60 °C. Results were taken from [9,15–17,42] and combined.

X35CrMo17 is less resistant against local corrosion at high pressure (100 bar) in the supercritical as well as the liquid phase [9,16,42] when compared to other possible injection pipe steel qualities (42CrMo4, X20Cr13, X46Cr13, X5CrNiCuNb16-4). X35CrMo17 is characterized by distinct pitting (pit per m^2) with a generally higher number of pits under supercritical CO_2 conditions [9,16,42] (Note that after 8000 h of exposure at ambient pressure, the number of pits per m^2 increases tremendously, exceeding that obtained in the liquid phase). Figure 4 shows initial pits in combination with the surface corrosion layer precipitated in the vapour phase at ambient pressure and therefore clearly states that the corrosion mechanism is initiated by the formation of the pits.

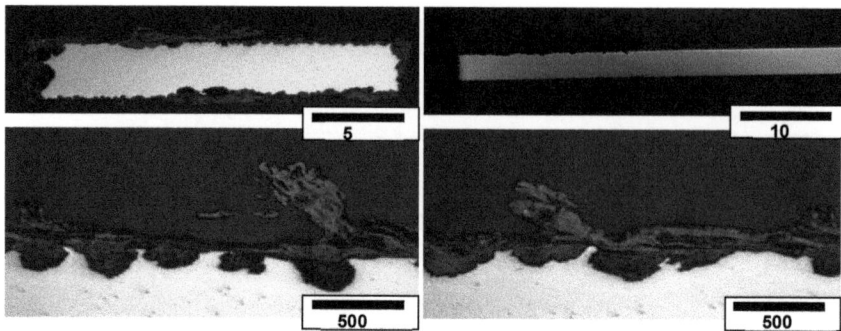

Figure 4. Depicted surface cross-sections with heavy local corrosive attack after 8000 h of exposure at 60 °C and 1 bar of X35CrMo17.

In general, higher nickel and chromium contents in heat-treated steels rectify the corrosion resistance [16,22,27]. For X35CrMo17 and X5CrNiCuNb16-4, the increased chromium content leads to passivation layers, producing lower surface corrosion rates but insufficient reliability, according to enhanced local corrosion phenomena. Hence, the influence of the heat treatment is less meaningful than the influence of chromium content and atmosphere. Both steel qualities may be considered as injection pipe steels regarding surface corrosion criteria but not regarding local corrosion.

3.2. Surface Morphology and Scale Precipitation

The three-phase boundary: water, steel, and supercritical CO_2 lead to the precipitation of thick corrosion layers in water-saturated CO_2 at ambient pressure (Figure 5) and a "leopard"-shaped corrosion layer (Figures 1, 5 and 6) typical for martensitic stainless steels with 17% Chromium X5CrNiCuNb16-4 [16,51,53] and X35CrMo17 [16,41,42]. This corrosion formation is present in supercritical water-saturated CO_2 and in CO_2-saturated brine clearly after 2000 h of exposure at 60 °C and 100 bar. On average, the thickness of the corrosion layer formed on X35CrMo17 is about 0.8 mm locally the magnitude of the outer and inner corrosion layer exceeds the average by a factor of four [42]. Sample surfaces reveal ellipsoidal regions. The centres of the ellipsoidal regions are light-coloured, indicating corrosion layers revealing siderite $FeCO_3$ and goethite alpha-FeOOH and also main precipitation phases [16,41,51]. The darker outer regions are not corroded at eyesight nor are they protected by a passivating layer.

Earlier phase analysis [9,42] for X35CrMo17 report various salts because alloying elements and iron from the base material react with the brine to form oxides, hydroxides and carbonates. The main phases of goethite α-FeOOH, mackinawite FeS and spinel phases of various compositions, for example, magnetite Fe_3O_4 and chromite $FeCr_2O_4$, arrange the complex multi-layer carbonate/oxide scale. Iron oxides are needle-shaped and halites NaCl precipitate in cubic habitus. Due to overlying peaks, siderite $FeCO_3$ could not be identified via XRD but EDX-Scans of cross-sections definitely analysed siderite to be the scale matrix phase [9,41]. Rhodocrosite $MnCO_3$, chromium iron oxide $Cr_{1.3}Fe_{0.7}O_3$ and akaganeite $Fe_8O_8(OH)_8Cl_{1.34}$ are minor phases.

"Ellipsoids" (Figures 1, 5 and 6) show increased oxygen content compared to the surrounding surface (Figure 7, measuring position two). This refers to the fast growth of siderite, $FeCO_3$. The oxygen content diminishes as a function of increasing distance from the centre of the ellipsoids. Therefore, only a thin passivating layer (possibly consisting of chromium iron oxide (Cr_2O_3 and $(Fe_x(Cr_{1-x}))_3O_4$)) remains between the homogeneous ellipsoids.

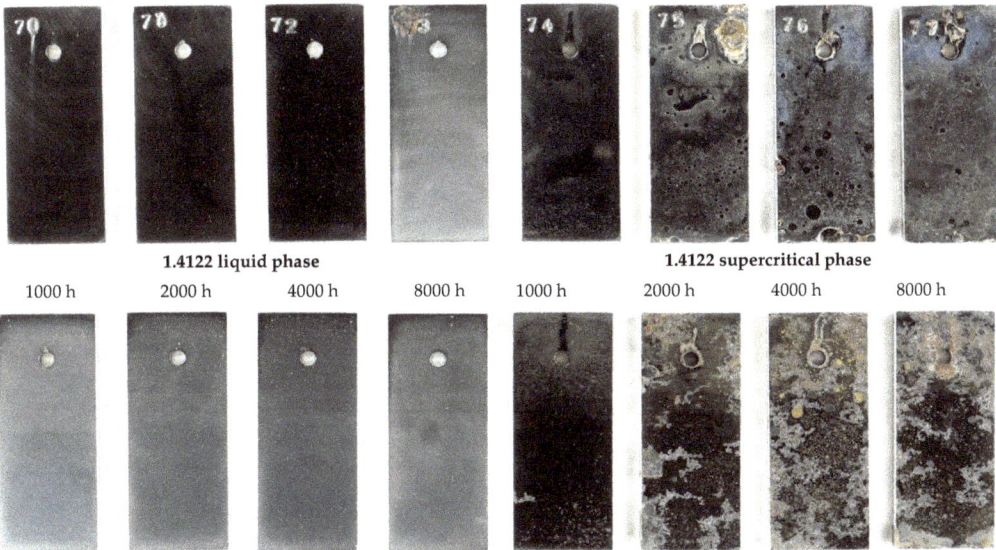

Figure 5. Surface images of X35CrMo17 after 1000 to 8000 h of exposure to CO_2–saturated aquifer water at 60 °C and 100 bar.

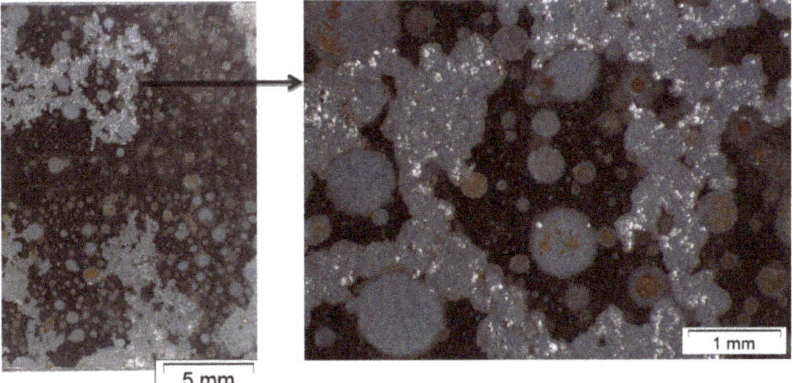

Figure 6. Corroded surfaces of X35CrMo17 after 8000 h of exposure to water-saturated supercritical CO_2 at 60 °C and 100 bar.

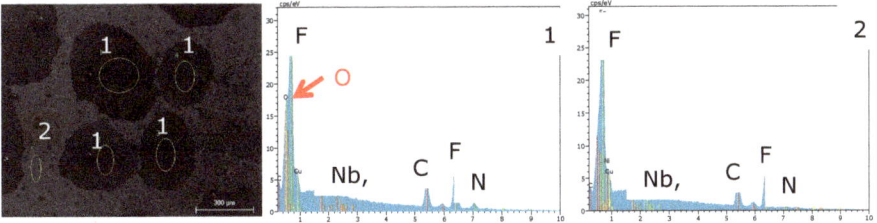

Figure 7. Scanning electron microscopy micrographs and elements distributed within the ellipsoids formed on the corroded surface of X5CrNiCuNb16-4 after hardening and tempering at 670 °C before being exposed for 8000 h at 60 °C and 100 bar to water-saturated supercritical CO_2.

In general, in contact with corrosive solutions (e.g., CO_2-saturated saline aquifer water) a passive film is formed on the surface of high-alloyed high chromium stainless steels. This acts as a reaction ion barrier between the metal surface and the aggressive environment. The passivating layer, mainly composed of chromium oxide Cr_2O_3, prevents the mutual diffusion of Fe from the base metal and O_2, C, S and other impurities from the CO_2-saturated brine. It therefore protects the metal from further dissolution and degradation. In a CCS environment a Cr_2O_3 passivating layer also precipitates on high chromium steels. However, this may either be destroyed locally after precipitation or precipitate discontinuously, probably because of inhomogeneous carbide distribution or local changes in pH due to the formation of carbonic acid in CO_2-saturated water or in water-saturated CO_2. In water-saturated CO_2 with pH 5.2–5.6, no stable chromium oxide film is formed (Figure 6) and local corrosion processes begin shortly after exposure. As a consequence, the leopard-shaped corrosion layer grows and reaches an equilibrium ellipsoid pattern with sufficient corrosion products (Figure 7, middle, indicated as measuring area (1) while the surrounding metal surface is still covered with the passivating layer (Figure 7, right, indicated as measuring area (2). (The oxygen and carbon content are too low for EDX analysis because the layer is less than 1-micrometer-thick.)

3.3. Corrosion Initiation in Water Saturated Supercritical CO_2 (SCC)

Because the "leopard" shape phenomenon is clearly visible at 100 bar (at ambient pressure, the corrosion rate is high due to surplus oxygen in the experimental system and the leopard structure is soon overgrown) and in water-saturated CO_2, the focus of this work is to describe the corrosion precipitation within this atmosphere.

Note, it may be assumed that the atmosphere (water-saturated supercritical CO_2 or CO_2-saturated brine) does not influence the corrosion mechanism because the leopard structure is present in both. Additionally, the corrosion phenomenon is assumedly independent of the microstructure of the steel because the "leopard" shape is found on coupons with ferrite-perlite microstructure as well as on coupons with martensitic or tempered martensitic microstructure (Figure 8). Mo and Ni do not seem to influence the corrosion mechanism either, because both steels show the same corrosion pattern, but one contains Ni and the other Mo. Because the steels' surface mainly being covered by a passivating chromium oxide Cr_2O_3 layer, it is most likely that the high chromium content of 16% and 17%, respectively, are the driving force for this particular corrosion phenomenon. Earlier studies presenting results of steels with lower chromium content (42CrMo4 (1% Cr) or X46Cr13 and X20Cr13 (each 13% Cr) [16] show pitting and discontinuous but layered corrosion precipitates.

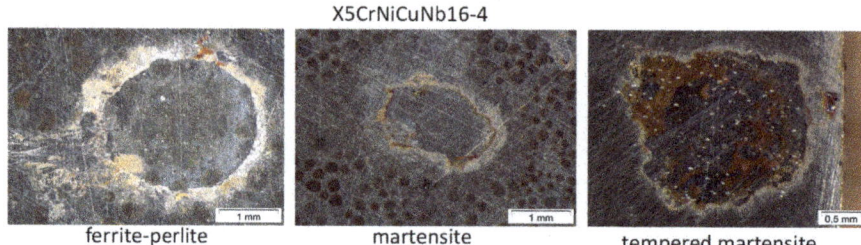

Figure 8. Sample surfaces (micro) of X5CrNiCuNb16-4 after 8000 h of exposure to water-saturated supercritical CO_2 at 60 °C and 100 bar.

The authors previously outlined the initiation of the typical "leopard" surface structure [41,51] and now present the most possible scenarios for corrosion in water-saturated supercritical CO_2:

(a) The passivating layer is locally destroyed, possibly due to locally very low pH as a consequence of the formation of carbonic acid in water-saturated supercritical CO_2 leading to anodic dissolution.
(b) The carbide distribution within the steels' microstructure is not homogeneous. Carbides located at the metal surface corrode locally because carbides are more susceptible to anodic dissolution [20]. Consequently, ellipsoids grow from the initial carbide dissolution leaving a newly exposed metal surface that is highly susceptible to the corrosive environment.
(c) Carbonic acid H_2CO_3 (as a reaction product from water and CO_2) is not soluted equally along the entire sample surfaces. Hence, a thin passivating layer is formed in the initial corrosion stage that then starts growing locally. Once a sufficient thickness of these corrosion islands is achieved, it detaches laterally, causing corrosion reactions.
(d) In general, raising the temperature accelerates the water solubility in supercritical CO_2. Choi et al. reported that the solubility of water in CO_2 decreases in the region 0 bar—50 bar and then slightly raises again [62]. Because the temperature was kept constant (60 °C) and the pressure was at a constant 100 bar, both, neither the temperature nor pressure influence the solubility of water in supercritical CO_2 over time. Furthermore, in this particular CCS environment, the solubility decreases overall. Consequently, at 100 bar and 60 °C, the metal surface that precipitated a passivating layer consisting of Cr_2O_3 and $(Fe_x(Cr_{1-x}))_3O_4$ is wetted by very thin and small water droplets. Distinct "leopard"-shaped corrosion layers form associated with initial droplets condensed on the surface. The residual water droplets can be seen in Figure 8, with bigger droplets in the middle and the small former droplets now being the "leopard" ellipsoids. At the metal–water–supercritical CO_2 phase boundary, the surface is locally depassivated, whereas the remaining surface is covered by thin passivating corrosion layers. This formation model will be described in detail below.

Note that even this unusual corrosion behaviour gives very low surface corrosion rates (<0.01 mm/year) for both steels. Therefore, ellipsoids and surrounding surfaces passivate the steel surfaces and prevent the metal from early degradation. Pitting is not taken into account here; the centres of the bigger droplets reveal pits (Figure 8), indicating that the passivating nature of the ellipsoids is highly dependent on their size.

3.4. Formation Mechanism in Water Saturated Supercritical CO_2 (SCC)

Contrary to our findings in Figures 2 and 3, Hassani et al. [63] found higher corrosion rates in supercritical CO_2 (in this study, this only accounts for pit corrosion [42]). They stated that the corrosion mechanisms in supercritical CO_2 as well as gaseous CO_2 are the same deriving from polarization curves [63]. Wei et al. [64] also state that the corrosion mechanisms at high pressure (supercritical CO_2 in liquid phase) are similar to those obtained at ambient pressure with low CO_2 partial pressure (liquid phase). This is contradicted by Liu et al. [65] who explain the difference of corrosion mechanism in water-saturated CO_2 and CO_2-saturated water by the distance of water chemistry.

A higher corrosion rate is mainly explained through increasing CO_2 partial pressure [9,15–17,32,34,41,49,51,52,58,59], resulting in a more acidic and reactive environment and more initially formed carbonic acid H_2CO_3, dissociating to H_3O^+ and HCO_3^- according to Equation (5). However, here the unusual corrosion pattern may contribute to the low corrosion rates in supercritical CO_2 saturated with aquifer water according to a geothermal CCS site.

The high chromium steel is passivated by Cr_2O_3 and $(Fe_x(Cr_{1-x}))_3O_4$ before being in contact with the CCS environment (Equation (2)).

$$4Cr + 3O_2 \rightarrow 2\,Cr_2O_3 \qquad (2)$$

Long exposure hours lead to high surface corrosion rates in the supercritical phase after 1000 h of exposure because the passivating layer decays exposing the newly formed metal surface to an electrolyte with high CO_2 partial pressure. As a consequence, after long exposure times, the base metal microstructure decomposes and internal corrosion processes accelerate in water-saturated supercritical CO_2.

Once the supercritical CO_2 (SCC) is saturated with water, droplets are formed on the metal surface due to the low solubility of water in SCC [62], even decreasing with time in this particular CCS environment, as described above. Here, carbonic acid H_2CO_3 is formed quickly, according to Equations (3) and (4):

$$H_2O \rightarrow H^+ + OH^- \tag{3}$$

$$CO_{2\,(SCC)} + H_2O_{(l)} \rightarrow H_2CO_3 \tag{4}$$

The cathodic reaction in the CO_2 corrosion process is driven by the formation of HCO_3, depending on the exchange of ionic species described in Equation (5), and by the CO_2 partial pressure in the encircling medium, leading to an increasing H_2CO_3 concentration [10,26]. The cathodic reactions consist of the reduction of H_2CO_3, $HCO_3^-{}_{(aq)}$ and H^+ (Equations (4–6)).

Cathodic reactions:

$$H_2CO_3 + e^- \rightarrow H^+ + HCO_3^-{}_{(aq)} \tag{5}$$

$$H_2CO_3 + H_2O \rightarrow H_3O^+ + HCO_3^- \tag{6}$$

$$2\,HCO_3^-{}_{(aq)} + 2\,e^- \rightarrow 2\,CO_3^{2-} + H_2 \tag{7}$$

$$2\,H^+ + 2\,e^- \rightarrow H_2 \tag{8}$$

According to Nesic et al. [3], the corrosion rate increases as the partial pressure of CO_2 increases for scale-free CO_2 corrosion processes. The environment becomes more acidic and reactive as a result of higher partial pressure of the CO_2 in the water-saturated supercritical CO_2 phase. It is well accepted that the concentration of carbonic acid H_2CO_3 increases with increasing CO_2 partial pressure that accelerate the cathodic reactions, consequently resulting in higher corrosion rates.

In the CO_2 corrosion process, the anodic reaction comprises of the dissolution of Fe (Equation (8) in the case of local depassivation or destruction of the C_2O_3 or $Fe_x(Cr_{1-x})_3O_4$ layer. After the CO_2 is dissipated to establish a corrosive environment (carbonic acid H_2CO_3), iron from the base metal is dissolved in the acidic water droplet. Because the solubility of $FeCO_3$ in water is low (pK_{sp} = 10.54 at 25 °C) [26,43] a siderite $FeCO_3$ corrosion layer expands on the alloy surface in the wake of the anodic iron dissolution [13,16,19–21], according to Equations (10)–(12) (Figure 9).

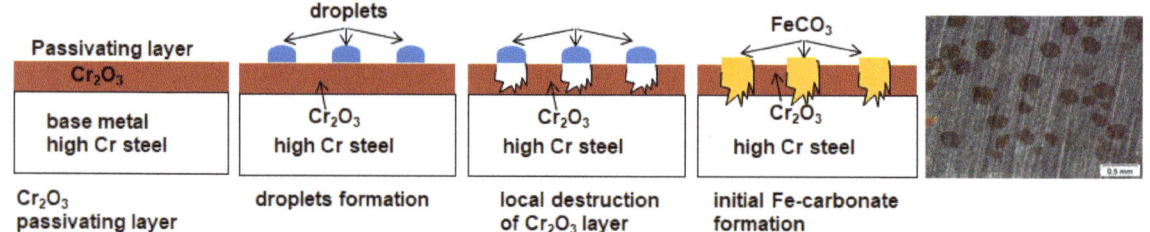

Figure 9. Schematic cross-section illustration of the corrosion procedure to form the leopard structured corrosion scale consisting of siderite $FeCO_3$ on 16–17% Cr high alloyed stainless steels X35CrMo17 and X5CrNi CuNb16-4. Reprintetd with permission from [16]. 2021 MDPI, A. Pfennig.

Anodic reactions:

$$Fe \rightarrow Fe^{2+} + 2e^- \tag{9}$$

$$Fe^{2+} + CO_3^{2-} \rightarrow FeCO_3 \qquad (10)$$

$$Fe^{2+} + 2\,HCO_3^- \rightarrow Fe(HCO_3)_2 \qquad (11)$$

$$Fe(HCO_3)_2 \rightarrow FeCO_3 + CO_2 + H_2O \qquad (12)$$

These reactions were discussed in detail by various authors [6,9]. CO_2 corrosion is mainly driven by the generation of carbonic acid and the existence of HCO_3 [17]. According to Han et al. [66] and Wei et al. [64], the corrosion takes place in a two-step reaction where an amorphous phase explains differences in the porous structure of the inner and outer layer of the corrosion layer. In the first stage, the steel is introduced to the corrosive environment, the water-saturated supercritical CO_2 (SCC). As soon as the solubility limit of water in SCC is exceeded, water droplets form on the steels' surface and the carbon dioxide forms carbonic acid H_2CO_3 within the droplets. An initial reaction step may be ascribed to the formation of Fe[II] compounds $Fe(OH)_2$ (Equation (13)), an amorphous metastable transient ferrous hydroxide passivating film [6,26], when $Fe(OH)_2$ exceeds its solubility limit. At the same time, the local pH near the hydroxide film increases locally (Figures 10 and 11).

$$Fe + 2H_2O \rightarrow [Fe(OH)_2]_{amorph} + 2H^+ + 2e^- \qquad (13)$$

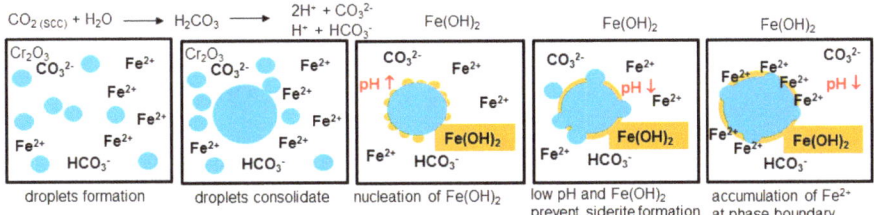

Figure 10. Schematic cross-section illustration of the first step of the corrosion procedure to form the leopard structured corrosion scale consisting of siderite $FeCO_3$ on 16–17% Cr high alloyed stainless steels X35CrMo17 and X5CrNi CuNb16-4.

Wei et al. [64] found that independent of the pressure, the CO_3^{2-} concentration was similar at high pressure and ambient pressure, but the pH in the liquid phases was much higher at high pressure. This also accounts for the initial water droplets forming on the steels' surface in supercritical water-saturated CO_2 and may be the result of the formation of the transient $Fe(OH)_2$ layer from the water droplet and not from SCC. Soon after the ferrous hydroxide is formed, the surrounding pH decreases again at high pressure, when it is exposed to fresh water-saturated SCC containing carbonic acid from the growing droplet. With pH being as low as 4.5, the solubility of siderite $FeCO_3$ increases, supersaturating the water droplets with CO_3^{2-}, H_3O^+ and HCO_3^- ions during the corrosion initiation period. As a consequence of the enhanced solubility of $FeCO_3$, the formation of a stable solid carbonate layer is impeded.

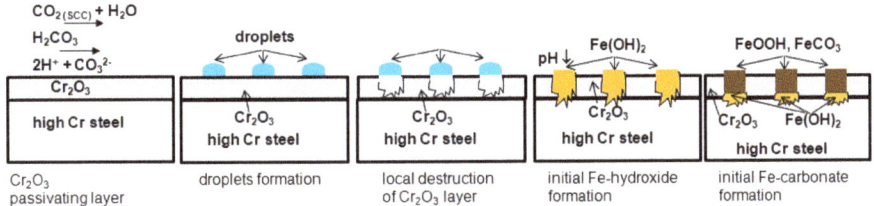

Figure 11. Schematic cross-section illustration of the second step of the corrosion procedure to form the leopard structured corrosion scale consisting of siderite $FeCO_3$ and goethite FeOOH on 16–17% Cr high alloyed stainless steels X35CrMo17 and X5CrNi CuNb16-4.

Additionally, because crystal growth is the dominating reaction at low supersaturation—nucleation dominates at high supersaturation [64]—crystal growth of siderite $FeCO_3$ is also prevented, leaving a transient nanocrystalline or amorphous hydroxide scale [6,17,26] on the steels' surface, according to Equation (13). In this initiation period, no continuous scale is formed in the CO_2-saturated droplet leading to the first decomposing reactions on the steel's surface. The formation of the amorphous or nanocrystalline scale prior to siderite precipitation reduces the corrosion rate and consequently, the concentration of iron ions Fe^+. Furthermore, it blocks the mutual diffusion of ionic species Fe^+, CO_3^{2-} and O^{2-} at the metal/amorphous phase boundary. Here the accumulation of Fe^+ species at the base metal–hydroxide interface favours reactions, according to a second reaction step (Equation (14)).

At the same time, the increased formation rate of Fe^{2+} ions (Equation (9)) enhances carbide precipitation close to the hydroxide/metal boundary at the metals' surface (Figure 10). Carbides are not only more susceptible to decomposing reactions they also affect the scale growth mechanism [64]. Growth of the carbonate layer will proceed internally and externally depending on the various carbon and oxygen partial pressures.

The following step refers to goethite $FeOOH$ and siderite $FeCO_3$ formation when carbon dioxide CO_2 and water consequently form when carbonic acid H_2CO_3 is present (Figure 11). $FeCO_3$ and goethite α-$FeOOH$ not only result from a rather low pH in CO_2-containing and its low solubility [26]; it may also form as a result from further reactions of the transient ferrous hydroxide phase.

$$[Fe(OH)_2]_{(aq)} + H_2O \rightarrow \alpha\text{-}FeOOH_{trans} + 3H^+ + 3e^- \quad (14)$$

$$[Fe(OH)_2]_{(aq)} + [H_2CO_3]_{(aq)} \rightarrow FeCO_3 + 2H_2O \quad (15)$$

The more acidic environment then leads to the complete formation of a discontinuous ferrous carbonate film in the area of former droplets, according to Equations (14) and (15). This is visible as centres of the ellipsoids after exposure to CCS environment. At high pressure with low CO_2 supersaturation, as found in the CO_2-saturated droplet phase, reactions kinetics are much slower than in SCC. Therefore, nucleation reactions are slow and stable crystal growth of siderite is then the dominating reaction mechanism. A stable and dense siderite layer is formed within the area of the droplets. The now passivating ellipsoids are surrounded by the passivating C_2O_3 layer giving low corrosion rates as stated in Figure 3.

When metastable hydroxides form before siderite precipitates the local arrangement of the phases at equilibration is changed [6]. The hydroxide/brine interface absorbs carbonate ions which react with oxygen vacancies and develop cation/oxygen vacancy pairs of the Mott–Schottky-type. At the same time, oxygen vacancies at the hydroxide/brine interface react in reverse with additional carbonate ions to form additional cation vacancies. The excess vacancies move and attach to the hydroxide/siderite interface, where they condense. As a negative result, the siderite detaches from the transient hydroxide film-enabling surface degradation and particularly pitting. However, after a long exposure time (8000 h), mechanical failure is assumed as well because of the different surface morphologies and because the thermal expansion coefficients most likely do not match. If a critical thickness is exceeded, the corrosion layer consequently detaches in a lateral direction [6,16,41].

3.5. Degradation of Carbonate and Hydroxide Layer

As mentioned before, the typical "leopard"-shaped corrosion layer forms, which indicates the initial small droplets on the metal surface. These grow in diameter with increasing exposure time. Here the surface is depassivated locally; first ferrous hydroxide was formed, then siderite $FeCO_3$ nucleated to build a passivating layer. Both reactions driven from HCO_3^- and CO_3^{2-} as well as a reaction via the amorphous/nanocrystalline transient $Fe(OH)_2$ take place. The resulting siderite is visible as darker ellipsoids in a grey-coloured metal surface (Figures 1 and 6–8). The remaining surface is covered by a thin passivating corrosion Cr_2O_3 layer. As a function of exposure time, new droplets condense on the metal surface, causing the pH to decrease (note, the precipitation of ferrous hydroxide causes an increase of pH, leading to a stable transient hydroxide layer). These droplets consolidate building a three-phase boundary (water, metal, SCC supercritical CO_2) at the outer area. The centres of the bigger droplets reveal pits, indicating that the passivating nature of the ellipsoids is highly dependent on their size. Once a critical size is exceeded, pitting is initiated (explaining the rather high number of pits precipitated on both steel qualities [41,42,51]). Degradation of the base material is initiated at the three-phase boundary because the thin passivating siderite $FeCO_3$ layer is destroyed locally (Figure 12). At the same time, the base metal is decomposed within the diameter of the condensed water droplets, whereas the outer regions remain covered by the Cr_2O_3 layer. Small pits surrounding the former droplet precipitate at the multiphase boundary as well as in the droplets' interior enhancing the corrosion processes (Figure 12). The flowing corrosive media removes the remaining film, causing the pit to grow wider and eventually cover larger parts of the surface. Because it takes much more time for pits to consolidate and grow wider than new droplets to form, water diffuses back into the supercritical CO_2. The consolidated droplets decrease inwards in size and reduce in total area (in which the siderite $FeCO_3$ is decomposed), leaving sulphates ($FeSO_4$) in the outer areas, whereas the centre shows goethite α-FeOOH as well as hematite Fe_2O_3 [16,51,52] as a result from oxidation reaction after the test periods.

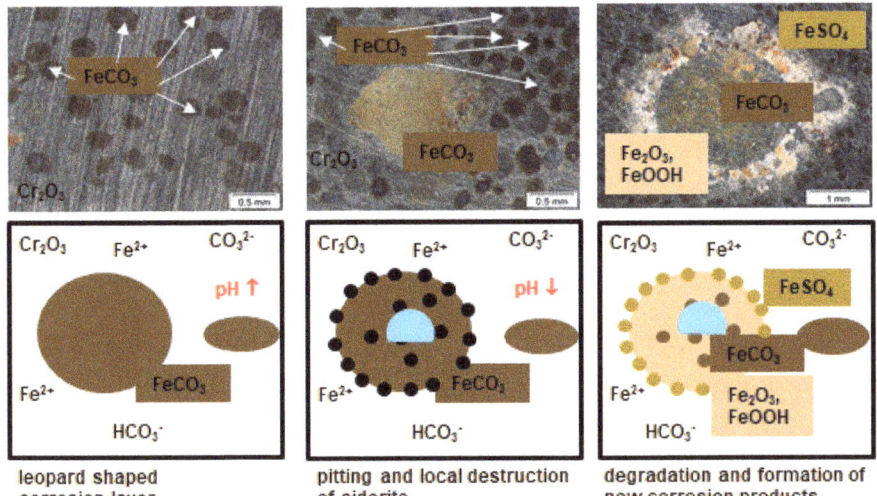

Figure 12. Schematic illustration of the degradation of the passivating siderite corrosion layer formed on 16–17% Cr high alloyed stainless steels X35CrMo17 and X5CrNi CuNb16-4.

4. Conclusions

The formation mechanism for elliptical corrosion layers on X35CrMo17 and X5CrNiCuNb16-4 exposed to a laboratory CCS atmosphere similar to the Northern German Basin was outlined and the assumed reaction mechanism was described. The corrosion scale is characterized by a "leopard"-shaped corrosion scale. Therefore, coupons of the steel quality X35CrMo17 and X5CrNiCuNb16-4 suitable as injection pipe with 17% and 16% Chromium were exposed up to approximately one year (8000 h) to supercritical CO_2 and saline aquifer water at 100 bar and 60 °C in laboratory experiments.

Both steel qualities passivate leading to the low surface corrosion rates on both steels (<0.012 mm/year). Due to excess oxygen in the open test circuit at ambient pressure, corrosion rates at ambient pressure exceed those measured after exposure at 100 bar by a factor of 50. In general, higher pressure induces pitting (pit per m^2). However, especially at

100 bar, the corrosion kinetics of X35CrMo17 are slower (max. 0.007 mm/year) compared to steel qualities 42CrMo4, X20Cr13, X46Cr13 and X5CrNiCuNb16-4 independent of the environment (water-saturated supercritical CO_2 or CO_2-saturated saline aquifer water). If the passivating FeOOH, α-$FeCO_3$ layer degrades severely, pitting corrosion is initiated, which results in ongoing local degradation of the base metal in a CCS environment.

At high pressure, a non-uniform corrosion layer ("leopard" shape) reveals products from carbonate corrosion on the surface comprising of α-$FeCO_3$ and FeOOH and more possibly $Cr_2(CO_3)_3$ and CrOOH due to the high chromium content. Inside the typical ellipsoids, Fe_2O_3 and Cr_2O_3 precipitate due to altering water solubility in supercritical CO_2 at high pressure and the dominating reaction mechanism changes from nucleation to crystal growth. It is assumed that in this particular CCS environment, the solubility of water in supercritical CO_2 decreases overall. Consequently, at 100 bar and 60 °C, the metal surface originally covered by a passivating layer consisting of Cr_2O_3 and $(Fe_x(Cr_{1-x}))_3O_4$ is wetted by very thin and small water droplets. The peculiar "leopard"-shaped corrosion layer is associated with these initial droplets on the surface. At the metal–water-supercritical CO_2 phase boundary, the surface is locally depassivated, whereas the remaining surface is covered by thin passivating corrosion layers. As a function of exposure time, regions of earlier droplets consolidate with former outer areas corroding the most at the three-phase boundary: metal–water–SCC. Small pits precipitate enhancing the corrosion processes. Because it takes more time for pits to consolidate than new droplets to form, the reverse process starts with water diffusing back into the supercritical CO_2, where it reduces the region of consolidated droplets from the outer area towards the centre. Consequently, sulphates ($FeSO_4$) remain in the outer areas whereas the centres show hematite Fe_2O_3 and goethite α-FeOOH.

Local corrosion is especially crucial in the decision process for suitable steels in CCS application. Steels are inoperable in pressure vessel applications if the surface corrosion rate exceeds 0.1 mm/year. Because X35CrMo17 and also X5CrNiCuNb16-4 stay way below this margin at high pressure, it may be considered safe in terms of surface corrosion. However, pitting corrosion—as an almost unpredictable statistical phenomenon—is not admitted in order to fulfil the regulations of DIN 6601 due to a rather high risk of notch effects on the surface. Notches may be the cause of fractures and the following failure of the component. Therefore, predicting the lifetime of steels susceptible to pit corrosion in CCS environment is not possible according to this study.

This paper comprises and compares data of previously published work: [9,15,17,32,34, 41,42,49–53,58–60].

Author Contributions: Conceptualization, A.P.; methodology, A.P.; software, validation, A.P., A.K.; formal analysis, A.P.; investigation, A.P.; resources, A.P., A.K.; data curation, A.P.; writing—original draft preparation, A.P.; writing—review and editing, A.P.; visualization, A.P.; supervision, A.P. and A.K.; project administration, A.P.; funding acquisition, A.P. All authors have read and agreed to the published version of the manuscript.

Funding: This research received no external funding.

Institutional Review Board Statement: Not applicable.

Informed Consent Statement: Not applicable.

Conflicts of Interest: The authors declare no conflict of interest.

References

1. Thomas, C. *Carbon Dioxide Capture for Storage in Deep Geologic Formations—Results from CO2 Capture Project*; Elsevier Ltd.: London, UK, 2005; ISBN 0080445748.
2. Broek, M.V.D.; Hoefnagels, R.; Rubin, E.; Turkenburg, W.; Faaij, A. Effects of technological learning on future cost and performance of power plants with CO2 capture. *Prog. Energy Combust. Sci.* **2009**, *35*, 457–480. [CrossRef]
3. Nešić, S. Key issues related to modelling of internal corrosion of oil and gas pipelines—A review. *Corros. Sci.* **2007**, *49*, 4308–4338. [CrossRef]

4. Hurter, S. Impact of Mutual Solubility of H_2O and CO_2 on Injection Operations for Geological Storage of CO_2. In Proceedings of the International Conference of the Properties of Water and Steam ICPWS, Berlin, Germany, 8–11 September 2012.
5. Zhang, L.; Yang, J.; Sun, J.S.; Lu, M. Effect of pressure on wet H2S/CO2 corrosion of pipeline steel. In Proceedings of the NACE Corrosion 2008 Conference and Expo, New Orleans, LA, USA, 16–20 March 2008. Paper No. 09565.
6. Mu, L.J.; Zhao, W.Z. Investigation on Carbon Dioxide Corrosion Behaviors of 13Cr Stainless Steel in Simulated Strum Water. *Corros. Sci.* **2010**, *2*, 82–89. [CrossRef]
7. Seiersten, M. Material selection for separation, transportation and disposal of CO2. In *NACE International, Houston, CORROSION/2001*; Paper No. 01042; NACE International: Houston, TX, USA, 2001.
8. Cui, Z.D.; Wu, S.L.; Zhu, S.L.; Yang, X.J. Study on corrosion properties of pipelines in simulated produced water saturat-ed with supercritical CO_2. *Appl. Surf. Sci.* **2006**, *252*, 2368–2374. [CrossRef]
9. Pfennig, A.; Kranzmann, A. Reliability of pipe steels with different amounts of C and Cr during onshore carbon dioxide injection. *Int. J. Greenh. Gas Control* **2011**, *5*, 757–769.
10. Zhang, H.; Zhao, Y.L.; Jiang, Z.D. Effects of temperature on the corrosion behavior of 13Cr martensitic stainless steel dur-ing exposure to CO2 and Cl environment. *Mater. Lett.* **2005**, *59*, 3370–3374. [CrossRef]
11. Alhajji, J.; Reda, M. The effect of alloying elements on the electrochemical corrosion of low residual carbon steels in stagnant CO2-saturated brine. *Corros. Sci.* **1993**, *34*, 1899–1911. [CrossRef]
12. Choi, Y.-S.; Nešić, S. Corrosion behavior of carbon steel in supercritical CO_2-water environments. In Proceedings of the NACE Corrosion 2008 Conference and Expo, New Orleans, LA, USA, 16–20 March 2008. Paper No. 09256.
13. Jiang, X.; Nešić, S.; Huet, F. The Effect of Electrode Size on Electrochemical Noise Measurements and the Role of Chloride on Localized CO_2 Corrosion of Mild Steel. In Proceedings of the NACE Corrosion 2008 Conference and Expo, New Orleans, LA, USA, 16–20 March 2008; p. 09575.
14. Ahmad, Z.; Allam, I.M.; Abdul, B.; Aleem, J. Effect of environmental factors on the atmospheric corrosion of mild steel in aggressive sea coastal environment. *Anti Corros. Methods Mater.* **2000**, *47*, 215–225. [CrossRef]
15. Pfennig, A.L.; Bäßler, R. Effect of CO_2 on the stability of steels with 1% and 13% Cr in saline water. *Corros. Sci.* **2009**, *51*, 931–940. [CrossRef]
16. Pfennig, A.; Wolf, M.; Kranzmann, A. Corrosion and Corrosion Fatigue of Steels in Downhole CCS Environment-A Summary. *Processes* **2021**, *9*, 594. [CrossRef]
17. Pfennig, A.; Zastrow, P.; Kranzmann, A. Influence of heat treatment on the corrosion behavior of stainless steels during CO_2-sequestration into saline aquifer. *Int. J. Greenh. Gas Control* **2013**, *15*, 213–224. [CrossRef]
18. Nyborg, R. Controlling Internal Corrosion in Oil and Gas Pipelines. Business Briefing: Exploration & Production. *Oil Gas Rev.* **2005**, *2*, 70–74.
19. Carvalho, D.S.; Joia, C.; Mattos, O. Corrosion rate of iron and iron–chromium alloys in CO_2 medium. *Corros. Sci.* **2005**, *47*, 2974–2986. [CrossRef]
20. Linter, B.R.; Burstein, G.T. Reactions of pipeline steels in carbon dioxide solutions. *Corros. Sci.* **1999**, *41*, 117–139. [CrossRef]
21. Wu, S.L.; Cui, Z.D.; Zhao, G.X.; Yan, M.L.; Zhu, S.L.; Yang, X.J. EIS study of the surface film on the surface of carbon steel form supercritical carbon dioxide corrosion. *Appl. Surf. Sci.* **2004**, *228*, 17–25. [CrossRef]
22. Bülbül, Ş.; Sun, Y. Corrosion behavior of high Cr-Ni cast steels in the HCl solution. *J. Alloys Compd.* **2010**, *598*, 143–147. [CrossRef]
23. Hou, B.; Li, Y.; Li, Y.; Zhang, J. Effect of alloy elements on the anti-corrosion properties of low alloy steel. *Bull. Mater. Sci* **2000**, *23*, 189–192. [CrossRef]
24. Cvijović, Z.; Radenković, G. Microstructure and pitting corrosion resistance of annealed duplex stainless steel. *Corros. Sci.* **2006**, *48*, 3887–3906. [CrossRef]
25. Park, J.-Y.; Park, Y.-S. The effects of heat-treatment parameters on corrosion resistance and phase transformations of 14Cr–3Mo martensitic stainless steel. *Mater. Sci. Eng. A* **2007**, *449–451*, 1131–1134. [CrossRef]
26. Banaś, J.; Lelek-Borkowska, U.; Mazurkiewicz, B.; Solarski, W. Effect of CO_2 and H_2S on the composition and stability of passive film on iron alloys in geothermal water. *Electrochim. Acta* **2007**, *52*, 5704–5714. [CrossRef]
27. Bilmes, P.; Llorente, C.; Méndez, C.; Gervasi, C. Microstructure, heat treatment and pitting corrosion of 13CrNiMo plate and weld metals. *Corros. Sci.* **2009**, *51*, 876–881. [CrossRef]
28. Zhang, L.; Zhang, W.; Jiang, Y.; Deng, B.; Sun, D.; Li, J. Influence of annealing treatment on the corrosion resistance of lean duplex stainless steel 2101. *Electrochimica Acta* **2009**, *54*, 5387–5392. [CrossRef]
29. Brown, B.; Parakala, S.R.; Nešić, S. CO_2 corrosion in the presence of trace amounts of H_2S. In Proceedings of the NACE International Corrosion Conference Series: Corrosion 2004, New Orleans, LA, USA, 28 March–1 April 2004. Paper No. 04736.
30. Isfahany, A.N.; Saghafian, H.; Borhani, G. The effect of heat treatment on mechanical properties and corrosion behavior of AISI420 martensitic stainless steel. *J. Alloys Compd.* **2011**, *509*, 3931–3936. [CrossRef]
31. Lucio-Garcia, M.; Gonzalez-Rodriguez, J.; Casales, M.; Martinez, L.; Chacon-Nava, J.; Neri-Flores, M.; Martinez-Villafañe, A. Effect of heat treatment on H_2S corrosion of a micro-alloyed C–Mn steel. *Corros. Sci.* **2009**, *51*, 2380–2386. [CrossRef]
32. Pfennig, A.; Wolthusen, H.; Wolf, M.; Kranzmann, A. Effect of heat Treatment of Injection Pipe Steels on the Reliability of a Saline Aquifer Water CCS-site in the Northern German Basin. *Energy Procedia* **2014**, *63*, 5762–5772. [CrossRef]
33. Thorbjörnsson, I. Corrosion fatigue testing of eight different steels in an Icelandic geothermal environment. *Mater. Des.* **1995**, *16*, 97–102. [CrossRef]

34. Pfennig, A.; Kranzmann, A. Borehole Integrity of Austenitized and Annealed Pipe Steels Suitable for Carbon Capture and Storage (CCS). *Int. J. Mater. Mech. Manuf.* **2017**, *5*, 213–218. [CrossRef]
35. Maranhão, J.P. Davim, finite element modelling of machining of AISI 316steel: Numerical simulation and experimental validation. *Simul. Modell. Pract. Theory* **2010**, *18*, 139–156.
36. Martin, M.; Weber, S.; Izawa, C.; Wagner, S.; Pundt, A.; Theisen, W. Influence of machining-induced martensite on hydrogen-assisted fracture of AISI type 304 austenitic stainless steel. *Int. J. Hydrog. Energy* **2011**, *36*, 11195–11206. [CrossRef]
37. Evgenya, B.; Hughesa, T.; Eskinba, D. Effect of surface roughness on corrosion behavior of low carbon steel in inhibited 4 M hydrochloric acid under laminar and turbulent flow conditions. *Corros. Sci.* **2016**, *103*, 196–205. [CrossRef]
38. Xu, M.; Zhang, Q.; Yang, X.X.; Wanga, Z.J.; Liub, J.; Li, Z. Impact of surface roughness and humidity on X70 steel corrosion in supercritical CO_2 mixture with SO_2, H_2O, and O_2. *J. Supercrit. Fluids* **2016**, *107*, 286–297. [CrossRef]
39. Llaneza, V.; Belzunce, F.J. Study of the effects produced by shot peening on the surface of quenched and tempered steels: Roughness, residual stresses and work. *Appl. Surf. Sci.* **2015**, *356*, 475–485. [CrossRef]
40. Lee, S.M.; Lee, W.G.; Kim, Y.H.; Jang, H. Surface roughness and the corrosion resistance of 21Cr ferritic stainless steel. *Corros. Sci.* **2012**, *63*, 404–409. [CrossRef]
41. Pfennig, A.; Kranzmann, A. Effect of CO_2 and pressure on the stability of steels with different amounts of chromium in saline water. *Corros. Sci.* **2012**, *65*, 441–452. [CrossRef]
42. Pfennig, A.; Kranzmann, A. Effect of CO_2, Atmosphere and Pressure on the Stability of X35CrMo17 Stainless Steel in Laboratory CCS-Environment. In Proceedings of the 14th Greenhouse Gas Control Technologies Conference, Melbourne, VIC, Australia, 21–26 October 2018.
43. Lopez, D.A.; Schreiner, W.H.; de Sánchez, S.R.; Simison, S.N. The influence of carbon steel microstructure on corrosion layers an XRS and SEM characterization. *Appl. Surf. Sci.* **2003**, *207*, 69–85. [CrossRef]
44. Akbari Mousavi, S.A.A.; Sufizadeh, A.R. Metallurgical investigations of pulsed Nd:YAG laser welding of AISI 321 and AISI 630 stainless steels. *Mater. Des.* **2009**, *30*, 3150–3157. [CrossRef]
45. Takemoto, M. *Study on the Failure Threshold Stress Criteria for the Prevention and Mechanism of Stress Corrosion Cracking*; Faculty of Science and Engineering, Aoyama Gakuin University: Tokyo, Japan, 1984.
46. Nor Asma, R.B.A.; Yuli, P.A.; Mokhtar, C.I. Study on the effect of surface finish on corrosion of carbon steel in CO_2 environment. *J. Appl. Sci.* **2011**, *11*, 2053–2057. [CrossRef]
47. Wang, J.; Zou, H. Relationship of microstructure transformation and hardening behavior of type 630 stainless steel. *J. Univ. Sci. Technol. Beijing* **2006**, *3*, 213–221. [CrossRef]
48. Islam, A.W.; Sun, A.Y. Corrosion model of CO_2 injection based on non-isothermal wellbore hydraulics. *Int. J. Greenh. Gas Control* **2016**, *54*, 219–227. [CrossRef]
49. Pfennig, A.; Wolthusen, H.; Zastrow, P.; Kranzmann, A. Evaluation of heat treatment performance of potential pipe steels in CCS-environment. In *Energy Technology 2015: Carbon Dioxide Management and Other Technologies*; Springer International Publishing: Berlin/Heidelberg, Germany, 2016; pp. 15–22. [CrossRef]
50. Pfennig, A.; Trenner, S.; Wolf, M.; Bork, C. *Vibration Tests for Determination of Mechanical Behavior in CO2-Containing Solutions in European Corrosion Congress EuroCorr 2013*; Estoril Congress Center: Estoril, Portugal, 2013.
51. Pfennig, A.; Wolthusen, H.; Kranzmann, A. Unusual Corrosion Behavior of 1.4542 Exposed a Laboratory Saline Aquifer Water CCS-environment. *Energy Procedia* **2017**, *114*, 5229–5240. [CrossRef]
52. Pfennig, A.; Kranzmann, A. Potential of martensitic stainless steel X5CrNiCuNb 16-4 as pipe steel in corrosive CCS environment. *Int. J. Environ. Sci. Dev.* **2017**, *8*, 466–473. [CrossRef]
53. Pfennig, A.; Kranzmann, A. Corrosion and Fatigue of Heat Treated Martensitic Stainless Steel 14542 Used For Geothermal Applications. *Matter Int. J. Sci. Technol.* **2019**, *5*, 138–158. [CrossRef]
54. Han, J.; Yang, Y.; Nešić, S.; Brown, N.B. Roles of passivation and galvanic effects in localized CO_2 corrosion of mild steel. In Proceedings of the NACE Corrosion 2008, New Orleans, LA, USA, 16–20 March 2008. Paper No. 08332.
55. Förster, A.; Norden, B.; Zinck-Jørgensen, K.; Frykman, P.; Kulenkampff, J.; Spangenberg, E.; Erzinger, J.; Zimmer, M.; Kopp, J.; Borm, G.; et al. Baseline characterization of the CO_2SINK geological storage site at Ketzin, Germany. *Environ. Geosci.* **2006**, *13*, 145–161. [CrossRef]
56. Forster, A.; Schoner, R.; Förster, H.-J.; Norden, B.; Blaschke, A.-W.; Luckert, J.; Beutler, G.; Gaupp, R.; Rhede, D. Reservoir characterization of a CO_2 storage aquifer: The Upper Triassic Stuttgart Formation in the Northeast German Basin. *Mar. Pet. Geol.* **2010**, *27*, 2156–2172. [CrossRef]
57. Bäßler, R.; Sobetzki, J.; Klapper, H.S. Corrosion Resistance of High-Alloyed Materials in Artificial Geothermal Fluids. In Proceedings of the NACE International Corrosion Conference Series: Corrosion 2013, New Orleans, LA, USA, 17–21 March 2013. Paper No. 2327.
58. Pfennig, A.; Wolf, M. Influence of geothermal environment on the corrosion fatigue behavior of standard duplex stainless steel X2CrNiMoN22-5-32. *J. Phys. Conf. Ser.* **2020**, *1425*, 012183.
59. Pfennig, A.; Wolf, M.; Kranzmann, A. Evaluating corrosion and corrosion fatigue behavior via laboratory testing techniques in highly corrosive CCS-environment. In Proceedings of the 15th Greenhouse Gas Control Technologies Conference 2020, Abu Dhabi, United Arab Emirates, 15–18 March 2021.

60. Pfennig, A.; Wolf, M.; Kranzmann, A. Effects of saline aquifer water on the corrosion behavior of martensitic stainless steels during exposure to CO_2 environment. In Proceedings of the 15th Greenhouse Gas Control Technologies Conference 2020, Abu Dhabi, United Arab Emirates, 15–18 March 2021.
61. Kraus, S.W.; Nolze, G. POWDER CELL—A program for the representation and manipulation of crystal structures and calculation of the resulting X-ray powder patterns. *J. Appl. Cryst.* **1996**, *29*, 301–303. [CrossRef]
62. Choi, Y.-Y.; Nešić, S. Determining the corrosive potential of CO_2 transport pipline in high pCO_2-water environments. *Int. J. Greenh. Gas Control* **2011**, *5*, 788–797. [CrossRef]
63. Hassani, S.; Vu, T.N.; Rosli, N.R.; Esmaeely, S.N.; Choi, Y.S.; Young, D.; Nešić, S. Wellbore integrinanoty and corrosion of low alloy steel and stainless steels in high pressure CO_2 geologic storage environment: An experimental study. *Int. J. Greenh. Gas Control* **2014**, *23*, 594–601. [CrossRef]
64. Wei, L.; Pang, X.; Liu, C.; Gao, K. Formation mechanism and protective property of corrosion product scale on X70 steel under supercritical CO_2 environment. *Corros. Sci.* **2015**, *100*, 404–420. [CrossRef]
65. Liu, Z.-G.; Gao, S.-H.; Du, S.-X.; Li, J.-P.; Yu, C.; Wang, Y.-X.; Wang, X.-N. Comparison of corrosion mechanism of low alloy pipeline steel used for flexible pipes at vapor-saturated CO_2 and CO_2-saturated brine conditions. *Mater. Corros.* **2017**, *68*, 1200–1211. [CrossRef]
66. Han, J.; Zhang, J.; Carey, J.W. Effect of bicarbonate on corrosion of carbon steel in CO_2-saturated brines. *Int. J. Greenh. Gas Control* **2011**, *5*, 1680–1683. [CrossRef]

Communication

Carbon Storage in Portland Cement Mortar: Influences of Hydration Stage, Carbonation Time and Aggregate Characteristics [†]

Luqman Kolawole Abidoye [1,*] and Diganta B. Das [2,*]

[1] Department of Process Engineering, International Maritime College, Suhar 322, Oman
[2] Chemical Engineering Department, Loughborough University, Loughborough, Leicestershire LE11 3TU, UK
* Correspondence: luqman@imco.edu.om (L.K.A.); d.b.das@lboro.ac.uk (D.B.D.); Tel.: +96894010081 (L.K.A.); +44-1509-222509 (D.B.D.)
[†] This paper is an extended version of our paper published in 1st International Conference on Engineering and Environmental Sciences (ICEES), Osun State University, Osogbo, Nigeria, 5–7 November 2019.

Abstract: This study elucidates the effects of the particle size, carbonation time, curing time and pressure on the efficiency of carbon storage in Portland cement mortar. Using pressure chamber experiments, our findings show how carbonation efficiency increases with a decrease in the particle size. Approximately 6.4% and 8.2% (w/w) carbonations were achieved in the coarse-sand and fine-sand based mortar samples, respectively. For the hydration/curing time of 7 h, up to 12% carbonation was achieved. This reduced to 8.2% at 40 h curing period. On the pressure effect, for comparable curing conditions, 2 bar at 7 h carbonation time gives 1.4% yield, and 8.2% at 5 bar. Furthermore, analysing the effect of the carbonation time, under comparable conditions, shows that 4 h of carbonation time gives up to 8.2% yield while 64 h of carbonation gives up to 18.5%. It can be reliably inferred that, under similar conditions, carbonation efficiency increases with lower-sized particles or higher-surface areas, increases with carbonation time and higher pressure but decreases with hydration/curing time. Microstructural analyses with X-ray diffraction (XRD) and scanning electron microscopy (SEM) further show the visual disappearance of calcium-silicate-hydrate (C-S-H) together with the inhibition of ettringite formation by the presence of CO_2 and $CaCO_3$ formation during carbonation.

Keywords: concrete; carbonation; hydration; particle; curing; size

1. Introduction

Global warming arising from climate change is a real global concern. The greenhouse gases, e.g., CO_2 and other gases like methane, have been identified as the worst culprits of this menace [1–3]. Several efforts have been made to mitigate global warming through the reduction of greenhouse gas emissions. Such efforts centre around carbon capture and storage (CCS) in deep geological aquifers, depleted oil reservoirs, ocean beds and abandoned coal seams [4–8]. Recent evidence reveals that the carbonation of Portland cement concrete is a more promising approach to carbon storage than the geological carbon storage approach. Cement can serve as carbon sink from the reactions of CO_2 with the hydrates of tri- and di-calcium silicates (C_3S, C_2S), which are generally present in cement (during hydration process), as well as the reaction of CO_2 with $Ca(OH)_2$, which is a by-product of cement hydration [9–11]. The process of carbonation in cement and/or concrete often takes place in various stages. However, the dominant reaction in fresh cement carbonations is as expressed in Equation (1) while Equations (2) and (3) show the early-stage carbonation of hydrated concrete [9].

$$2(3CaO.SiO_2) + 3CO_2 + 3H_2O \rightarrow 3CaO.2SiO_2.3H_2O + 3CaCO_3 \quad (1)$$

$$Ca(OH)_2 + CO_2 \rightarrow CaCO_3 + H_2O \tag{2}$$

$$3CaO.2SiO_2.3H_2O + 3CO_2 \rightarrow 3CaCO_3 + 2SiO_2 + 3H_2O \tag{3}$$

Further, the presence of AFm phases in cement offers additional routes to carbonate formation with myriads of other products. AFm are a family of hydrated calcium aluminates based on the hydrocalumite-like structure [12], $Ca_4Al_2(OH)_2.SO_4.6H_20$ [13]. Yaseen et al. [13] express the decomposition of AFm phases during carbonation with graphene oxide (GO) as follows:

$$\begin{aligned}&nCaO.Al_2O_3.nCaSO_4.mH_2O + nGO + nH_2O \\ &\rightarrow nCaCO_{3(s)} + n(CaSO_4.2H_2O) + Al_2O_3.mH_2O + (n-m)H_2O\end{aligned} \tag{4}$$

Equation (4) shows that, in addition to carbonates from calcium hydroxide (CH) and calcium-silicate-hydrate (C-S-H) phases, the carbonation of AFm phases generates more carbonates and produces gypsum and alumina gel. This increased productivity of carbonates contributes substantially to $CaCO_3$ polymorph formation, and it is frequently mentioned that the formed $CaCO_3$ amount far exceeded that which could be obtained from the entire CH dissolution [14,15].

The manufacturing of cement follows a process whereby CO_2 emissions are generated into the atmosphere. Besides the emissions from fuel combustion in the cement manufacturing process, the chemistry of the reactions is responsible for almost two-thirds (64%) of the CO_2 emissions emanating from the Portland cement industry [16,17]. Eventually, the cement is used in the making of concrete for building houses, bridges, road constructions, etc. However, these structures made with cement can reabsorb some of those emissions over time, reversing a portion of the calcination reaction in a process referred to as carbonation [18].

Among the factors expatiated in the literature, the quantity of carbonation in concrete is strongly connected to the available microstructural space in the matrix of the concrete. Galan et al. [19] state that available pore space decreases as the carbonation process continues over time, owing to the formation of calcium carbonate, which contains larger molecules than calcium hydroxide, thus decreasing the available pore space with time. Additionally, if the humidity is too high, the porous system can be blocked off by condensed water, so the carbonation could be obstructed [20]. Zhang [21] reports that the fastest carbonization rate will be reached in the relative humidity of 50–70%, while carbonization will stop when the relative humidity reaches 100% (or in water) or when it is less than 25% (or in the dry environment). Wang et al. [22] also acknowledge the importance of moisture in efficient carbonation. Thus, an effective management of concrete humidity and microstructural space will enhance the carbonization of concrete.

Apart from humidity, the pore size of the concrete can also be affected by the precipitation of the carbonation reaction products owing to the fact that calcium carbonate is formed, which is a larger molecule than calcium hydroxide, thereby reducing the available pore microstructure, making it difficult for the diffusion of carbon dioxide into the sample core as time progresses [19]. This behaviour makes carbonization a surface-based reaction that discontinues after a layer of calcium carbonate has covered the concrete, while calcium hydroxide lies underneath [21]. Therefore, the task of improving carbonation efficiency in concrete involves exposing the available calcium hydroxide as well as the hydrates of tri- and di-calcium silicates (C_3S, C_2S) for further reaction with incoming CO_2. This can be achieved in various ways, including the flow mixing of the concrete with CO_2 before the concrete is set. In a semi-batch system, this involves turning the concrete continuously while CO_2 is being poured in. Increasing the surface area of the cement aggregate phase is another viable means. Since the reactions take place at the exposed surface, the use of particles with larger surface areas can bring about an improvement in carbonation efficiency.

Furthermore, concrete being exposed to a higher CO_2 concentration in the environment might accelerate the carbonation process [23]. Besides cement variety, the water–

cement ratio and the relative humidity, an important external factor that can limit carbonization is the concentration of CO_2 [21]. Haselbach and Thomas [21] investigated the carbonation of decades-old concrete sidewalk samples. They found the carbonation level decreasing with increasing thickness from the surface of the sample. Up to 80% carbonation was calculated for surface samples. The effects of curing and the time of carbonation were demonstrated on carbonation efficiency by [24], with up to 24% carbonation found for an initial curing time of 18 h, but 8.5% efficiency observed in the absence of initial curing. The authors reported up to 35% efficiency for 4-day carbonation time, when using slag in the concrete. For 2 h of carbonation after 1 h of cooling, [9] recorded only 15% efficiency in hollow-core concrete slab.

Furthermore, the carbonation process improves the mechanical properties of concrete [25,26]. This indicates that mankind will benefit immensely from the practice of concrete carbonation. The quality of calcium carbonate generated by the carbonation process and the developed mechanical strength are both affected by the internal water content [27].

On the strength of the above reports, it can be argued that the simultaneous influences of particle sizes and operating pressures have not been involved in the earlier investigations. The particle sizes, distribution and arrangement will definitely influence the effective microstructural spaces in concrete. This will further determine the pore-scale diffusivity of the CO_2, in order to improve the carbonation yield. This work presents a scientific effort to distinguish the effects of various particles sizes on carbonation efficiency under various operating parameters of pressure, carbonation time, the hydration stage and so on. Gravimetric and pyrolysis analyses were used to determine carbonation levels, while the microstructures of the carbonated and uncarbonated concrete samples were examined with the X-ray diffraction (XRD) instrument.

2. Materials and Methods

2.1. Sample Preparation

A pressure chamber made of stainless steel was used for the experiment. This has been previously described in [28]. The gas was 99.9% pure CO_2 (BOC Gases, Leicester, UK) and the mortar was prepared from Portland cement (Lafarge Nigeria Limited, CEM 11/B-L 32.5N, Ewekoro, Ogun State, Nigeria) with silica aggregates of two different sizes (4.9 mm and 0.9 mm mean particle diameters, respectively) together with tap water. The mortar samples prepared with a 0.9 mm particle size are referred to as the fine-sand sample while those prepared with a 4.9 mm particle size are referred to as the coarse-sand sample. The silica aggregates were industry-sorted into the above sizes. Therefore, no further sieving process was conducted in our laboratory.

The mix proportion recommended in Building Materials in Civil Engineering [21] was used for the sample preparation, i.e., cement: sand, stone, water, at the ratio, 1:2.21:4.09:0.60. The only modification made in this work to the mix proportion of [21] was that the sand and stone ratios were merged for either fine sand or coarse sand. This modification was necessary because this work uses either fine sand or coarse sand, but not both, in each concrete sample. This is to enable the determination of the particle size/surface area effect in the carbonation of concrete. The concrete samples were prepared in a mould of 1 cm thickness and a 5 cm diameter. To ensure uniformity of weight in each sample, 40 g of concrete mixture was weighed into the mould and then compacted uniformly using a tamping rod for each sample made. In order to investigate the effect of particle sizes, uniform size aggregates were used in each concrete sample, i.e., in every sample, either coarse or fine sand was used, but not both. As a result, for each concrete sample, the combined ratios of both aggregates were used as either fine or coarse sand, i.e., a 6.3 ratio of aggregate (i.e., 2.21 + 4.09) was used in each sample. The cement content in concrete was approximately 12.7%, and the water–cement ratio was 0.60. Both the coarse and fine aggregates were mainly of a silicate origin.

The concrete was prepared by pouring a well-mixed concrete mixture into the 1 cm high and 5 cm diameter metallic mould. This was followed by light tamping with a rod to ensure the compactness of the particles. In order to ensure the different samples had similar weight characteristics, an equal amount (40 g) of concrete was weighed into the mould every time. Figure 1 shows samples of the concrete and the sample holder for the experiment.

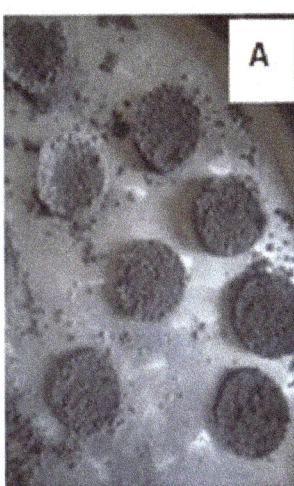
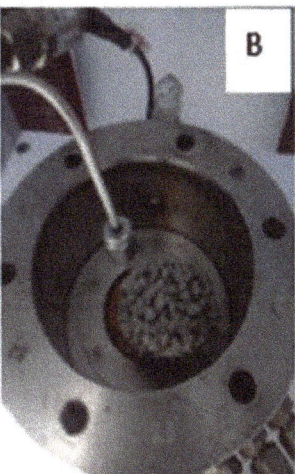

Figure 1. (**A**) Mortar samples of a 5 cm diameter and 1 cm thickness. (**B**) Sample holder with the mortar sample, ready for carbonation experiment [28].

2.2. Experimental Set-Up and Carbonation Quantification

The carbonation experiment was performed in a CO_2 pressure chamber at the Chemical Engineering Laboratory of the Loughborough University, Loughborough, Leicestershire, UK. The schematic representation of the experimental set-up is shown in Figure 2. It consists of a sample holder (pressure chamber), which is a steel cell measuring 4 cm high and 10 cm in diameter. It has stainless steel end-pieces at the top and bottom. The top end-piece of the sample holder was connected to the supercritical fluid pump (Model 260D, Isco Teledyne, city, State abbreviation, USA) via a steel tube. The pump was also connected to a pressurised CO_2 cylinder. For the purpose of the pressure chamber experiment, the bottom end-piece of the sample holder was plugged to prevent the outflow of gas. The experimental rig was located in a heating cabinet with electric heaters to regulate the system temperature. The instrument used for the temperature regulation was a PID temperature controller (West Control Solutions, Brighton, UK).

A constant temperature of 25 °C was used in all experiments. The pump was filled with CO_2 gas at the beginning of the experiment and then set at constant pressure. Three pressure regimes were used: 2, 5 and 20 bars, in order to determine the effect of pressure on the carbonation efficiency. To begin the experiment, connecting valves between the pump and chamber were opened to let gas into the chamber. The system equilibrated readily to maintain pressure and temperature. The experiment continued for different durations, e.g., 4 h, 12 h, 64 h, etc. This was done in order to investigate the influence of time duration on the carbonation efficiency.

Similarly, the effect of the hydration stage or curing time was investigated by subjecting the samples to carbonation after a different curing period or different hydration stages, e.g., 12 h, 24 h, 100 h, etc. Curing is the process in which the concrete is preserved in conditions that prevent the excessive loss of moisture to promote hydration reaction and for the concrete to gain strength. In this work, the hydration time, for the experimental concrete samples, was taken to be the time between concrete preparation and the start of the carbonation experiment for the selected samples.

Gravimetric analysis was primarily used to detect the extent of carbonation in the concrete. The experimental concrete samples were weighed before the beginning of each experiment. They were then carefully placed in the chamber to avoid crumbling or scratching, which might reduce their weights. After the set duration of carbon injection in the pressure chamber, the experiment was stopped, and the carbonated sample weighed again. The differences in the weights of the samples were primarily taken as evidence of percentages of carbonation. The quantification of carbonation was based on the mass of dry cement in the concrete according to Equation (5) [9]:

$$\text{CO}_2 \text{ Uptake } (\%) = \frac{\text{Mass after carbonation } - \text{ Mass before carbonation}}{\text{Mass of dry cement}} \times 100\% \quad (5)$$

The cement mass in concrete was determined by two means: (1) by determining the cement percentage applied in the original mix proportion, and (2) by drying selected samples of concrete at 105 °C for 24 h to drive off the water content and to determine the cement and sand content. This was followed by scraping the cement on a sand surface with a metal sponge in hot water with soap. After noting that most of the cement was scraped off, the sand particles were dried again to determine the actual amount of cement in the samples of concrete. This gravimetry method was later corroborated by the pyrolysis technique, using a few samples. The pyrolysis procedure is described below. This approach of determining the carbonation yield essentially relates to weight gain in the gravimetry technique and weight loss in the pyrolysis procedure. Martín et al. [29] relate the measurements of elemental carbon and thermo-gravimetrical weight loss to the amount of captured CO_2.

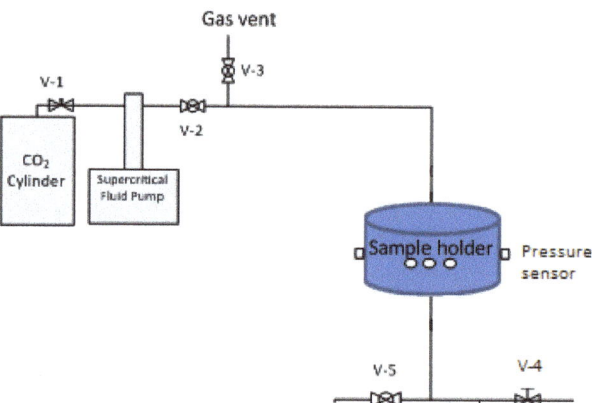

Figure 2. Experimental set-up for concrete carbonation [28].

Table 1 presents the summary of the experimental conditions for the different tests performed in this study.

Table 1. Summary of the experimental conditions of the different tests.

Exp. No	Pressure (Bar)	Temp (°C)	Curing (h)	Carbonation Duration (h)	Sand Type
1	5	25	40	4	Coarse
					Fine
2	5	25	40	64	Coarse
					Fine
3	5	25	104	6	Coarse
					Fine
4	2	25	110	7	Fine
			120	24	Fine
5	5	25	164	20	Fine
			164	26	Coarse
6	20	25	200	6	Coarse
					Fine
					Fine
7	5	25	7	4	Fine

2.3. Pyrolysis Analysis

Pyrolysis analysis was conducted on carbonated and uncarbonated concrete samples by ramping temperature between 550 °C and 1000 °C in a furnace (Carbolite, Eurothermal, Essex, UK). Before pyrolysis, the samples were first dried in an oven at 105 °C to a constant weight. Following the assumption of Leber and Blakey [30], the weight loss between 550 °C and 1000 °C was attributed to the decomposition of carbonates and was used as a direct measure for the CO_2 content with respect to the mass of cement [31,32]. Pyrolysis was only conducted on selected samples of carbonated and uncarbonated concrete. To get the final carbonation percentage, the average CO_2 content in the reference/uncarbonated concrete samples was subtracted from that in the carbonated concretes to obtain the CO_2 uptake only due to carbonation curing. The CO_2 content in the reference samples might have come from limestone used in cement or might have been from CO_2 absorbed from the atmosphere [19]. Leber and Blakey [30] estimate the carbonation degree in mortars and concretes based on the assumption that all absorbed CO_2 reacts with the limestone to form calcium carbonate. A similar assumption was made in this work

2.4. Microstructural Analysis

X-ray diffraction (XRD) and scanning electron microscopy (SEM) were performed on the carbonated and fresh (reference) concrete samples. The instrument used for the XRD was D2-Phaser Bruker (BX00412, Bruker, Coventry, UK LTD) while that of the SEM was FEGSEM (JSM-7100F, JEOL, Tokyo, Japan). In most cases, the concrete samples were split into halves and one half was carbonated. The two halves were then examined microstructurally for evidence of carbonation.

For the XRD analysis, a powdery layer of cement was extracted from selected samples of the concrete by lightly crushing the sample in a laboratory mortar by pestle. The crushed particles were then sieved in a 80 μm sieve. Wide-angle XRD patterns were obtained using a Bruker D2 Phaser diffractometer fitted with a one-dimensional LynxEye detector. A copper X-ray source ($K\alpha$ = 1.54184 Å) maintained at 30 kV and 10 mA was used, with the $K\beta$ radiation suppressed by a 0.5 mm thick nickel filter. Patterns were recorded over a 2θ range of 10–90° with a step size of 0.02° and an equivalent step time of 49.2 s. Sample rotation was set at 15 rpm. Bruker's proprietary Eva 2.0 software (version 5.2) was used to obtain the spectra. The phase identification database used in this work involved the use of the International Centre for Diffraction Data (ICDD) PDF-2.

3. Results

The results of various investigations described above are reported in this section. The results presented here are that of the CO_2 involved in the carbonation process, since the average CO_2 of the reference (uncarbonated) mortar (1.28%) has been subtracted from the results presented in the following discussion. The average CO_2 of 1.28% in the uncarbonated concrete was detected in the pyrolysis analysis. Additionally, the issue of water loss during carbonation, expounded in the work of [9], does not hold here since the carbonation was done in a sealed pressure chamber where gas and moisture did not leave the chamber until the end of the experiment, unlike in [9], where a flow-through process was used and the injected gas, together with the entrained moisture, escaped out of the chamber during the experiment. The CO_2 quantity found in reference (uncarbonated) samples might have come from the manufacturing process of cement.

We present the results of the experiments in this section, while a more detailed discussion of the results is presented in the next section (Section 4, discussion).

Gravimetric/Pyrolytic Analysis

Figure 3 shows the comparisons between the analyses conducted on similar samples using gravimetric and pyrolysis methods.

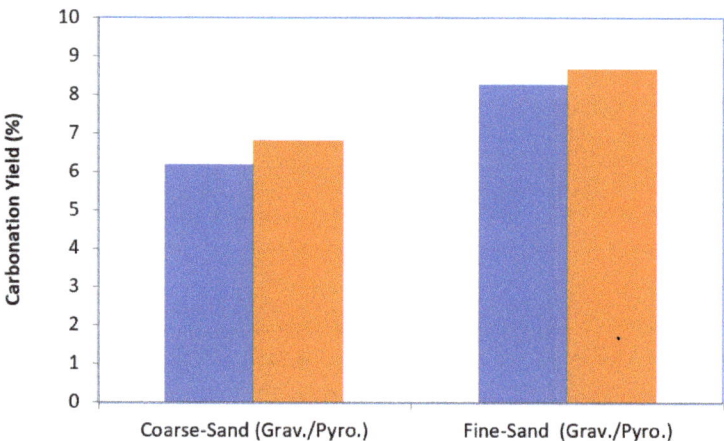

Figure 3. Comparisons of carbonation yields obtained from gravimetric and pyrolytic analysis (5 bar pressure, 104 h curing time and 6 h carbonation time).

Figure 4 shows the carbonation yield at 5 bar pressure, 4 h carbonation time and 40 h curing time, while Figure 5 shows the carbonation yield at 5 bar pressure, 64 h carbonation time and 40 h curing time. Figure 6 shows the carbonation yield at 5 bar pressure, 6 h carbonation time for 104 h curing time, while Figure 7 shows the carbonation yield at 2 bar pressure, 7 and 24 h carbonation times for 110 and 120 h curing times.

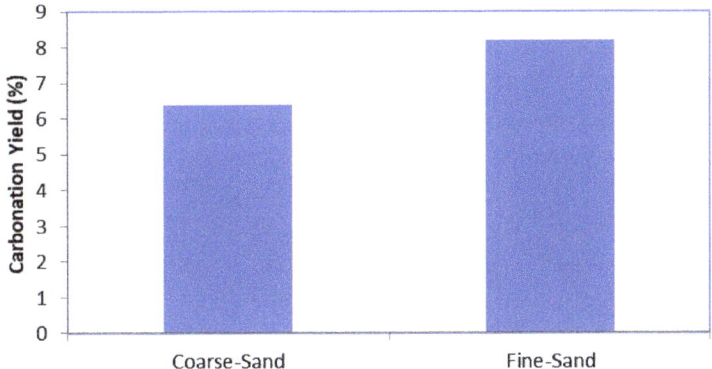

Figure 4. Carbonation yield at 5 bar pressure, 4 h carbonation time and 40 h curing time.

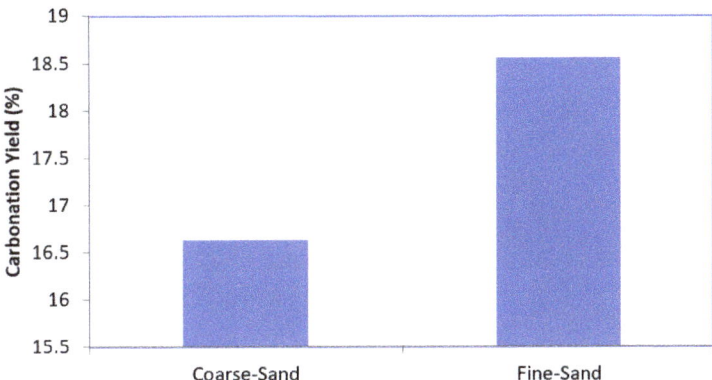

Figure 5. Carbonation yield at 5 bar pressure, 64 h carbonation time and 40 h curing time.

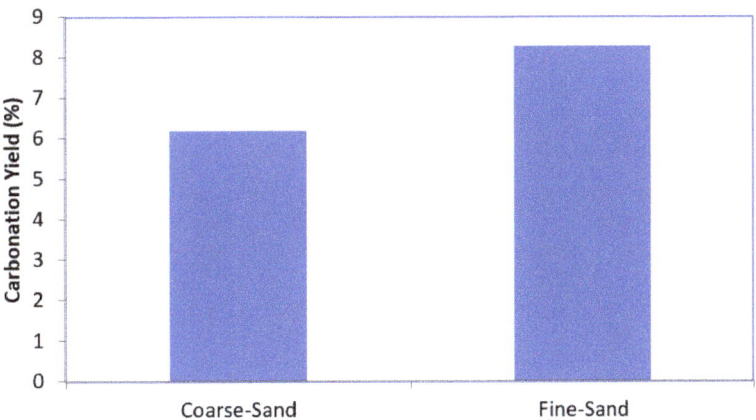

Figure 6. Carbonation yield at 5 bar pressure, 6 h carbonation time and 104 h curing time.

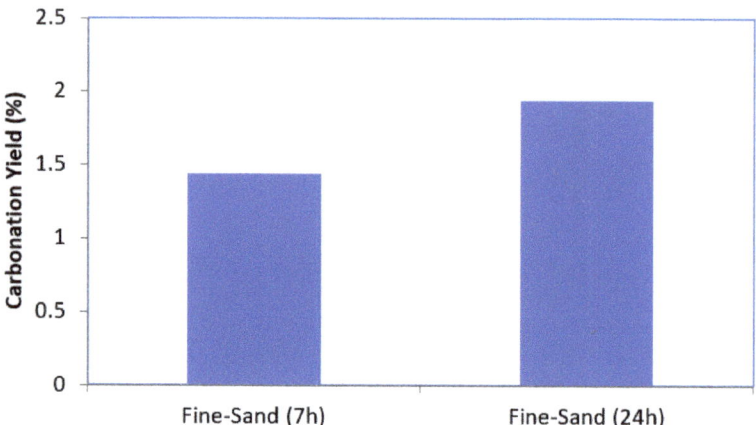

Figure 7. Carbonation yield at 2 bar pressure, 7 and 24 h carbonation times, 110 and 120 h curing times.

Figure 8 shows the carbonation yields in fine sand mortar at different pressures, namely, 5 bar (104 h curing, 6 h carbonation), and 2 bar (110 h curing, 7 h carbonation). Figure 9 shows the carbonation yield at 5 bar, carbonation/curing times for coarse (26/164 mesh) and fine sand (20/164 mesh) based mortar.

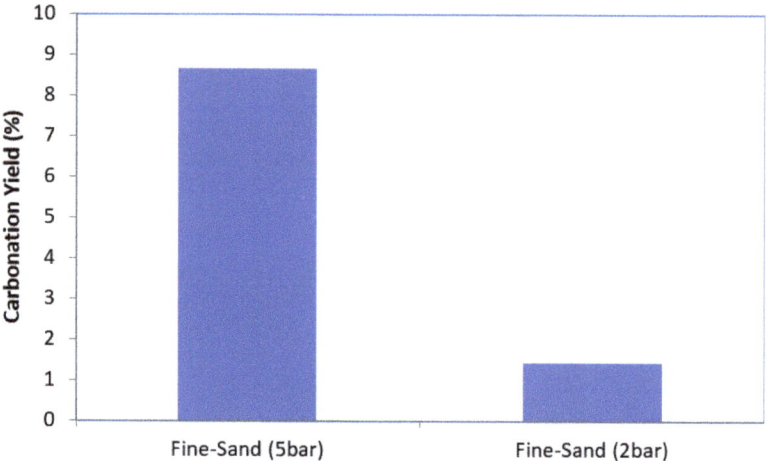

Figure 8. Carbonation yields in fine sand mortar at different pressures: 5 bar (104 h curing time, 6 h carbonation time) and 2 bar (110 h curing time, 7 h carbonation time).

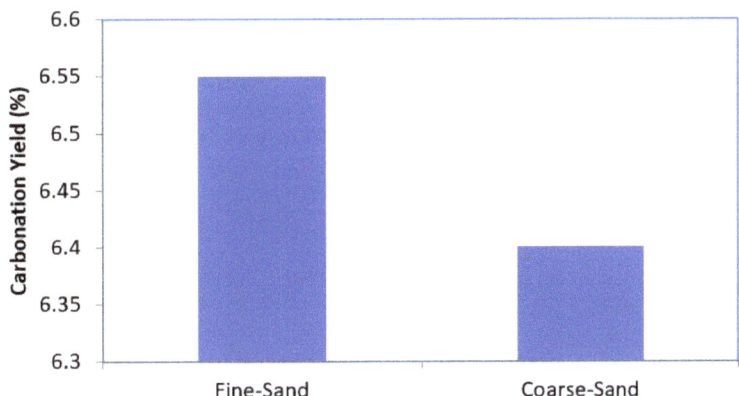

Figure 9. Carbonation yield at 5 bar, carbonation/curing times for coarse (26/164 mesh) and fine (20/164 mesh) sand mortar.

Figure 10 shows the carbonation yield at 20 bar, 6 h carbonation time, 200 h curing time, while Figure 11 shows the influence of curing time on carbonation yield.

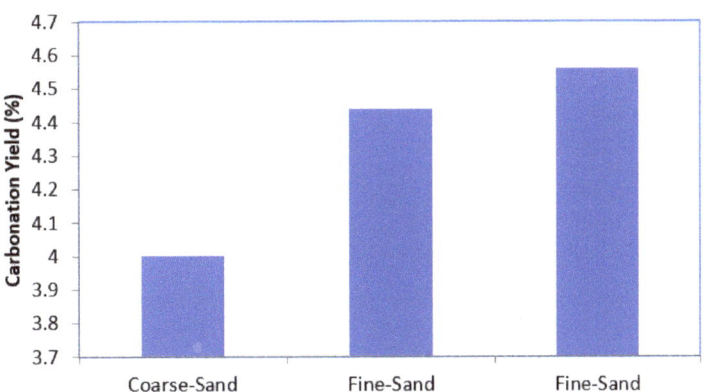

Figure 10. Carbonation yield at 20 bar, 6 h carbonation time, 200 h curing time.

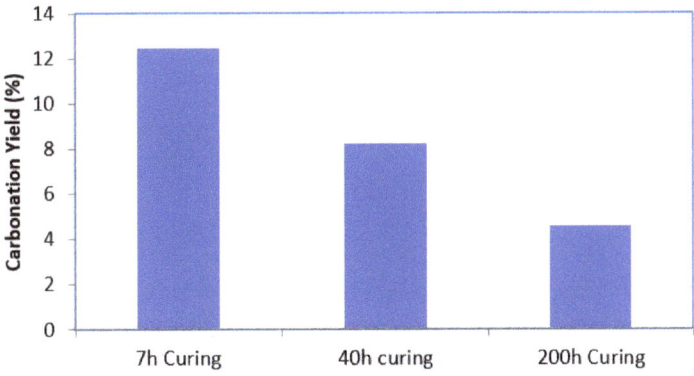

Figure 11. Influence of curing time on carbonation yield.

Figure 12 is the XRD analysis showing the tricalcium silicate (C_3S), $Ca(OH)_2$ and $CaCO_3$ patterns in the carbonated and uncarbonated mortar samples. Carbonation took place at 2 bar for 24 h following 120 h of curing. On the other hand, Figure 13 is the XRD analysis showing tricalcium silicate (C_3S), $Ca(OH)_2$ and $CaCO_3$ patterns in carbonated and uncarbonated mortar samples. In this case, carbonation took place at 5 bar for 24 h. In Figure 14 we show the XRD analysis of $CaCO_3$ peaks in carbonated mortar samples. Here, the sample 1 (red line) is 6.55% carbonated, while sample 2 (black line) is 1.94% carbonated.

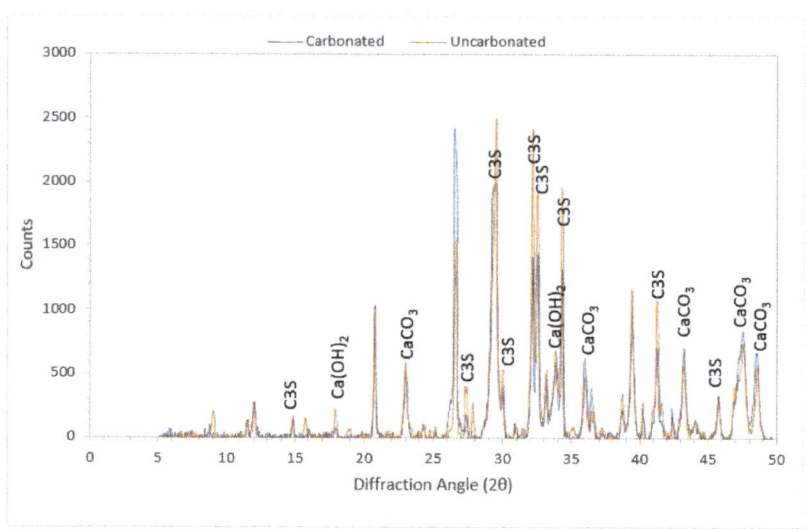

Figure 12. XRD showing tricalcium silicate (C_3S), $Ca(OH)_2$ and $CaCO_3$ patterns in the carbonated and uncarbonated mortar samples. Carbonation took place at 2 bar for 24 h following 120 h of curing [28].

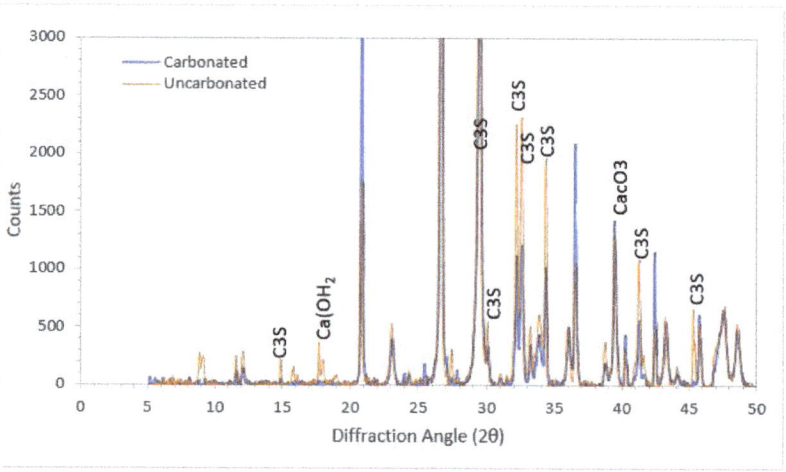

Figure 13. XRD analysis showing tricalcium silicate (C_3S), $Ca(OH)_2$ and $CaCO_3$ patterns in carbonated and uncarbonated mortar samples. Carbonation took place at 5 bar for 24 h [28].

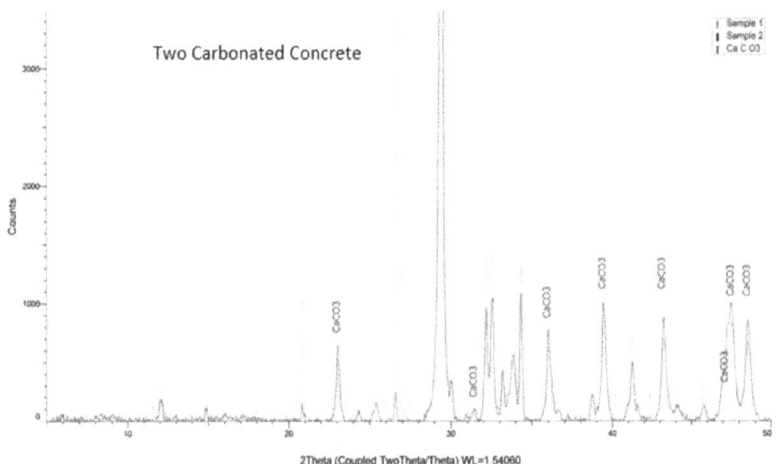

Figure 14. XRD analysis showing CaCO₃ peaks in carbonated mortar samples. Sample 1 (red line) is 6.55% carbonated, while sample 2 (black line) is 1.94% carbonated [28].

Figure 15 shows the SEM images of uncarbonated and carbonated samples. Carbonation took place at 2 bar for 24 h following 120 h of curing. Figure 16 shows the SEM images of uncarbonated and carbonated samples. Carbonation took place at 2 bar for 24 h.

Figure 15. SEM of (**A**) uncarbonated and (**B**) carbonated samples. Carbonation took place at 2 bar for 24 h following 120 h of curing.

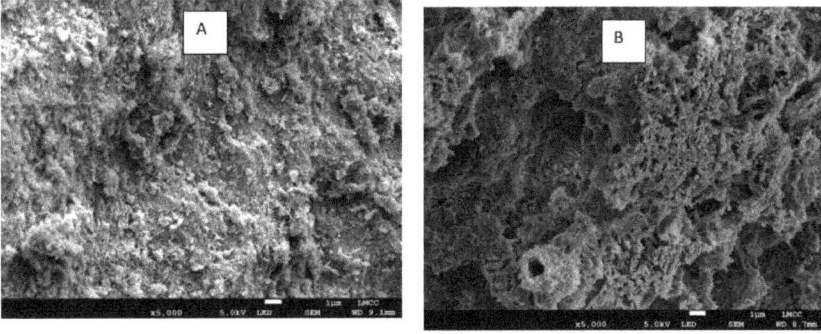

Figure 16. SEM of (**A**) uncarbonated and (**B**) carbonated samples. (2 bar for 24 h).

4. Discussion

The graphical representations of the results from the experiments were described earlier. As a way of affirming the reliability and repeatability of the results, Figure 3 shows the comparisons between the analyses conducted on similar samples using gravimetric and pyrolysis methods. Under similar conditions, the results from the two analyses (gravimetric and pyrolysis (Grav./Pyro.)) were comparable. The figure shows coarse-sand concrete with 6.19% carbonation by gravimetric analysis, while it shows 6.81% carbonation by pyrolysis. In fine-sand concrete, the results show 8.27% carbonation by gravimetric analysis while it shows 8.66% carbonation by pyrolysis. These are close results, and the trends show the reliability of either of the methods.

As shown in Figure 4, following 4 h of carbonation at 5 bar and 25 °C as well as 40 h of curing, approximately 6.4 and 8.2% carbonations were achieved in the coarse-sand and fine-sand mortar samples, respectively. The results showed higher carbonation in fine-sand mortar. Ordinarily, owing to its higher porosity, it is expected that the coarse-sand based mortar will have more interstitial pore spaces to allow CO_2 penetration and carbonation reaction to take place. However, the results show that the higher surface area present in fine-sand based concrete is a stronger factor to be considered.

Under similar conditions to that in Figure 4, but with 64 h of carbonation, the carbonation of the mortar increased to 16.6% and 18.3% for coarse-sand and fine-sand based mortar samples, respectively. This also gives higher carbonation in fine-sand based mortar than the coarse-sand type, thereby confirming the influence of particle sizes on the carbonation reactions in mortar. It further shows the influence of carbonation time on the reaction. Thus, under suitable conditions, longer exposure of concrete to CO_2 plume will improve the carbonation reaction. The plot of these results is shown in Figure 5.

After 104 h of curing and 6 h of carbonation time, similar levels of carbonation were observed as after 4 h of carbonation (Figure 4), while other conditions were similar. The results show that carbonation efficiency is reduced as curing time increases. This is depicted in Figure 6. In the figure, there are 6.2% and 8.3% carbonations in coarse-sand and fine-sand based concrete samples, respectively, which, again, shows higher carbonation potential in fine-sand mortar.

The results show that the carbonation time and particle sizes are stronger factors in carbonation efficiency while the curing time has a reverse influence. However, the slightly longer carbonation time (6 h) with the slightly lower carbonation yield and longer curing time compared to Figure 4 implies that the carbonation potential reduces as the curing of the mortar progresses. This means that as silicates are consumed during the hydration of concrete, the carbonation potential reduces [9,32]. Early-age carbonation of the concrete has been reported to improve the mechanical strength of concrete composite [25]. Thus, there is a higher advantage in concrete carbonation at a shorter curing period. Further, since the concrete samples used here are not rewetted with water after preparation, the result might imply that as carbonation progresses, concrete loses water, which reduces its potential for reaction with CO_2 under dry conditions [21]. It should be noted that the works of [9] as well as [24] employed rewetting of the mortar, which might have enhanced carbonation, while [24] compensates for the water loss during initial curing and carbonation curing with water spray, which was applied to restore the original water content. The early carbonation approach was also employed by [9] by conducting the process after about 2 h of curing. This approach of compensating water loss is also expected to improve carbonation efficiency by making use of the moderate moisture in the mortar. This practice of compensating water loss or rewetting obviously explains the differences in the results of the current work and earlier investigations by [24] as well as by [9]. Adding liquid water into samples at appropriate time intervals could enhance carbonation reactions effectively, with a maximum improvement of 34.1% previously recorded [22].

Figure 7 shows the results of experiments conducted at a lower pressure of 2 bar in order to test the influence of pressure on the carbonation yield in the mortar. The results show that the carbonation yield became lower than before, ostensibly owing to the lower pressure of the experiment. Similar to the earlier observations, the influence of carbonation time is reflected in the results, with the yield at 24 h carbonation time (1.9%) obviously higher than that at 7 h (1.4%). Figure 8 shows the effect of pressure on the carbonation yields in fine sand samples under the similar conditions of the curing and carbonation periods. The results show the clear positive influence of pressure on the carbonation efficiency under similar conditions.

Figure 9 shows another interesting dimension to the previous discussions. In this case, experiments were conducted at 5 bar but at a longer curing period of 164 h. These experiments took place for longer carbonation times of 20 h and 26 h. Despite the higher carbonation time, the yields here (6.5% for fine and 6.4% for coarse) are even lower than those for fine-sand mortar at 4 h carbonation time (8.2% in Figure 4). These results defy the influence of the carbonation time, as propounded earlier. However, the ostensible reason for this can be traced to the longer curing time of 164 h. Owing to this lengthy period of curing, much of the original silicates in the mortar have been consumed by the hydration reaction and much moisture was lost as the hydration progressed. Thus, the hydration stage accounts for the degree of carbonation in concrete. Furthermore, the influence of particle size is evident in the results. The coarse-sand concrete had a lower carbonation yield than fine-sand concrete. This was despite the longer number of hours involved for the carbonation of coarse-sand concrete (26 h) compared to the time involved for the fine-sand concrete (20 h). As stated earlier, a longer curing time often leads to reduced water content in the sample, which hampers the efficiency of the carbonation process. Carbonization will stop when the relative humidity reaches 100% (or in water) or is less than 25% (or in the dry environment) [21].

Figure 10 shows the experimental results obtained at a higher pressure in concrete using 20 bar gas pressure in the chamber. Approximately 4.0% carbonation was achieved in the coarse-sand mortar, while the first and second fine-sand mortars had 4.4% and 4.6% carbonation yields, respectively. This again shows the influence of particle size in the carbonation efficiency of concrete. However, compared to the earlier results (see, e.g., Figure 6), the expected influence of higher pressures is missing in this case (Figure 10). In Figure 6, a more than 8% carbonation yield was obtained at 5 bar, 104 h curing time and 6 h carbonation time. The low carbonation yield in Figure 10 (e.g., 4.6%), was ostensibly as a result of a higher curing period of 200 h, which reduces the carbonation efficiency. Thus, it can be inferred that the curing or hydration stage has great influence on the carbonation in the mortar. After days of hydration, the carbonation effectiveness reduces. The stage of hydration at which this effectiveness starts to decline needs to be investigated in the future. However, rewetting the mortar can reduce the effect of moisture loss on the carbonation potential after a long period of curing. Furthermore, as shown in Figure 10, the results of the repeat experiments for the two fine samples (4.4% and 4.6%) show the repeatability of the results in these investigations. The expected influence of higher pressures can be seen between the performances of the carbonation processes in Figures 6 and 7, under similar curing conditions. In Figure 6, at a pressure of 5 bar, fine-sand concrete recorded 8.27% carbonation at a 104 h curing period and 6 h of carbonation time, while in Figure 7, at a pressure of 2 bar, fine-sand concrete recorded 1.4% carbonation at a 110 h curing period and 7 h of carbonation time. This performance, despite the higher carbonation time in the latter, shows the obvious influence of pressure (5 bar in the former and 2 bar in the latter).

The influence of the curing period on the carbonation yield is comparatively shown in Figure 11, with yields depicted for 7 h of curing (5 bar, 4 h carbonation), 40 h of curing (5 bar, 4 h carbonation) and 200 h of curing (20 bar pressure, 6 h carbonation). After 4 h carbonation time, 12.5% carbonation was achieved in the fine-sand mortar cured for 7 h, while around one-third of the yield (4.56%) was obtained for 200 h-cured concrete. This clearly depicts the strong influence of the curing period on the carbonation efficiency in concrete.

Summary analysis of the gravimetric results shows that for a curing period of 7 h (Figure 11), more than 12% carbonation was achieved. The carbonation reduced to 8.2% at the 40 h curing period (Figures 4 and 11). On the pressure effect, for comparable curing conditions, 2 bar at 7 h carbonation time gave a 1.4% yield (Figure 5) while 5 bar at 6 h carbonation time gave more than 8.2% (Figure 6). Furthermore, analysing the effect of carbonation time for comparable conditions shows that 4 h of carbonation time gave up to a 8.2% yield, while 64 h of carbonation gave up to 18.5%. The carbonation time is effective in ensuring that CO_2 diffusion takes place at the surface and core parts of the concrete. Thus, the longer the carbonation time, the further the reach of the gas in the concrete structure.

In comparison, [9] record up to 14.5% carbonation in 2 h of carbonation following 2 h of heat curing. This carbonation yield is comparable to the 64 h of carbonation and 40 h of curing in this work (Figure 5). It may appear that the efficiency is higher in the work of [9], but a consideration of the different experimental processes and conditions explains some underlying factors. In [9], they used a flow-through experiment where CO_2 was made to pass through the microstructural pores in the mortar, unlike in this work, where CO_2 had to overcome surface inhibition or pore blockage to diffuse further into the mortar matrix. That explains why the authors were able to report the carbonation yields at the top, core and bottom of the carbonated mortar. Furthermore, the carbonation by [9] took place immediately following 2 h of curing. This enhances the performance of the process, unlike in the current case, where carbonation took place after 40 h of curing, which resulted in more loss of moisture. Furthermore, the cement content in the mortar used by [9] was 15% compared to 12.7% used in this work. They used a water content of 0.3 compared to 0.6 in this work. Thus, the mortar samples in [9] had more pore space owing to the lower water content and more binder cement, both of which contributed to their better carbonation performance.

Ref [24] reported up to 24% carbonation with initial curing of up to 18 h. The effect of carbonation time was also reported, with up to 35% carbonation based on 4-day carbonation time (96 h) recorded. Again, this seems to show a better carbonation than recorded in this work. However, it should be noted that the authors used slag in their concrete mixture, which might have enhanced the carbonation efficiency. Pozzolana cement and blast-furnace slag cement readily undergo carbonization [21]. Similar to the current findings, their work corroborated the fact that carbonation increases with CO_2 exposure time (carbonation time). The observation that carbonation increases with time is similar to the findings of [18] on the carbonation of decades-old concrete sidewalk samples.

In Figure 12, at 2θ =14.9, 27.5, 29.6, 32.3, 32.6, 34.3 and 41.3, the peaks of tricalcium silicate (C_3S) were conspicuously stronger in the uncarbonated mortar samples than in the carbonated ones. The reason for this is that the carbonation reactions reduced the amount of remaining silicates in the concrete [19]. At 2θ = 36.1, 43.3, 47.5 and 48.6, $CaCO_3$ peaks were stronger in the carbonated mortar samples, which shows evidence for carbonation. At 2θ = 18.1 and 34.2, stronger peaks appear for $Ca(OH)_2$ in uncarbonated samples than in the carbonated ones. This observation is similar to the findings by [19]. This observation may be accounted for by the fact that $Ca(OH)_2$ growth is inhibited by carbonation in carbonated mortar while its growth is less-restrained in uncarbonated mortar samples.

Figure 13 shows further evidence of carbonation in mortar samples. In the figure, peaks for $Ca(OH)_2$ and C_3S are conspicuously stronger in the uncarbonated samples than in the carbonated ones. This implies that these compounds were consumed by the carbonation process, leading to their reduced quantities in the carbonated sample. Evidence of carbonation is further indicated in the carbonated sample at 2θ = 39.6, with a stronger peak for $CaCO_3$ in the carbonated mortar. Meanwhile, it should be noted that the reaction of CO_2 with $Ca(OH)_2$ during the carbonation process leads to reduced pH. Thus, acidity increases in carbonated mortar, which may have a detrimental effect on the steel in reinforced concrete. Therefore, carbonation may be more suitable for unreinforced concrete.

Furthermore, Figure 14 shows relative carbonation in two carbonated mortar samples with $CaCO_3$ peaks. Sample 1 (red line) was 6.55% carbonated while sample 2 (black line) was 1.94% carbonated. At every point considered, there are relatively stronger peaks in sample 1 (6.55%) than sample 2 (1.94%). This indicates the increase of carbonate content in carbonated samples and may be taken as evidence of relative percentages of carbonation.

Micrographs of scanning electron microscopy (SEM) for the carbonated and uncarbonated concrete samples are shown in Figure 15. The uncarbonated sample (Figure 15A) displays the amorphous C-S-H phase with $Ca(OH)_2$ background. The ettringite needles are hardly visible in the uncarbonated sample owing to the early stage curing of the samples (120 h or 5 days). In [9], they did not observe ettringite needles in heat-cured concrete until after 28 days. Therefore, the faint and scarce needle-like structures at the background of the samples may indicate the early formation of the product. In their work, [33] observed a small amount of ettringite at the early stage but more at a later age.

Figure 15B shows a uniform denser and cloudy mass of carbonated concrete with a crispy look and without the elements of connecting needles or a nebulous connection as in Figure 15A. The absence of the needle-like connection in the carbonated concrete can be attributed to the early stage of curing as well as the inhibition of ettringite formation by the presence of CO_2 and $CaCO_3$ formation [19,34]. According to [34], early-age carbonation curing decomposed this ettringite into calcium carbonates. In the figure, the $CaCO_3$ has been seamlessly integrated into the C-S-H phase to give a more solid and denser structure. As concrete is carbonated, it becomes denser, because $CaCO_3$ occupies a greater volume than the $Ca(OH)_2$, which it replaces in the concrete [35,36]. Thus, Figure 15B reflects the evidence of carbonation. The darker appearance of the micrograph in Figure 15B is also evidence of carbonation in the sample. The production of crystalline $CaCO_3$ under carbonation had a filling effect that refined the pore sizes [34], thus leading to a uniform denser and cloudy mass of carbonated concrete with a crispy look.

Figure 16 shows another set of carbonated and counterpart uncarbonated concrete samples. In Figure 16A, widespread distribution of the needle form of C-S-H (calcium-silicate-hydrate) is visible together with the background $Ca(OH)_2$ phase. These patterns disappeared in the carbonated concrete, leading to denser-looking structure (Figure 16B). This dense look can mean increased strength in the carbonated concrete. Carbon dioxide reacts with tricalcium silicate (C_3S) and this accelerates the setting and early strength development [37].

Similarly, a change in appearance is also noticeable. The carbonated concrete has a darker appearance than the uncarbonated sample. This is likely the effect of carbonation on the sample. The un-hydrated cement particles appear brightest [38], while the carbonated concrete has more grey to dark grey areas [39].

5. Conclusions

Laboratory experiments and micro-structural analyses were carried out to show the performance of Portland cement mortar for storage of carbon dioxide. Particle size, carbonation time, curing time and carbonation pressure exhibited various effects on the efficiency of carbon storage in Portland cement mortar. The carbonation efficiency increased with an increased surface area, which resulted in higher carbonation in fine-sand based mortar samples. The results showed about 28% higher carbonation in fine-sand based mortar samples as compared to coarse-sand based mortar samples. The hydration time had a reversed effect on carbonation, with a reduction in the carbonation level by about 31% recorded, owing to a difference of about 33 h hydration time in the mortar samples. The carbonation efficiency increased with pressure, from 1.4% to 8.2% for a pressure rise from 2 bar to 5 bar. Similarly, the duration of carbonation shows a positive effect, with the carbonation efficiency rising from 8.2% to 18.5% for a change in carbonation duration from 4 to 64 h. It can be reliably inferred that, under similar conditions, the carbonation efficiency will increase with lower-sized particles or a higher-surface area, increase with carbonation time, increase with higher pressure but decrease with hydration time. Microstructural analyses with X-ray diffraction (XRD) and scanning electron microscopy (SEM) further show the consumption of calcium-silicate-hydrate (C-S-H) together with the inhibition of ettringite formation by the presence of CO_2 and $CaCO_3$ formation during carbonation.

Owing to the consumption of $Ca(OH)_2$, carbonation leads to reduced pH, which engenders corrosion in reinforced concrete. Therefore, carbonation should be limited to unreinforced concrete.

Author Contributions: Conceptualization, L.K.A. and D.B.D.; methodology, L.K.A.; software, D.B.D.; validation, D.B.D.; formal analysis, L.K.A.; investigation, L.K.A.; resources, D.B.D.; data curation, L.K.A.; writing—original draft preparation, L.K.A.; writing—review and editing, L.K.A. and D.B.D.; visualization, L.K.A.; supervision, D.B.D.; project administration, D.B.D.; funding acquisition, D.B.D. All authors have read and agreed to the published version of the manuscript.

Funding: This research received no external funding.

Institutional Review Board Statement: Not applicable.

Informed Consent Statement: Not applicable.

Data Availability Statement: The corresponding author can be contacted for further information and access to raw data.

Acknowledgments: The authors acknowledge Loughborough University, Loughborough, UK, especially the Department of Chemical Engineering for granting Abidoye, L.K. the use of its laboratory in which this work was successfully executed. The authors are also grateful to the technologists in the Department of Chemical Engineering, Loughborough University, especially Sean Creedon and Tony Eyre, for their efforts in the successful academic visit of Abidoye.

Conflicts of Interest: The authors declare no conflict of interest.

References

1. Petvipusit, K.R.; Elsheikh, A.H.; Laforce, T.C.; King, P.R.; Blunt, M.J. Robust optimisation of CO_2 sequestration strategies under geological uncertainty using adaptive sparse grid surrogates. *Comput. Geosci.* **2014**, *18*, 1–16. [CrossRef]
2. Stephens, J.C. Growing interest in carbon capture and storage (CCS) for climate change mitigation. *Sustain. Sci. Pract. Policy* **2006**, *2*, 4–13. [CrossRef]
3. Rabiu, K.O.; Abidoye, L.K.; Das, D.B. Geo-electrical characterization for CO2 sequestration in porous media. *Environ. Process.* **2017**. [CrossRef]
4. Abidoye, L.K.; Das, D.B. Tracking CO_2 Migration in Storage Aquifer. In *Carbon Capture, Utilization and Sequestration*; Agarwal, R.K., Ed.; IntechOpen: London, UK, 2018. Available online: https://www.intechopen.com/books/carbon-capture-utilization-and-sequestration/tracking-co2-migration-in-storage-aquifer (accessed on 12 January 2019). [CrossRef]
5. Goel, G.; Abidoye, L.K.; Das, D.B.; Chahar, B.R.; Singh, R. Scale dependency of dynamic relative permeability-saturation curves in relation with fluid viscosity and dynamic capillary pressure effect. *Environ. Fluid Mech.* **2016**, *16*, 945–963. [CrossRef]
6. Wang, Z. Effects of Impurities on CO_2 Geological Storage. Master's Thesis, University of Ottawa, Ottawa, ON, Canada, 2015.
7. Abidoye, L.K.; Das, D.B.; Khudaida, K. Geological carbon sequestration in the context of two-phase flow in porous media: A review. *J. Crit. Rev. Environ. Sci. Technol.* **2015**, *45*, 1105–1147. [CrossRef]
8. Mathieson, A.; Wright, I.; Roberts, D.; Ringrose, P. Satellite imaging to monitor CO_2 movement at Krechba, Algeria. *Energy Procedia* **2009**, *1*, 2201–2209. [CrossRef]
9. Shao, Y.; Morshed, A.Z. Early carbonation for hollow-core concrete slab curingand carbon dioxide recycling. *Mater. Struct.* **2015**, *48*, 307–319. [CrossRef]
10. Adedokun, D.A.; Ndambuki, J.M.; Salim, R.W. Improving Carbon Sequestration in Concrete: A Literature Review. *Int. Sch. Sci. Res. Innov.* **2013**, *7*, 269–271.
11. Lagerblad, B. *Carbon Dioxide Uptake during Concrete Life Cycle, State of the Art*; Swedish Cement and Concrete Research Institute—CBI: Gothenburg, Sweden, 2005; pp. 1–47. ISBN 91-976070-0-2.
12. Matschei, T.; Lothenbach, B.; Glasser, F. The AFM phase in Portland cement. *Cem. Concr. Res.* **2007**, *37*, 118–130. [CrossRef]
13. Yaseen, S.A.; Yiseen, G.A.; Li, Z. Elucidation of Calcite Structure of Calcium Carbonate Formation Based on Hydrated Cement Mixed with Graphene Oxide and Reduced Graphene Oxide. *ACS Omega* **2019**, *4*, 10160–10170. [CrossRef]
14. Šavija, B.; Lukovic, M. Carbonation of Cement Paste: Understanding, Challenges, and Opportunities. *Constr. Build. Mater.* **2016**, *117*, 285–301. [CrossRef]
15. Stepkowska, E.T.; Perez-Rodríguez, J.L.; Sayagues, M.J.; Martínez-Blanes, J.M. Calcite, Vaterite and Aragonite Forming on Cement Hydration from Liquid and Gaseous Phase. *J. Therm. Anal. Calorim.* **2003**, *73*, 247–269.
16. Hills, T.; Leeson, D.; Florin, N.; Fennell, P. Carbon Capture in the Cement Industry: Technologies, Progress, and Retrofitting. *Environ. Sci. Technol.* **2016**, *50*, 368–377. [CrossRef]
17. Elzinga, D.; Bennett, S.; Best, D.; Burnard, K.; Cazzola, P.; D'Ambrosio, D.; Dulac, J.; Fernandez Pales, A.; Hood, C.; LaFrance, M.; et al. *Energy Technology Perspectives 2015: Mobilising Innovation to Accelerate Climate Action*; International Energy Agency: Paris, France, 2015.

18. Haselbach, L.; Thomas, A. Carbon sequestration in concrete sidewalk samples. *Constr. Build. Mater.* **2014**, *54*, 47–52. [CrossRef]
19. Galan, I.; Andrade, C.; Mora, P.; Sanjuan, M.A. Sequestration of CO_2 by concrete carbonation. *Environ. Sci. Technol.* **2010**, *44*, 3181–3186. [CrossRef] [PubMed]
20. Pade, C.; Guimaraes, M. The CO_2 uptake of concrete in a 100 year perspective. *Cem. Concr. Res.* **2007**, *37*, 1348–1356. [CrossRef]
21. Zhang, H. *Building Materials in Civil Engineering*; Woodhead Publishing Limited: Cambridge, UK, 2011; ISBN 978-1-84569-955-0.
22. Wang, D.; Noguchi, T.; Nozaki, T. Increasing efficiency of carbon dioxide sequestration through high temperature carbonation of cement-based materials. *J. Clean. Prod.* **2019**, *238*, 117980. [CrossRef]
23. Monkman, S.; Shao, Y. Assessing the carbonation behavior of cementitious materials. *J. Mater. Civ. Eng.* **2006**, *18*, 768–776. [CrossRef]
24. El-Hassan, H.; Shao, Y. Carbon Storage through Concrete Block Carbonation Curing. *J. Clean Energy Technol.* **2014**, *2*, 287–291. [CrossRef]
25. Junior, A.N.; Ferreira, S.R.; Filho, R.D.T.; Fairbairn, E.M.R.; Dweck, J. Effect of early age curing carbonation on the mechanical properties and durability of high initial strength Portland cement and lime-pozolan composites reinforced with long sisal fibres. *Compos. Part B Eng.* **2019**, *163*, 351–362. [CrossRef]
26. Ashraf, W.; Olek, J.; Sahu, S. Phase evolution and strength development during carbonation of low-lime calcium silicate cement (CSC). *Constr. Build. Mater.* **2019**, *210*, 473–482. [CrossRef]
27. Ho, L.S.; Nakarai, K.; Ogawa, Y.; Sasaki, T.; Morioka, M. Effect of internal water content on carbonation progress in cement-treated sand and effect of carbonation on compressive strength. *Cem. Concr. Compos.* **2018**, *85*, 9–21. [CrossRef]
28. Abidoye, L.K.; Das, D.B. Carbon storage in concrete: Influences of hydration stage, carbonation time and aggregate characteristics. In Proceedings of the 1st International Conference on Engineering and Environmental Sciences, Osun State University, Osogbo, Nigeria, 5–7 November 2019; pp. 212–224.
29. Martin, D.; Flores-Alés, V.; Aparicio, P. Proposed Methodology to Evaluate CO_2 Capture Using Construction and Demolition Waste. *Minerals* **2019**, *9*, 612. [CrossRef]
30. Leber, I.; Blakey, F.A. Some effects of carbon dioxide on mortars and concrete. *J. Am. Concr. Inst.* **1956**, *53*, 295–308. [CrossRef]
31. Young, J.F.; Berger, R.L.; Breese, J. Accelerated curing of compacted calcium silicate mortars on exposure to CO_2. *J. Am. Ceram. Soc.* **1974**, *57*, 394–397. [CrossRef]
32. Khoshnazar, R.; Shao, Y. Characterization of carbonation-cured cement paste using X-ray photoelectron spectroscopy. *Constr. Build. Mater.* **2018**, *168*, 598–605. [CrossRef]
33. Siedel, H.; Hempel, S.; Hempel, R. Secondary ettringite formation in heat treated portland cement concrete: Influence of different W/C ratios and heat treatment temperatures. *Cem. Concr. Res.* **1993**, *23*, 453–461. [CrossRef]
34. Liu, Y.; Zhuge, Y.; Chow, C.-W.K.; Keegan, A.; Li, D.; Pham, P.N.; Huang, J.; Siddique, R. Properties and microstructure of concrete blocks incorporating drinking water treatment sludge exposed to early-age carbonation curing. *J. Clean. Prod.* **2020**, *261*, 121257. [CrossRef]
35. You, K.; Jeong, H.; Hyung, W. Effects of Accelerated Carbonation on Physical Properties of Mortar. *J. Asian Archit. Build. Eng.* **2014**, *13*, 217–221. [CrossRef]
36. Mors, R.; Jonkers, H. Effect on Concrete Surface Water Absorption upon Addition of Lactate Derived Agent. *Coatings* **2017**, *7*, 51. [CrossRef]
37. De Weerdt, K.; Ben Haha, M.; Le Saout, G.; Kjellsen, K.O.; Justnes, H.; Lothenbach, B. Hydration mechanisms of ternary Portland cementscontaining limestone powder and fly ash. *Cem. Concr. Res.* **2011**, *41*, 279–291. [CrossRef]
38. Zhao, H.; Darwin, D. Quantitative backscattered electron analysis of cement paste. *Cem. Concr. Res.* **1992**, *22*, 695–706. [CrossRef]
39. Supit, S.W.M.; Shaikh, F.U.A. Effect of Nano-$CaCO_3$ on Compressive Strength Development of High Volume Fly Ash Mortars and Concretes. *J. Adv. Concr. Technol.* **2014**, *12*, 178–186. [CrossRef]

Article

Carbon Capture from Biogas by Deep Eutectic Solvents: A COSMO Study to Evaluate the Effect of Impurities on Solubility and Selectivity

Thomas Quaid and M. Toufiq Reza *

Department of Biomedical and Chemical Engineering and Sciences, Florida Institute of Technology, 150 West University Boulevard, Melbourne, FL 32901, USA; tquaid2018@my.fit.edu
* Correspondence: treza@fit.edu; Tel.: +1-321-674-8578

Abstract: Deep eutectic solvents (DES) are compounds of a hydrogen bond donor (HBD) and a hydrogen bond acceptor (HBA) that contain a depressed melting point compared to their individual constituents. DES have been studied for their use as carbon capture media and biogas upgrading. However, contaminants' presence in biogas might affect the carbon capture by DES. In this study, conductor-like screening model for real solvents (COSMO-RS) was used to determine the effect of temperature, pressure, and selective contaminants on five DES' namely, choline chloride-urea, choline chloride-ethylene glycol, tetra butyl ammonium chloride-ethylene glycol, tetra butyl ammonium bromide-decanoic acid, and tetra octyl ammonium chloride-decanoic acid. Impurities studied in this paper are hydrogen sulfide, ammonia, water, nitrogen, octamethyltrisiloxane, and decamethylcyclopentasiloxane. At infinite dilution, CO_2 solubility dependence upon temperature in each DES was examined by means of Henry's Law constants. Next, the systems were modeled from infinite dilution to equilibrium using the modified Raoults' Law, where CO_2 solubility dependence upon pressure was examined. Finally, solubility of CO_2 and CH_4 in the various DES were explored with the presence of varying mole percent of selective contaminants. Among the parameters studied, it was found that the HBD of the solvent is the most determinant factor for the effectiveness of CO_2 solubility. Other factors affecting the solubility are alkyl chain length of the HBA, the associated halogen, and the resulting polarity of the DES. It was also found that choline chloride-urea is the most selective to CO_2, but has the lowest CO_2 solubility, and is the most polar among other solvents. On the other hand, tetraoctylammonium chloride-decanoic acid is the least selective, has the highest maximum CO_2 solubility, is the least polar, and is the least affected by its environment.

Keywords: biogas; carbon capture; deep eutectic solvents; Henry's Law; Raoult's Law; selectivity; solubility

Citation: Quaid, T.; Reza, M.T. Carbon Capture from Biogas by Deep Eutectic Solvents: A COSMO Study to Evaluate the Effect of Impurities on Solubility and Selectivity. *Clean Technol.* **2021**, *3*, 490–502. https://doi.org/10.3390/cleantechnol3020029

Academic Editor: Diganta B. Das

Received: 25 March 2021
Accepted: 18 May 2021
Published: 1 June 2021

Publisher's Note: MDPI stays neutral with regard to jurisdictional claims in published maps and institutional affiliations.

Copyright: © 2021 by the authors. Licensee MDPI, Basel, Switzerland. This article is an open access article distributed under the terms and conditions of the Creative Commons Attribution (CC BY) license (https://creativecommons.org/licenses/by/4.0/).

1. Introduction

Anaerobic digestion (AD) is the process of breaking down organic substances in anoxic conditions by bacteria [1]. Organic macro-molecules such as fats, carbohydrates, and proteins are digested into micro-molecules during AD, which results in a nutrient-rich solid for plants (fertilizer) and biogas [2]. This process occurs naturally in landfills, but also in a controlled environment in equipment called anaerobic digestors. The feedstock for AD are materials that are otherwise considered waste, such as agricultural waste, manure, organic waste from animal processing plants, food waste, and many others [3,4]. The growing adoption of AD offers a new approach to these waste streams which supports a recycle economy that increases market efficiency and bolsters the renewable energy industry as the globe shifts towards green fuel.

During AD, several reactions occur, but the process can be categorized into four stages: hydrolysis, acidogenesis, acetogenesis, and methanogenesis. During hydrolysis, long-chain polymers like cellulose are hydrolyzed into fermentable forms like glucose.

Acidogenesis and acetogenesis are characterized by the generation of hydrogen gas and carbon dioxide from monomers and glucose. The final stage, methanogenesis, is the stage where most of the methane is produced. Apart from CH_4 and CO_2, several other impurities are formed dependent upon the feed, such as ammonia, hydrogen sulfide, water, nitrogen, and siloxanes. The presence of CO_2 and the impurities lower the overall energy content of the biogas and can cause premature failure of point-of-use equipment [5]. For these reasons, carbon capture and biogas upgrading are often required prior to biogas application. Currently, biogas upgrading is conventionally performed through amine-based ionic liquid absorption or water scrubbing [6]. Ionic liquid (IL) amine-based absorption is desirable due to the solvents having a high selectivity for CO_2 over CH_4, which can achieve ~99% CH_4 purity [6–9]. However, the high viscosity, high cost, and toxicity of these solvents suggest the need for an alternative [9–11]. Water scrubbing has a high efficiency (~97% CH_4 purity achieved), but it has been associated with bacterial growth issues, massive water consumption, and its necessity for additional processes in series to remove feed impurities [6,9]. Other processes have also been developed for CO_2 removal, such as solid sorbents. These solid-based sorbents are found to have a large range of CO_2 capacity that reach up to 80 weight percent but have high operating temperatures that exceed 500 °C [12,13]. However, due to the low combustive properties of some impurities, low-temperature solid adsorbents like zeolites are the only feasible option, which have significantly lower capacities [14,15].

Deep eutectic solvents (DES) are a relatively new material that is being studied as a carbon capture media [16–18]. DES are made from a hydrogen bond donor (HBD) and a hydrogen bond acceptor (HBA) [16,19]. The melting point of DES is decreased significantly compared to individual HBA and HBD due to charge delocalization from hydrogen bonding [20–22]. Studies have proven DES to exhibit desirable traits for use as a CO_2 absorbent, such as thermal stability, tunability, reversibility, and reasonable CO_2 solubility [14,17,23], with Zhang et al. [15] reporting a 1:1 mol CO_2 per mol solvent solubility ratio [24], Bi et al. [25] reporting a 0.25 g/g of CO_2 per solvent solubility, and Ren et al. [25] reporting 0.4 mol CO_2 per mol solvent solubility. The literature often uses experimental methods to develop CO_2 capture on DES. However, the use of computational software with highly accurate determinations may make the down-selection of DES easier. Therefore, conductor-like screening model for real solvents (COSMO-RS), which is a thermodynamic property prediction software that relies on the generation of sigma profiles rather than databases of functional group interactions, was used in this study. COSMO has been used by several authors to model CO_2 capture, such as Song et al. [26], who was able to screen a database containing thousands of HBD and HBA combinations for potential CO_2-capturing solvents. Of the various DES, quaternary ammonium salts have garnered a significant amount of attention for their ability to solvate CO_2 [18,27]. The accuracy of COSMO was also studied by Liu et al. [28] by testing hundreds of DES for CO_2 absorption, and they found a maximum of 10.3% error after tuning the program across the studied samples. Several studies have been performed on the solubility of CO_2 in DES [22,29], however, to the best of the authors' knowledge, none was conducted on understanding how various impurities in biogas affect the carbon capture by DES. This knowledge is essential to design an absorption system for biogas upgrading since solubility and selectivity of a solvent can be adversely affected by contaminants, especially when accounting for accumulation during repeated use.

This study focuses on evaluating the affinity various DES have for selected contaminants and how their presence in various amounts affects the affinity for CO_2 in these solvents. This will be performed using COSMO by first modeling the DES and contaminants not found in the software library, then generating thermophysical properties of Henry's Law constants and activity. Selectivity of CO_2 over CH_4 and solubility of CO_2 changes in a selected group of DES were studied here for both infinite dilution and partial pressure at various temperature ranges. Finally, effects of impurities ranging from 0 to 5 mole % on CO_2 solubility in various DES were evaluated.

2. Materials and Methods

2.1. Composition of Biogas

The standard percent ranges of biogas composition used in this study have been listed in Table 1. The variance of the composition depends upon several factors surrounding the AD process, such as temperature, retention time, kinetics, and feed stock composition [30]. Table 1 shows the components studied with their respective abbreviations for the investigation and their industrial compositions.

Table 1. Pre-treatment biogas components and composition for studied molecules.

Molecule	Abbreviation	Composition Volume %	PPM	References
Hydrogen Sulfide	H_2S	0–2	0–10,000	[31,32]
Ammonia	NH_3	0–1	0–100	[31,33]
Nitrogen	N_2	0–15	-	[31]
Water	H_2O	5–10	-	[32]
Propanone	Acetone	-	0–15	[34]
Octamethyltrisiloxane	Octa	-	0–41.35	[35]
Decamethylcyclopentasiloxane	Deca	-	0–5.17	[33]
Carbon Dioxide	CO_2	15–47	-	[31]
Methane	CH_4	35–70	-	[31]

2.2. Deep Eutectic Solvents

Table 2 lists the five common DES considered for this study, including choline chloride-urea, choline chloride-ethylene glycol, tetra butyl ammonium chloride-ethylene glycol, tetra butyl ammonium bromide-decanoic acid, and tetra octyl ammonium chloride-decanoic acid, along with their components and component mixing ratios. The solvents studied are termed quaternary ammonium salts due to the structure of the HBD. The quaternary ammonium salts are relatively cheap, safe for the environment, and naturally derived [16,36,37]. The specific solvents were chosen as an attempt to represent a large range of their class by means of carbon chain length of the quaternary ammonium salts and commonly paired HBDs.

Table 2. Selected deep eutectic solvents for biogas upgrading and their abbreviations.

DES	Abbreviation	HBA	HBD	Molar Ratio	Molar Mass (g/g mol)
$N_{8888}Br$:Decanoic Acid	N8Br:DA	$N_{8888}Br$	Decanoic Acid	1:3	1019.08
$N_{4444}Br$:Decanoic Acid	N4Br:DA	$N_{4444}Br$	Decanoic Acid	1:3	839.15
$N_{4444}Cl$:Ethylene Glycol	N4Cl:EG	$N_{4444}Cl$	Ethylene Glycol	1:3	464.11
ChCl:Ethylene glycol	ChCl:EG	ChCl	Ethylene Glycol	1:3	325.83
ChCl:Urea	ChCl:U	ChCl	Urea	1:2	259.74

2.3. COSMO Simulation

COSMO is a quantum modeling software that determines thermodynamic properties using density functional theory (DFT). To determine the thermodynamic properties, the HBAs and HBDs are modeled using TurboMoleX software. The impurities are selected from the COSMO library. HBAs and HBDs are then mathematically evaluated for their natural geometrical lowest energy state and conformers. COSMO was then used for all thermophysical property calculations. TurboMoleX® was used to generate all molecular sigma profiles, conformers, and data not already found in the included database. TZVP (tri-zeta-valence-polarized) settings were used with default numerical grid of m3 and BP86 functions. COSMOThermX® was used for all thermodynamic property calculations. These properties were used to calculate sigma profile of the molecules, where charge density is plotted with charge of the molecule. Here, the molecule is differentiated into charge density segments, with each segment representing areas with charge density ranging from -0.3 to $+0.3$ e/Å^2. The charge density segments are plotted to form the sigma profiles. The data from the sigma profiles are used to model microscopic molecular surface charge interactions between analytes, then a statistical thermodynamic procedure is carried out to derive macroscopic thermodynamic properties from the generated information [38]. The base values generated are chemical potentials of the systems' constituents, these are then applied to thermodynamic calculations of Henry's Law coefficient and activity coefficients. Determination of the solubility and selectivity of the systems was carried out by COSMO-RS, whose results are based upon the chemical potential generated by COSMO-RS.

3. Results and Discussion

3.1. Sigma Profiles of DES's, Polar, and Non-Polar Molecules

A sigma profile is a distribution function that relates the surface area of a molecule to the charge density of the surface [39]. In this study, sigma profiles are used to understand the electrostatic interactions between DES and selected polar and nonpolar molecules. The sigma profiles explain the trends of solubility and selectivity for a DES-based extraction. To generate these profiles, COSMO creates incremental segments of the studied molecule, which are then organized based upon surface charge density. The area under these sigma profile curves gives the total surface area of the studied molecule. Peaks between ± 0.0082 e/Å^2 charge density indicate that the molecule readily undergoes van der Waals interactions [39,40]. Peaks outside of this range indicate hydrogen bonding as the preferred interaction due to polarity [40].

Sigma profiles are useful for determining how molecules will interact in a solvent-solute system. From a range of sigma profiles, appropriate solvents may be identified for a given molecule based on how the charge densities between the two profiles align. A highly polar solvent that has significant charge density in the HBA region (-0.0082 e/Å^2) could be expected to have a high affinity for a solute that shows a significant charge density in the HBD region ($+0.0082$ e/Å^2). The same is true for two molecules that have significant charge densities in the non-polar region of the sigma profile (± 0.0082 e/Å^2). This logic can be used to determine if an impurity will have a lesser or higher affinity than a solute, giving rise to competition for the solvents' binding sites.

In Figure 1, the sigma profiles of each DES are displayed. The order of the solvents from the most to the least polar and, therefore, most available for hydrogen bonding to least available, are as follows: ChCl:U > ChCl:EG > N4Cl:EG > N4Br:DA > N8Br:DA. The peaks between 0.015 and 0.002 e/Å^2 are from the halogens associated with each solvent. It is observed that by changing the HBD groups as with the tetrabutylammonium variants, the sigma structure is significantly altered, which lends to the notion of DES properties being highly tunable [16,41].

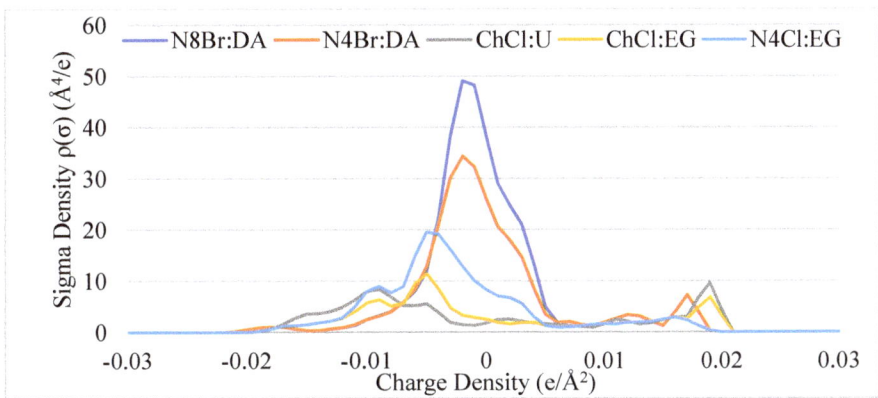

Figure 1. Sigma potential profiles of DES with respect to charge density.

Sigma profiles of non-polar gases can be seen in Figure 2. For the non-polar gases, the key difference in the sigma profiles of the molecules is the charge density distribution of CO_2 vs. N_2 vs. CH_4. N_2 and CH_4 have most of their area concentrated around the zero-x-axis compared to CO_2. CO_2 is considered a non-polar gas, since the distribution of the charges for CO_2 are weighted between ± 0.0082 e/Å^2. However, CO_2 can be influenced by its environment to make it behave more like a polar molecule and participate in hydrogen bonding or behave more like a non-polar molecule and participate in van der Waals interactions. The potential for this behavior can be seen in the sigma profile as the charge density is concentrated closely to the ± 0.0082 e/Å^2 boundary. It is also understood that CO_2 contains two polar bonds, but the linear structure of the molecule creates a net-zero dipole moment. However, in a polar environment such as CO_2 in water, it behaves as an acid gas.

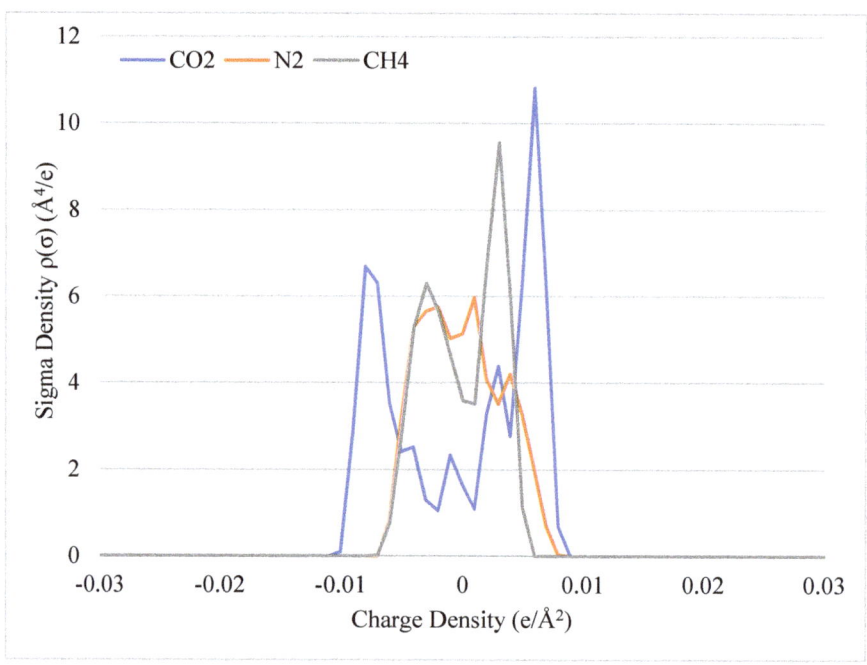

Figure 2. Sigma potential profiles of non-polar molecules with respect to charge density.

Regarding polar gases, only gases reported as impurities of biogas are selected for this study. There exist large variations in profiles among this group, as seen in Figure 3. The most notable impurity is water, which reaches the farthest among the other gases on the charge density and is relatively symmetric, which concludes its adaptability in assuming the roles as a Lewis acid or base. Acetone has a large peak near the 0 e/Å^2 yet behaves as a Lewis base due to the considerable peak beyond 0.01 e/Å^2. H_2S is relatively evenly dispersed along the x-axis, suggesting it can participate in both van der Waals interactions and hydrogen bonding depending upon its environment. SO_2 is heavily concentrated around the boundaries of ± 0.0082 e/Å^2, and as such, would be expected to have lower solubility among the less polar DES. Ammonia is a weak base, and this is indicated in the large peaks near the HBA region (-0.0082 e/Å^2) but is capable of hydrogen donating interactions, as seen in the trailing area in the positive region of the plot as it extends to nearly 0.03 e/Å^2.

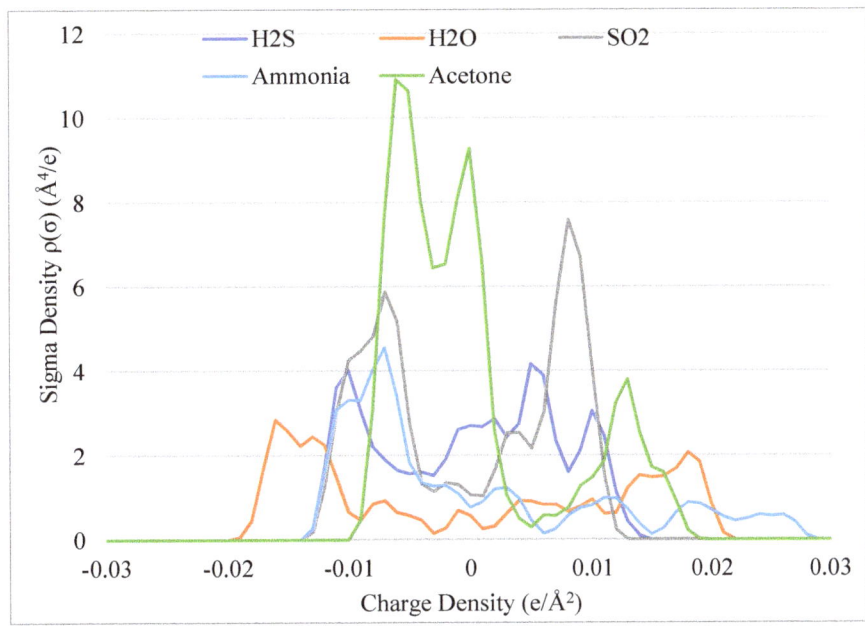

Figure 3. Sigma potential profiles of polar molecules with respect to charge density.

As discussed previously, CO_2 can be influenced by its environment to partake in hydrogen bonding or van der Waals interactions. Due to this and the generated sigma profiles, it stands to reason that a DES containing significant amounts of a polar or non-polar contaminant may change the level of solubility of CO_2 within that system. For example, when considering the relatively polar profile of ChCl:U, it could be reasoned that if it were to accumulate strong polar molecules like water then the effect of hydrogen bond affinity for CO_2 would be enhanced. Thus, resulting in a higher selectivity for CO_2 than CH_4 in this particular solvent.

3.2. Selectivity for CO_2 over CH_4 by DES in Infinite Dilution

Considering the valuable product of biogas upgrading is methane, the selectivity of a solvent to solvate is of significant importance. The selectivity of CO_2 over CH_4 was first studied for various DES at infinite dilution by Henry's Law calculations and presented in Figure 4. Henry's Law constants are used to study the solubility of CO_2 vs. CH_4 for a pure DES regarding the first molecules of gas and how they selectively enter the DES and are only valid at low concentrations of gases in the DES. At room temperature and at infinite dilution, the largest selectivity of 4.7 can be observed in ChCl:U. Here, approximately 4.7 moles of CO_2 are expected to be absorbed per mole of CH_4. The least selective solvent in this model is N8Br:DA at approximately 1.75. The remaining solvents show a slight trend up from N8Br:DA. The data follows a rational trend of selectivity to size, with the smallest DES molecular constituents displaying the highest selectivity. However, this does not explain the dramatic increase in selectivity between ChCl:EG and ChCl:U, considering they are nearly the same mass (Table 2) and considering the selectivity is molar-based. This behavior could be explained from sigma profiles. Figure 1 shows ChCl:U as being the most likely to participate in hydrogen bonding of the five solvents and N8Br:DA as most likely to participate in van der Waals interactions. As previously mentioned, CO_2 can become polarized in a polar environment, which makes it much more likely to bind with ChCl:U than methane. In a relatively non-polar environment like N8Br:DA, both molecules will behave non-polar and bind closer to a 1:1 ratio. The values for simulated vs. experimental solubilities of CO_2 in ChCl:U at 5.6 MPa and 303.15 K are reported as

5.7 and 3.56 (mol/kg), respectively. The difference was reported to be caused by poorly optimized DES structures [42]. Xie et al. and Ji et al. report experimental solubilities of CO_2 in ChCl:U at 308.2 K and 0.651 and 0.678 p/MPa respectively, as 0.05 and 0.045 mole fraction, respectively. The solubility parameters were studied in this paper at a highest-pressure condition of 0.6 MPa and 25 °C, and for ChCl:U, the solubility of CO_2 at these conditions is 0.074. The discrepancies between experimental and calculated values could be attributed by the limitations of COSMO to fully model all solvation phenomena that occur, such as hole theory, induced polarity of solutes, and induced conformers of analytes. The selectivity appears to be mostly influenced by the polarity of the DES at room temperature. Similar observation was found in the literature, where Slupek et al. [10] compared the sigma profiles of their studied DES with solutes and determined that the overlapping regions between the two plots suggested interaction compatibility.

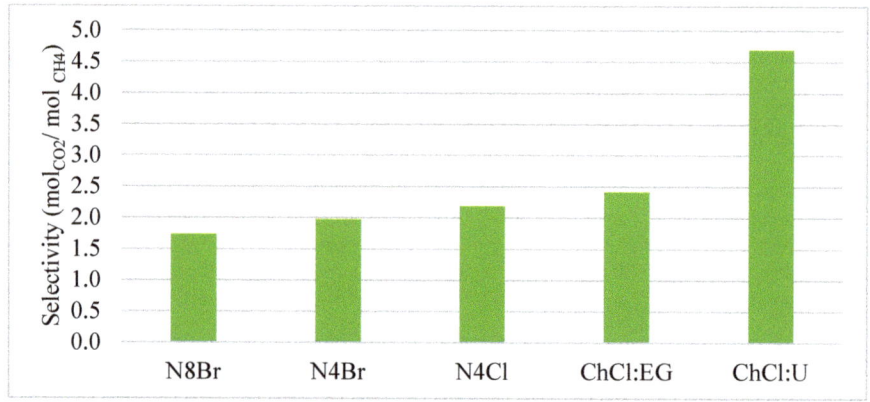

Figure 4. Selectivity of CO_2 vs. CH_4 at STP and infinite dilution calculated from Henry's Law coefficients for each DES.

The selectivity thus far has been discussed at 25 °C, however, temperature of the biogas could be as high as 55 °C depending on mesophilic or thermophilic microorganisms. Therefore, the effect of temperature on selectivity at infinite dilution is of practical interest. Figure 5 has shown the effect of temperature on Henry's Law constant, which is analogous to selectivity. Due to the unit of the Henry constant, the lower values are associated with higher solubility. With the increase of temperature, the Henry's Law constant increases. Interestingly, for the same HBA (e.g., ChCl), exceptional deviations in Henry's Law constant can be found for different HBD (e.g., urea versus ethylene glycol). This is probably due to the smaller HBA chain lengths that might have a naturally smaller affinity for CO_2 [41]. However, the induced polarity phenomena have a stronger impact on the solubility outcome. This is due to CO_2 being naturally non-polar, as seen in Figure 2. Thus, the magnitude of the dipole moment of a solvent will determine the affinity CO_2 will have for it.

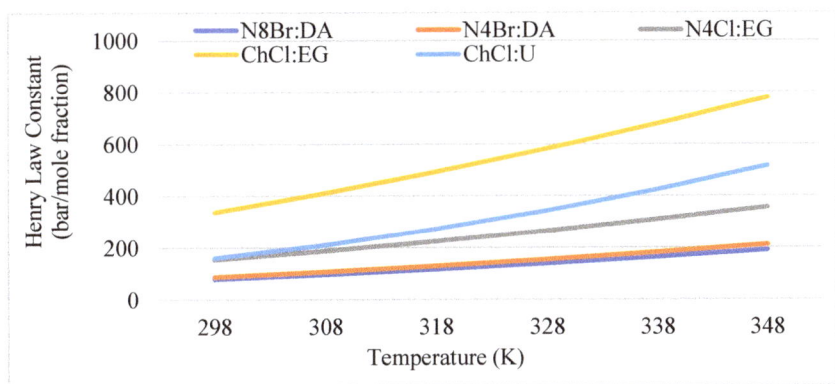

Figure 5. Effect of temperature on solubility of CO_2 at infinite dilution calculated with Henry's Law coefficients for each DES.

3.3. Effect of Pressure on Selectivity and Solubility of CO_2 in Various DES

Selectivity of CO_2 over CH_4 in various DES at infinite dilution provides valuable information on how polarity of DES and solute affect the selectivity. However, Henry's Law is only valid for infinite dilution, which might be misleading for carbon capture from biogas, as CO_2 concentration in biogas is often high. Therefore, Raoult's Law might provide more accurate information of the solubility and selectivity. In this study, modified Raoult's Law calculations are used to determine the maximum solubilities for a pure solvent by studying the last molecules to enter the system at any concentration. Understanding the effect pressure has on a system and how its constituents behave away from ideality is crucial to its design parameters. Figure 6 investigates the last molecules entering the system at equilibrium. It provides total saturation values for CO_2 on the left axis and selectivity of CO_2 vs. CH_4 on the right axis at varying partial pressures in 40% increments, since this falls within the composition range for both CO_2 and CH_4, as shown in Table 1. The first observation in this Figure 6 is the increase in solubility of CO_2 with increased pressure, regardless of solvent. The next is the same trend being seen in Figure 4 with respect to the solvent ordering of selectivity. This trend becomes significantly more pronounced when the system is closer to saturation. For example, the selectivity of ChCl:U at 1 bar is nearly 25 in Figure 6 compared to the Henry's Law calculations which were 4.7 in Figure 4. A possible explanation for this could be due to the solvent matrix becoming more of a polar environment as the holes fill with CO_2 and CH_4 has to squeeze into the smaller polarized spaces in order to occupy the solvent, which is not energetically favorable. The negative slopes of the selectivity analysis are due to the increase in pressure, as the molecules are forced into solvent, they become less selective. The more drastic change occurs within ChCl:U as the influence of polarity is overcome by the force of pressure, resulting in a non-linear relationship unlike the other less acidic solvents. The total capacity for CO_2 varies significantly between pressures, and the resulting trends of the bars suggest that the effect on the solvents also vary significantly. As discussed previously, the order of solvents in their ability to solvate CO_2 and the gaps in capacities are explained through alkyl-chain lengths [16], HBD selection, and the resulting polarity of these combinations with little effect from the halogens. The results here further confirm this by segregating the solvents into 3 visible groupings regarding solubility of CO_2 of N8Br:DA and N4Br:DA, N4Cl:EG and ChCl:U, and ChCl:EG. The most significant finding from this grouping is the relative effects on solubility between HBA chain length and associated HBD. N4Br:DA and N8Br:DA have relatively similar capacities for CO_2 that are significantly higher compared to N4Cl:EG. N4Br:DA finds a maximum ratio of approximately 1.9 over the CO_2 solubility of N4Cl:EG, where the alkyl chain lengths are the same but the HBD are different. However,

N8Br:DA only finds a maximum approximate ratio of 1.08 over the CO_2 solubility of N4Br:DA, which displays a difference in alkyl chain length but the same HBD.

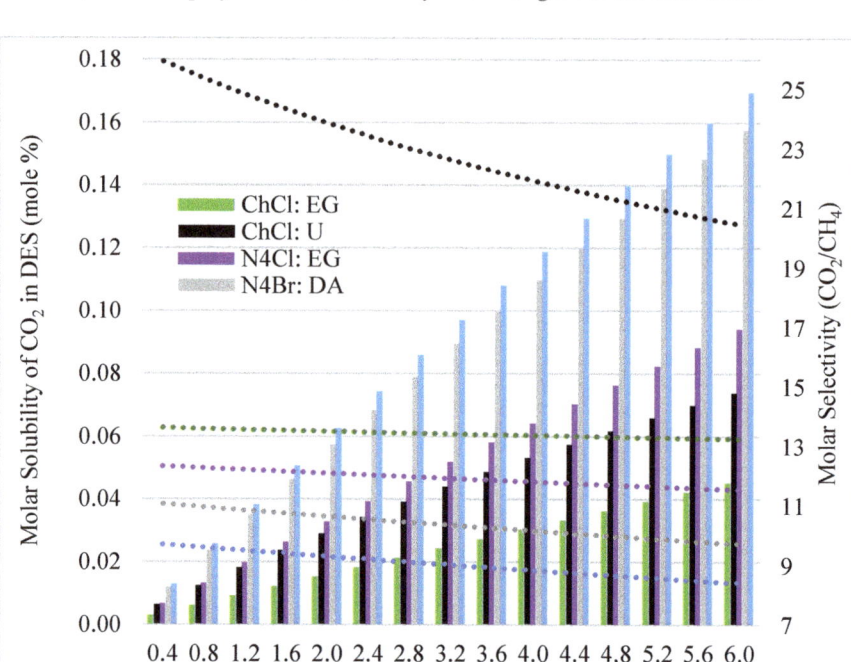

Figure 6. Effect of pressure on selectivity of CO_2 vs. CH_4 and solubility of CO_2 at equilibrium and 25 °C for each DES. Y-axis 1 is the solubility and y-axis 2 is the selectivity. The dotted lines coincide with y-axis 2 and the bars coincide with y-axis 1. The partial pressure is the same for CO_2 and CH_4.

3.4. Effect of Impurities on CO_2 Solubility in Various DES

Effect of selected impurities on CO_2 solubility of various DES at different temperatures under 3.6 bar pressure conditions are studied by solubilities. Analysis was performed on each DES to determine how the presence of contaminants within the feed gas, captured by the solvent, would affect the absorptive capacity for CH_4 and CO_2. This was performed on a wide range of contaminants found in Table 1 over three temperatures (25, 37, and 55 °C) at ambient pressure and three mole fractions of contaminant within the solvent (1, 3, and 5 mol%). The solubilties were normalized to show the deviation from the maximum solubility of CO_2 and CH_4 at DES, with no contaminants.

Of the five DES, ChCl:U is the most affected to the presence of all the impurities within biogas, as can be seen in Table 3. With the increase of ammonia in biogas, the maximum solubility of CO_2 and CH_4 increase in ChCl:U. For instance, the values for CO_2 at 37 °C are 1.01 and 1.03 for ammonia in ChCl:U at 1 and 5 mol%, respectively. However, the presence of all other contaminants decrease the maximum solubility of both CO_2 and CH_4 in ChCl:U. All contaminants produce a change greater than 5% from the base case, with the octa and deca siloxane compounds inciting the greatest changes. This finding is significant, as Jiang et al. [43] report an average concentration of siloxanes in untreated biogas reaching up to 2000 $\frac{mg}{m^3}$. It is observed that change in temperature produces minimal effect on how the impurities in ChCl:U alter the maximum solubility of CO_2. Although, there is a significant change on the solubility of CO_2. For example, the presence of propanone at 5 mole percent in CH_4 shows a deviation from the baseline of 1 as the values 0.89 and 0.92 for temperatures of 25 and 55 °C respectively, while the same conditions provide a range of 0.93 to 0.94 for CO_2.

Table 3. Normalized values for solubility of CO_2 at various mole percentages in ChCl:U and at varying temperatures. The values are normalized to fresh solvent solubilities of respective CO_2 and CH_4.

Temp (°C)	25			37			55		
Mol%	1%	3%	5%	1%	3%	5%	1%	3%	5%
H_2O	1.01	0.99	0.96	1.00	0.98	0.96	1.00	0.98	0.95
CO_2	-	-	-	-	-	-	-	-	-
CH_4	1.00	0.96	0.92	1.00	0.95	0.91	0.99	0.95	0.91
Octa	0.92	0.84	0.76	0.99	0.82	0.76	0.96	0.83	0.78
Deca	0.91	0.81	0.74	0.99	0.80	0.75	0.95	0.81	0.78
H_2S	1.01	0.98	0.95	1.00	0.98	0.95	1.00	0.97	0.95
NH_3	1.03	1.01	1.03	1.01	1.02	1.03	1.01	1.02	1.03
N_2	1.02	0.99	0.99	1.00	0.99	0.98	1.00	0.99	0.98
Acetone	1.01	0.97	0.94	1.00	0.96	0.93	1.00	0.96	0.93
SO_2	1.00	0.97	0.92	1.00	0.96	0.92	0.99	0.95	0.92

The solvents with the HBD of ethylene glycol (in Supplementary Information Tables S1 and S2) show a positive effect from every contaminant except H_2O, H_2S, and SO_2. The other contaminants show asymmetry with a weighted area around the HBA region, whereas H_2O, H_2S, and SO_2 are significantly more symmetrical regarding sigma profiles. A notable difference between the two DES with these HBD groups is the response to the contaminants at varying concentrations. At lower concentrations of the contaminants (1 mol%), N4Cl:EG is much more affected in terms of maximum CO_2 and CH_4 solubility compared to its ChCl:EG counterpart, but the opposite is true at higher concentrations. For example, at 25 °C, the CO_2 maximum solubility increases by 4% when octa makes up 1 mole percent of N4Cl:EG, however there is virtually no change when these same conditions are met for ChCl:EG as a value of 1 is reported. The trend found in ChCl:U between the temperature change and solubility change is not present in either of these DES.

The solvents with the HBD decanoic acid (N4Br:DA and N8Br:DA, Tables S3 and S4, respectively) show negative effects from all contaminants except siloxanes. Here, CH_4 solubility increases with the presence of octa and deca but CO_2 decreases with their presence. For these two DES, another similar trend follows regarding CO_2 and CH_4 solubility. The solubility varies little with contaminant mole percent, with nearly all changes being within 2%, with the exception of H_2O and ammonia for N8Br:DA and H_2O, ammonia, deca, and octa for N4Br:DA. At 1% contamination presence, the solubility of CO_2 in both DES start above 1 with higher solubility and decrease with increasing percentages of contaminant. Another trend to note is the slightly less negative effect the contaminants have upon N4Br:DA than N8Br:DA, whose main difference is their alkyl chain length.

4. Conclusions

The results of this study contain important preliminary data regarding the implementation of DES in biogas upgrading systems. The fundamental understanding of the solvents and their behavior under various temperatures, pressures, and influences from contaminants show that a complex web of variables exists that must be considered when choosing a DES for any application. It has been shown that the polarity of a solvent, its size, and its constituents are factors contributing to solubility, but the main determinant is the HBD selection. The significance of the varied contaminant concentrations is providing a method to model the accumulation that occurs within recycled solvent, where not all contaminants will be purged through the regeneration process. This study is a glimpse into the potential lifetime of the solvent, and how each solvent will be suited for a specific feed gas composition. The results show that the DES are affected by these contaminants in varying degrees in order of most to least, as follows: ChCl:U, ChCl:EG, N4Cl:EG, N4Br:DA, and N8Br:DA. This trend is the same for polarity and the reverse of alkyl chain length, and also suggests the order in which the length of time the solvents will be able to operate before regeneration is necessary, from least to most. The pressure study suggests the ideal operating environment is closer to atmospheric pressure considering selectivity but not for solubility. The selectivity at ambient temperature and pressure (STP) and infinite dilution are 4.7, 2.4, 2.2, 2.0, and 1.7 mol CO_2/mol CH_4 for ChCl:U, ChCl:EG, N4Cl:EG, N4Br:DA, and N8Br:DA, respectively. However, the selectivity at STP and finite dilution conditions are 25.9, 13.6, 12.3, 11.1, and 9.7 mol CO_2/mol CH_4. For ChCl:U, the absorbance was decreased by the presence of deca at STP and 1, 3, and 5 mole % by 0.91, 0.81, and 0.74 respectively, from a normalized value of 1. The changes in the presence of CH_4 at STP and 1, 3, and 5 mole % are 1.00, 0.96, and 0.92, respectively. These solvents have been shown to behave differently to each other when subjected to differing environmental factors such as temperature and pressure. All of these factors point to high tunability and complexity for these solvents.

Supplementary Materials: The following are available online at https://www.mdpi.com/article/10.3390/cleantechnol3020029/s1, Table S1: Normalized values for solubility of CO_2 and CH_4 at various mole percentages in ChCl: EG and at varying temperatures. The values are normalized to fresh solvent solubilities of respective CO_2 and CH_4., Table S2: Normalized values for solubility of CO_2 and CH_4 at various mole percentages in N4Cl: EG and at varying temperatures. The values are normalized to fresh solvent solubilities of respective CO_2 and CH_4., Table S3: Normalized values for solubility of CO_2 and CH_4 at various mole percentages in N4Br: DA and at varying temperatures. The values are normalized to fresh solvent solubilities of respective CO_2 and CH_4., Table S4: Normalized values for solubility of CO_2 and CH_4 at various mole percentages in N8Br: DA and at varying temperatures. The values are normalized to fresh solvent solubilities of respective CO_2 and CH_4.

Author Contributions: Conceptualization, T.Q. and M.T.R..; methodology, T.Q.; software, T.Q.; validation, T.Q.; formal analysis, T.Q.; investigation, T.Q.; resources, M.T.R.; data curation, T.Q.: writing—original draft preparation, T.Q.; writing—review and editing, T.Q. and M.T.R.; visualization, T.Q.; supervision, M.T.R.; project administration, M.T.R.; funding acquisition M.T.R. All authors have read and agreed to the published version of the manuscript.

Funding: This material is based upon work supported by the National Science Foundation under Grant No. 1856058 and American Chemical Society Petroleum Research Fund Grant No: PRF # 60342-DNI9.

Institutional Review Board Statement: Not applicable.

Informed Consent Statement: Not applicable.

Acknowledgments: We especially wish to acknowledge Kyle McGaughy for his assistance with COSMO-RS simulations.

Conflicts of Interest: The authors declare no conflict of interest.

References

1. Zhang, Q.; Hu, J.; Lee, D.-J. Biogas from anaerobic digestion processes: Research updates. *Renew. Energy* **2016**, *98*, 108–119. [CrossRef]
2. Khalid, A.; Arshad, M.; Anjum, M.; Mahmood, T.; Dawson, L. The anaerobic digestion of solid organic waste. *Waste Manag.* **2011**, *31*, 1737–1744. [CrossRef] [PubMed]
3. Holm-Nielsen, J.; Al Seadi, T.; Oleskowicz-Popiel, P. The future of anaerobic digestion and biogas utilization. *Bioresour. Technol.* **2009**, *100*, 5478–5484. [CrossRef] [PubMed]
4. Mao, C.; Feng, Y.; Wang, X.; Ren, G. Review on research achievements of biogas from anaerobic digestion. *Renew. Sustain. Energy Rev.* **2015**, *45*, 540–555. [CrossRef]
5. Santiago, R.; Moya, C.; Palomar, J. Siloxanes capture by ionic liquids: Solvent selection and process evaluation. *Chem. Eng. J.* **2020**, *401*, 126078. [CrossRef]
6. Xu, Y.; Huang, Y.; Wu, B.; Zhang, X.; Zhang, S. Biogas upgrading technologies: Energetic analysis and environmental impact assessment. *Chin. J. Chem. Eng.* **2015**, *23*, 247–254. [CrossRef]
7. García-Gutiérrez, P.; Jacquemin, J.; McCrellis, C.; Dimitriou, I.; Taylor, S.F.R.; Hardacre, C.; Allen, R.W.K. Techno-economic feasibility of selective CO_2 capture processes from biogas streams using ionic liquids as physical absorbents. *Energy Fuels* **2016**, *30*, 5052–5064. [CrossRef]

8. Xie, Y.; Björkmalm, J.; Ma, C.; Willquist, K.; Yngvesson, J.; Wallberg, O.; Ji, X. Techno-economic evaluation of biogas upgrading using ionic liquids in comparison with industrially used technology in Scandinavian anaerobic digestion plants. *Appl. Energy* **2018**, *227*, 742–750. [CrossRef]
9. Ryckebosch, E.; Drouillon, M.; Vervaeren, H. Techniques for transformation of biogas to biomethane. *Biomass Bioenergy* **2011**, *35*, 1633–1645. [CrossRef]
10. Słupek, E.; Makoś, P.; Gębicki, J. Theoretical and economic evaluation of low-cost deep eutectic solvents for effective biogas upgrading to bio-methane. *Energies* **2020**, *13*, 3379. [CrossRef]
11. Pham, T.P.T.; Cho, C.-W.; Yun, Y.-S. Environmental fate and toxicity of ionic liquids: A review. *Water Res.* **2010**, *44*, 352–372. [CrossRef]
12. Bhatta, L.K.G.; Subramanyam, S.; Chengala, M.D.; Olivera, S.; Venkatesh, K. Progress in hydrotalcite like compounds and metal-based oxides for CO_2 capture: A review. *J. Clean. Prod.* **2015**, *103*, 171–196. [CrossRef]
13. Dou, B.; Wang, C.; Song, Y.; Chen, H.; Jiang, B.; Yang, M.; Xu, Y. Solid sorbents for in-situ CO_2 removal during sorption-enhanced steam reforming process: A review. *Renew. Sustain. Energy Rev.* **2016**, *53*, 536–546. [CrossRef]
14. Siongco, K.R.; Leron, R.B.; Li, M.-H. Densities, refractive indices, and viscosities of N,N-diethylethanol ammonium chloride-glycerol or -ethylene glycol deep eutectic solvents and their aqueous solutions. *J. Chem. Thermodyn.* **2013**, *65*, 65–72. [CrossRef]
15. Bi, Y.; Hu, Z.; Lin, X.; Ahmad, N.; Xu, J.; Xu, X. Efficient CO_2 capture by a novel deep eutectic solvent through facile, one-pot synthesis with low energy consumption and feasible regeneration. *Sci. Total Environ.* **2020**, *705*, 135798. [CrossRef] [PubMed]
16. Sarmad, S.; Nikjoo, D.; Mikkola, J.-P. Amine functionalized deep eutectic solvent for CO_2 capture: Measurements and modeling. *J. Mol. Liq.* **2020**, *309*, 113159. [CrossRef]
17. Ghaedi, H.; Ayoub, M.; Sufian, S.; Shariff, A.M.; Hailegiorgis, S.M.; Khan, S.N. CO_2 capture with the help of Phosphonium-based deep eutectic solvents. *J. Mol. Liq.* **2017**, *243*, 564–571. [CrossRef]
18. Leron, R.B.; Li, M.-H. Solubility of carbon dioxide in a choline chloride-ethylene glycol based deep eutectic solvent. *Thermochim. Acta* **2013**, *551*, 14–19. [CrossRef]
19. Abbott, A.P.; Capper, G.; Davies, D.L.; Rasheed, R.K.; Tambyrajah, V. Novel solvent properties of choline chloride/urea mixtures. *Chem. Commun.* **2003**, 70–71. [CrossRef]
20. Smith, E.L.; Abbott, A.P.; Ryder, K.S. Deep Eutectic Solvents (DESs) and their applications. *Chem. Rev.* **2014**, *114*, 11060–11082. [CrossRef]
21. Perna, F.M.; Vitale, P.; Capriati, V. Deep eutectic solvents and their applications as green solvents. *Curr. Opin. Green Sustain. Chem.* **2020**, *21*, 27–33. [CrossRef]
22. Deep Eutectic Solvents Formed between Choline Chloride and Carboxylic Acids: Versatile Alternatives to Ionic Liquids | Journal of the American Chemical Society. Available online: https://pubs-acs-org.portal.lib.fit.edu/doi/10.1021/ja048266j (accessed on 15 March 2021).
23. Figueroa, J.D.; Fout, T.; Plasynski, S.; McIlvried, H.; Srivastava, R.D. Advances in CO_2 capture technology—The U.S. Department of Energy's carbon sequestration program. *Int. J. Greenh. Gas Control.* **2008**, *2*, 9–20. [CrossRef]
24. Zhang, N.; Huang, Z.; Zhang, H.; Ma, J.; Jiang, B.; Zhang, L. Highly efficient and reversible CO_2 capture by task-specific deep eutectic solvents. *Ind. Eng. Chem. Res.* **2019**, *58*, 13321–13329. [CrossRef]
25. Ren, H.; Lian, S.; Wang, X.; Zhang, Y.; Duan, E. Exploiting the hydrophilic role of natural deep eutectic solvents for greening CO_2 capture. *J. Clean. Prod.* **2018**, *193*, 802–810. [CrossRef]
26. Song, Z.; Hu, X.; Wu, H.; Mei, M.; Linke, S.; Zhou, T.; Qi, Z.; Sundmacher, K. Systematic Screening of deep eutectic solvents as sustainable separation media exemplified by the CO_2 capture process. *ACS Sustain. Chem. Eng.* **2020**, *8*, 8741–8751. [CrossRef]
27. McGaughy, K.; Reza, M.T. Systems analysis of SO_2-CO_2 Co-capture from a post-combustion coal-fired power plant in deep eutectic solvents. *Energies* **2020**, *13*, 438. [CrossRef]
28. Liu, Y.; Yu, H.; Sun, Y.; Zeng, S.; Zhang, X.; Nie, Y.; Zhang, S.; Ji, X. Screening deep eutectic solvents for CO_2 capture with COSMO-RS. *Front. Chem.* **2020**, *8*. [CrossRef]
29. Adeyemi, I.; Abu-Zahra, M.R.; Alnashef, I. Experimental study of the solubility of CO_2 in novel amine based deep eutectic solvents. *Energy Procedia* **2017**, *105*, 1394–1400. [CrossRef]
30. Anukam, A.; Mohammadi, A.; Naqvi, M.; Granström, K. A review of the chemistry of anaerobic digestion: Methods of accelerating and optimizing process efficiency. *Processes* **2019**, *7*, 504. [CrossRef]
31. Hagen, M.; Polman, E.; Jensen, J.K.; Myken, A.; Joensson, O.; Dahl, A. Adding Gas from Biomass to the Gas Grid. 2001. Available online: https://www.osti.gov/etdeweb/biblio/20235595 (accessed on 14 March 2021).
32. Al Mamun, M.R.; Torii, S. Enhancement of methane concentration by removing contaminants from biogas mixtures using combined method of absorption and adsorption. *Int. J. Chem. Eng.* **2017**, *2017*, 7906859. [CrossRef]
33. Nyamukamba, P.; Mukumba, P.; Chikukwa, E.S.; Makaka, G. Biogas upgrading approaches with special focus on siloxane removal—A review. *Energies* **2020**, *13*, 6088. [CrossRef]
34. Zhang, Y.; Zhu, Z.; Zheng, Y.; Chen, Y.; Yin, F.; Zhang, W.; Dong, H.; Xin, H. Characterization of Volatile Organic Compound (VOC) emissions from swine manure biogas digestate storage. *Atmosphere* **2019**, *10*, 411. [CrossRef]
35. Shen, M.; Zhang, Y.; Hu, D.; Fan, J.; Zeng, G. A review on removal of siloxanes from biogas: With a special focus on volatile methylsiloxanes. *Environ. Sci. Pollut. Res.* **2018**, *25*, 30847–30862. [CrossRef]

36. Zhekenov, T.; Toksanbayev, N.; Kazakbayeva, Z.; Shah, D.; Mjalli, F.S. Formation of type III deep eutectic solvents and effect of water on their intermolecular interactions. *Fluid Phase Equilibria* **2017**, *441*, 43–48. [CrossRef]
37. Hsu, Y.-H.; Leron, R.B.; Li, M.-H. Solubility of carbon dioxide in aqueous mixtures of (reline + monoethanolamine) at T = (313.2 to 353.2)K. *J. Chem. Thermodyn.* **2014**, *72*, 94–99. [CrossRef]
38. Wichmann, K. *COSMOthermX User Guide*; COSMOlogic GmbH & Co.: Leverkusen, Germany, 2019; p. 131.
39. Palmelund, H.; Andersson, M.P.; Asgreen, C.J.; Boyd, B.J.; Rantanen, J.; Löbmann, K. Tailor-made solvents for pharmaceutical use? Experimental and computational approach for determining solubility in deep eutectic solvents (DES). *Int. J. Pharm. X* **2019**, *1*, 100034. [CrossRef]
40. McGaughy, K.; Reza, M.T. Liquid—Liquid extraction of furfural from water by hydrophobic deep eutectic solvents: Improvement of density function theory modeling with experimental validations. *ACS Omega* **2020**, *5*, 22305–22313. [CrossRef]
41. García, G.; Aparicio, S.; Ullah, R.; Atilhan, M. Deep eutectic solvents: Physicochemical properties and gas separation applications. *Energy Fuels* **2015**, *29*, 2616–2644. [CrossRef]
42. Alioui, O.; Benguerba, Y.; Alnashef, I.M. Investigation of the CO_2-solubility in deep eutectic solvents using COSMO-RS and molecular dynamics methods. *J. Mol. Liq.* **2020**, *307*, 113005. [CrossRef]
43. Jiang, T.; Zhong, W.; Jafari, T.; Du, S.; He, J.; Fu, Y.-J.; Singh, P.; Suib, S.L. Siloxane D4 adsorption by mesoporous aluminosilicates. *Chem. Eng. J.* **2016**, *289*, 356–364. [CrossRef]

Article

Integrated and Metal Free Synthesis of Dimethyl Carbonate and Glycidol from Glycerol Derived 1,3-Dichloro-2-propanol via CO_2 Capture

Santosh Khokarale [1,*], Ganesh Shelke [1] and Jyri-Pekka Mikkola [1,2,*]

[1] Technical Chemistry, Chemical-Biological Centre, Department of Chemistry, Umeå University, SE-90187 Umeå, Sweden; ganesh.shelke@umu.se
[2] Industrial Chemistry & Reaction Engineering, Johan Gadolin Process Chemistry Centre, Department of Chemical Engineering, Åbo Akademi University, FI-20500 Åbo-Turku, Finland
* Correspondence: santosh.khokarale@umu.se (S.K.); jyri-pekka.mikkola@umu.se (J.-P.M.)

Abstract: Dimethyl carbonate (DMC) and glycidol are considered industrially important chemical entities and there is a great benefit if these moieties can be synthesized from biomass-derived feedstocks such as glycerol or its derivatives. In this report, both DMC and glycidol were synthesized in an integrated process from glycerol derived 1,3-dichloro-2-propanol and CO_2 through a metal-free reaction approach and at mild reaction conditions. Initially, the chlorinated cyclic carbonate, i.e., 3-chloro-1,2-propylenecarbonate was synthesized using the equivalent interaction of organic superbase 1,8-diazabicyclo [5.4.0] undec-7-ene (DBU) and 1,3-dichloro-2-propanol with CO_2 at room temperature. Further, DMC and glycidol were synthesized by the base-catalyzed transesterification of 3-chloro-1,2-propylenecarbonate using DBU in methanol. The synthesis of 3-chloro-1,2-propylenecarbonate was performed in different solvents such as dimethyl sulfoxide (DMSO) and 2-methyltetrahydrofuran (2-Me-THF). In this case, 2-Me-THF further facilitated an easy separation of the product where a 97% recovery of the 3-chloro-1,2-propylenecarbonate was obtained compared to 63% with DMSO. The use of DBU as the base in the transformation of 3-chloro-1,2-propylenecarbonate further facilitates the conversion of the 3-chloro-1,2 propandiol that forms in situ during the transesterification process. Hence, in this synthetic approach, DBU not only eased the CO_2 capture and served as a base catalyst in the transesterification process, but it also performed as a reservoir for chloride ions, which further facilitates the synthesis of 3-chloro-1,2-propylenecarbonate and glycidol in the overall process. The separation of the reaction components proceeded through the solvent extraction technique where a 93 and 89% recovery of the DMC and glycidol, respectively, were obtained. The DBU superbase was recovered from its chlorinated salt, [DBUH][Cl], via a neutralization technique. The progress of the reactions as well as the purity of the recovered chemical species was confirmed by means of the NMR analysis technique. Hence, a single base, as well as a renewable solvent comprising an integrated process approach was carried out under mild reaction conditions where CO_2 sequestration along with industrially important chemicals such as dimethyl carbonate and glycidol were synthesized.

Keywords: carbon dioxide; dimethyl carbonate; glycidol; organic superbase; integrated synthesis

Citation: Khokarale, S.; Shelke, G.; Mikkola, J.-P. Integrated and Metal Free Synthesis of Dimethyl Carbonate and Glycidol from Glycerol Derived 1,3-Dichloro-2-propanol via CO_2 Capture. *Clean Technol.* 2021, 3, 685–698. https://doi.org/10.3390/cleantechnol3040041

Academic Editor: Diganta B. Das

Received: 27 August 2021
Accepted: 17 September 2021
Published: 24 September 2021

Publisher's Note: MDPI stays neutral with regard to jurisdictional claims in published maps and institutional affiliations.

Copyright: © 2021 by the authors. Licensee MDPI, Basel, Switzerland. This article is an open access article distributed under the terms and conditions of the Creative Commons Attribution (CC BY) license (https:// creativecommons.org/licenses/by/ 4.0/).

1. Introduction

Glycerol is a highly precious and industrially important biomass-derived molecule since it has numerous applications in the pharmaceuticals, cosmetics, and food industries [1]. Besides that, it is also serving a vital role in the production of various low molecular weight commodity chemicals, e.g., ethylene glycol, 1,2- and 1,3-propanediol, acrylic acid, glycerol carbonate, glyceraldehyde, dihydroxyacetone, etc. [2–4]. Glycerol is a co-product of the biodiesel synthesis process and since the production of biodiesel increased tremendously, the glycerol is also produced in huge amounts simultaneously [3].

Albeit, excess glycerol is often disposed of as a waste, it is not economically beneficial for the biodiesel industries considering the overall cost of the process as well as the negligible value addition to such a vital and renewable chemical. Hence, it is necessary to establish more efficient and alternative pathways to utilize glycerol, for example in the synthesis of value-added chemicals, fuel additives, etc. Considering the excellent source of C_3 carbon backbone, glycerol is used to produce lactic acid, carbonates (linear and cyclic), diols, esters, and epichlorohydrin (ECH), which further reduces the dependency on fossil-derived routes upon their production [5,6].

The chlorination of the liquid glycerol to di-chlorinated analogies such as 2,3-dichloro-1-propanol and 1,3-dichloro-2-propanol is a well industrially applied process [5–7] (Scheme 1). This process is a part of Solvay's Epicerol process, which is applied for the synthesis of industrially important ECH (epoxy resin monomer) where the annual production has reached more than 100 kt [8,9]. This process not only replaced the traditional method for the synthesis of ECH, such as the chlorination of propene at high temperatures, but also increased the renewable nature of ECH since the processes use glycerol as one of the reagents in the synthesis.

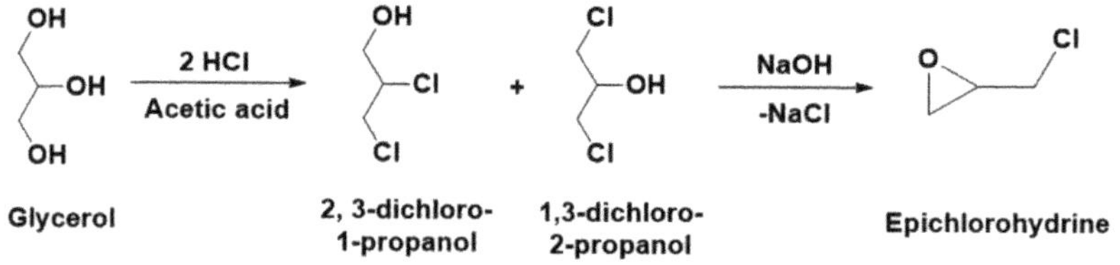

Scheme 1. Epichlorohydrin synthesis from glycerol (Solvay's Epicerol process).

As shown in Scheme 1, the chlorination of glycerol is carried out with two moles of hydrochloric acid using Lewis acid catalysts such as carboxylic acid (e.g., acetic acid) to form a mixture of 2, 3-dichloro-1-propanol and 1,3-dichloro-2-propanol. Further, out of these chlorinated derivatives of glycerol, 1,3-dichloro-2-propanol is converted to ECH following the alkaline hydrolysis process [7]. The synthesis of these chlorinated analogues of the glycerol and their further application is only limited to the ECH synthesis, whereas these analogues are not explored for other fruitful applications. 2-chloro-1,3-propanol, one of the mono-chlorinated analogues of glycerol, is considered as a waste in the Epicerol process. In this case, 2-chloro-1,3-propanol cannot be converted to the di-chlorinated species because the chlorine at the beta position (β form) inhibits further chlorination. Proto et al. proposed that 2-chloro-1,3-propanol can be converted to glycidol, which is also considered a highly active and vital chemical entity in polymer, rubber, as well as in dye industries [7]. In other words, identical to the synthesis of ECH from glycerol, the processing of 2-chloro-1,3-propanol for glycidol synthesis can emerge as a new alternative for the existing allyl alcohol epoxidation using an H_2O_2 precursor and titanium silicate catalyst, TS-1 [10]. Hence, this integrated approach for the synthesis of ECH and glycidol can increase the atom economy as well as the overall sustainability of the Epicerol process. However, besides synthesis of the ECH and glycidol, it is necessary to implement more available applications of the chlorinated analogs of the glycerol to enhance the applicability of a surplus amount of glycerol from the biodiesel industries.

In this work, we report the integrated method for the valorization of 1,3-dichloro-2-propanol to dimethyl carbonate (DMC) and glycidol through an organic superbase involving CO_2 capture and a base-catalyzed process. Being less toxic as well as having versatile reactivity, the use of DMC as a reagent as well as a solvent in various organic transformations is considered a green, sustainable, and environmentally friendly approach. Besides having combined the functionality of CO_2 and the methyl group, DMC is successfully utilized for the valorization of bio-based building blocks to value-added chemicals and fuels as well as for the derivatization of cellulose-to-cellulose methyl carbonate [11,12]. Considering the vital role of DMC in synthetic chemistry, several catalytic processes with and without the use of CO_2 have been developed for the DMC synthesis where some of the methods have been commercialized [13,14].

In this reaction approach, the CO_2 molecule was initially activated through the equivalent interaction between 1,3-dichloro-2-propanol and organic superbase diazabicyclo [5.4.0] undec-7-ene (DBU), where the resultant 3-chloro-1,2-propylenecarbonate was further transesterified with methanol to form DMC and glycidol (Scheme 2).

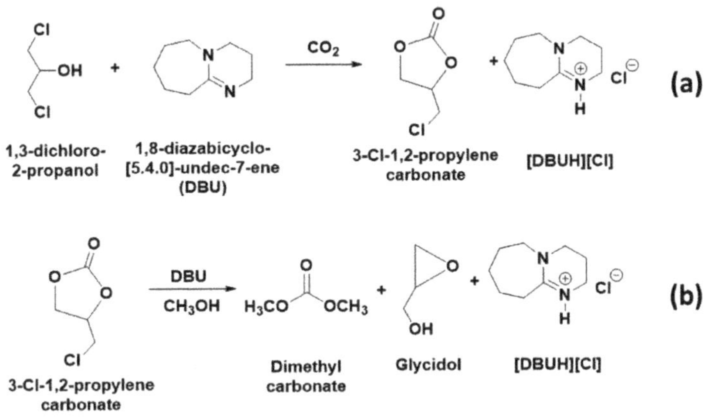

Scheme 2. (a) Synthesis of 3-chloro-1,2-propylenecarbonate from 1,3-dichloro-2-propanol and, (b) Transesterification of 3-chloro-1,2-propylenecarbonate to dimethyl carbonate and glycidol.

The DBU superbase (pk_a = 23)-mediated activation of CO_2 is a well-studied process where it not only emerged as a new and greener pathway for the up-gradation of CO_2, but it also introduced new reversible solvent media called switchable ionic liquid (SIL), which was efficiently used for the processing of lignocellulosic biomass such as wood and cellulose esters synthesis [15,16]. In this actual case, DBU initially deprotonates alcohols (R-OH) where the resultant alkoxide anion equivalently reacts with CO_2 to form [DBUH][ROCO$_2$] salt in the form of an SIL [17,18]. Besides the CO_2 capture, the synthesis of SIL has also been further explored upon the synthesis of linear as well as cyclic carbonates, methyl formate, as well as acrylic plastic precursors synthesis [19–22]. In this regard, the synthesis of cyclic carbonates using 1,2 chlorohydrins has been also previously reported for the synthesis of various cyclic carbonates [20]. Similar work of the synthesis of cyclic carbonates now mimicked in this report in the case of the synthesis of the 3-chloro-1,2-propylenecarbonate where 1,3-dichloro-2-propanol is assumed as the 1,2 chlorohydrin. After the complete synthesis, the recovery method has also been further set up for the separation of DMC and glycidol following solvent extraction techniques. In addition, as shown in Scheme 2, the overall process of the synthesis of DMC and glycidol was also accompanied by the formation of [DBUH][Cl] salt, which was further separated from the reaction mixture and further used for the recovery of DBU. The progress of the reaction as well as the purity of the recovered chemical species was confirmed by means of NMR analysis techniques.

2. Materials and Methods
2.1. Chemicals and Methods
2.1.1. Chemicals

1,3-dichloro-2-propanol (98%), 1,8-Diazabicyclo[5.4.0]undec-7-ene (DBU, 98%), dimethyl carbonate (DMC), glycidol (96%), D_2O (99.9 atom % D), and CDCl3 (99.8 atom % D) were purchased from Sigma Aldrich (Saint Louis, MO, USA), whereas methanol ($\geq$99.0%), dimethyl sulfoxide ($\geq$99.0%) and 2-Methyltetrahydrofuran (2-Me-THF, biorenewable, anhydrous, $\geq$99%, Inhibitor-free) were purchased from VWR chemicals and used without further purification. The CO_2 gas bottle was supplied by AGA AB (Linde Group) and used without further purification.

2.1.2. NMR Analysis

The progress during the synthesis of DMC and glycidol in the process was confirmed by means of NMR analysis using Bruker Avance 400 MHz instrument (Billerica, MA, USA). The $CDCl_3$ or capillary filled with D_2O was used as an internal standard during the analysis. The obtained data were further processed with TopSpin 4.0.7 software (Billerica, MA, USA). After NMR analysis, the types of chemical species observed during the synthesis are shown in Figure 1.

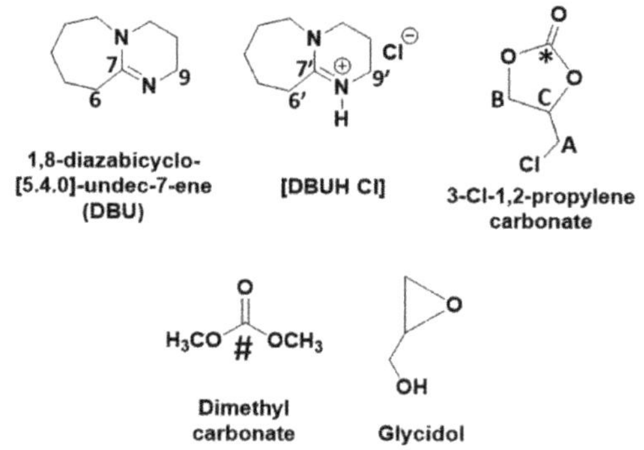

Figure 1. Chemical species formed during the synthesis. (The * and # signs used to highlight carbonyl carbon in 3-chloro-1,2-propylenecarbonate and dimethyl carbonate, respectively in NMR spectra).

2.2. Synthesis of 3-Chloro-1,2-propylenecarbonate in DMSO or 2-Me-THF

The synthesis of 3-chloro-1,2-propylenecarbonate was carried out either in DMSO or 2-Me-THF. In this case, initially, 0.63 g (4.9 mmol) of 1,3-dichloro-2-propanol was mixed with 4 mL of DMSO solvent under stirring and the reaction flask was kept in a water bath (21 °C) for 10 min. Then, the CO_2 gas (100 mL/min) was bubbled at room temperature in the reaction mixture for 5 min followed by dropwise addition of 0.75 mL (4.9 mmol) of DBU carried out at water bath temperature. The CO_2 gas was bubbled for a further 10 min in the reaction mixture to ensure complete interaction between added reagents. In the case of 2-Me-THF as a solvent, a similar process was applied during the synthesis of 3-chloro-1,2-propylenecarbonate where similar amounts of all the reagents have been used.

2.3. Recovery of 3-Chloro-1,2-propylenecarbonate

Solvent extraction was used to separate the various reaction components formed during the 3-chloro-1,2-propylenecarbonate synthesis. In this case, before applying the recovery method, the complete conversion of the reagents was initially confirmed with NMR analysis. The DMSO solvent comprised reaction mixture was added to 20 mL of ethyl acetate under stirring where white solid belonged to the chloride salt of DBU, i.e., [DBUH][Cl] was precipitated out. The [DBUH][Cl] salt was separated by filtration from the reaction mixture and washed with 15 mL of ethyl acetate. The DBU salt was further vacuum dried at 40 °C and stored in a desiccator before its purity confirmation using NMR analysis. The DMSO containing organic phase was added to 20 mL of water where DMSO was extracted with water while 3-chloro-1,2-propylenecarbonate remained in the ethyl acetate phase. Both phases were separated using a separating funnel. Further, the organic phase was concentrated on a rotary evaporator to obtain 3-chloro-1,2-propylene carbonate after drying over anhydrous sodium sulfate. In the case of 2-Me-THF solvent containing reaction mixture, the solvent was removed from the reaction mixture by a rotary evaporator. Further, ethyl acetate was added to the reaction mixture where precipitated [DBUH][Cl] salt and 3-chloro-1,2-propylenecarbonate with ethyl acetate were separated using filtration, rotatory evaporator, and vacuum drying methods. Recovery of the 3-chloro-1,2-propylenecarbonate with various solvents was calculated using equation 1, where the theoretical amount of the 3-chloro-1,2-propylenecarbonate was calculated based on the initial moles of 1,3-dichloro-2-propanol used in the synthesis.

$$\% \text{ Recovery of } 3-\text{Cl}-1,2-\text{propylenecarbonate} = \frac{\text{Recovered } 3-\text{Cl}-1,2-\text{propylenecarbpnate (moles)} \times 100}{\text{Theriotical amount of } 3-\text{Cl}-1,2-\text{propylenecarbonate (moles)}} \qquad (1)$$

2.4. Transesterification of 3-Chloro-1,2-propylenecarbonate to Synthesis DMC and Glycidol and Their Recoveries

The synthesis of DMC and glycidol from 3-chloro-1,2-propylenecarbonate was performed in methanol and studied at various temperatures. Equivalent amounts of 3-chloro-1,2-propylenecarbonate and DBU were mixed with methanol and the reaction mixture was stirred at room temperature (22 °C), 35 or 50 °C for various reaction times. At end of the reaction, methanol with DMC was separated from the reaction mixture by high vacuum distillation at 40 °C. The amount of DMC recovered with methanol was confirmed using gas chromatography where the calibration method and Equation number (2) were used to calculate the actual amount of DMC recovered (supporting information Figure S1). Glycidol and [DBUH][Cl] salt were separated from each other using 2-Me-THF and brine solution (NaCl saturated aqueous solution) as extracting solvents. In this case, 2-Me-THF and brine solution were added to the [DBUH][Cl] salt and glycidol mixture where the organic phase was separated from the aqueous phase using a separating funnel. For the glycidol recovery, the organic phase was dried over anhydrous sodium sulfate and concentrated by rotary evaporation. The recovery of glycidol was calculated using Equation (3). The aqueous phase was concentrated by rotation evaporation where dry methanol was added further in [DBUH][Cl] salt and NaCl mixture to precipitate the NaCl. The NaCl salt was separated from the alcoholic solution by filtration, whereas the [DBUH][Cl] salt was recovered after the alcoholic solution by rotary evaporation. The purity of the recovered [DBUH][Cl] salt and glycidol was confirmed using NMR analysis.

$$\% \text{ Recovery of dimethyl carbonate (DMC)} = \frac{\text{Recovered DMC (moles)} \times 100}{\text{Theortical amount of DMC (moles)}} \qquad (2)$$

$$\% \text{ Recovery of Glycidol} = \frac{\text{Recovered glycidol (moles)} \times 100}{\text{Theoretical amount of glycidol (moles)}} \qquad (3)$$

2.5. Recovery of DBU from [DBU][Cl] Salt Using a Neutralization Method

A total of 0.12 g (3 mmol) of NaOH was added to 10 mL of dry methanol, the reaction mixture was stirred at 50 °C for 1 h, and a transparent solution was obtained. Further, 0.47 g (2.5 mmol) of [DBUH][Cl] salt was added to this alkaline methanol solution where the reaction mixture was stirred at 50 °C for 1 h. As the reaction progressed, a white and crystalline precipitate of NaCl separated and settled at the bottom of the reaction flask. The NaCl salt was separated from the reaction mixture by filtration and DBU was recovered from the alcoholic solution by rotary evaporation. The recovery degree of DBU was calculated by using Equation (4), whereas the purity was confirmed using NMR analysis. To calculate the amount of recovered DBU, the theoretical amount of DBU was calculated based on the initial amount of [DBUH][Cl] salt (moles) that was used in the recovery process.

$$\% \text{ Recovery of DBU} = \frac{\text{Recovered DBU (moles)} \times 100}{\text{Theoretical amount of DBU (moles)}} \qquad (4)$$

3. Results

DMC and glycidol synthesis proceeded via the integrated two-step process approach. In this case, initially, the synthesis of 3-chloro-1,2-propylenecarbonate was prepared through the equivalent interaction of 1,3-dichloro-2-propanol, DBU, and CO_2, at room temperature. Further, the DMC along with glycidol were synthesized via a base-catalyzed transesterification of 3-chloro-1,2-propylenecarbonate in methanol. The synthesis of 3-chloro-1,2-propylenecarbonate was initially carried out in DMSO as a solvent and a similar synthesis was further studied in other solvents such as 2-Me-THF. After the complete addition of DBU in the reaction mixture containing DMSO and 1,3-dichloro-2-propanol under CO_2 bubbling, the composition of the resultant reaction mixture was confirmed by one- as well as two-dimensional NMR analysis. The ^1H and ^{13}C NMR spectra of the reaction mixture are shown in Figure 2 and supporting information Figure S2, respectively. As shown in Figure 2, after the interaction between 1,3-dichloro-2-propanol, DBU, and CO_2, DBU as well as 1,3-dichloro-2-propanol were completely consumed in the reaction mixture as their corresponding signals for the carbon atoms disappeared. In this case, the characteristics signals for the carbon atoms at positions six, seven, and nine, respectively, in the molecular DBU disappeared and new shielded signals for the carbon atoms at position six and seven (C-6′ and C-7′) as well as a de-shielded signal for carbon atom at position nine (C-9′), respectively, were observed.

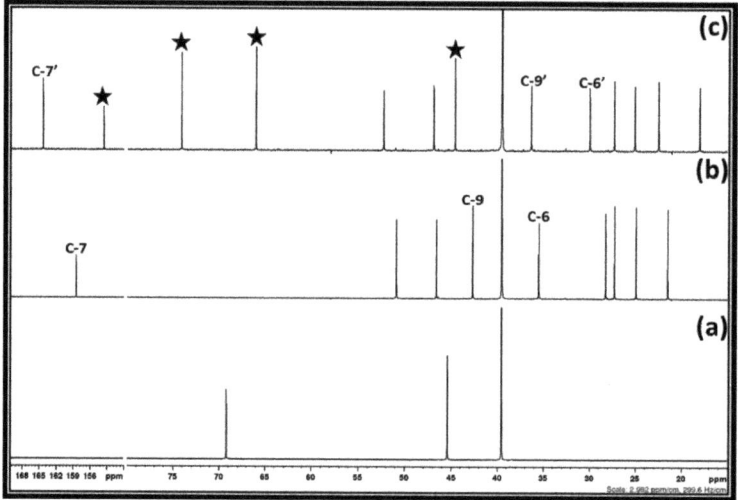

Figure 2. ^{13}C NMR spectra of the (**a**) 1,3-dichloro-2-propanol, (**b**) DBU, and (**c**) Reaction mixture after equivalent interaction of 1,3-dichloro-2-propanol, DBU, and CO_2 in DMSO (NMR analysis with capillary filled with D_2O).

This observation represents that the sp^2-N atom in the DBU molecule became protonated, which is in agreement with the previous reports [20]. Besides the signals for the protonated DBU, the signals for the unknown chemical species were also observed in the ^{13}C NMR analysis (shown by a filled star). ^1H NMR spectra also depict that the characteristics signals for the protons in both 1,3-dichloro-2-propanol and DBU molecules, respectively, disappeared, while signals for the unknown chemical species as well as protonated DBU appeared. As described previously, the DBU molecule is popularly known as a superbase to activate the CO_2 molecule through the formation of SIL in the presence of proton sources such as water or alcohol. Besides that, it was also previously confirmed that the equivalent interaction between 1, 2-halohydrin and DBU molecule in the presence of CO_2 results in the formation of cyclic carbonate [20]. Since 1,3-dichloro-2-propanol molecule has a similar structure to the 1,2-halohydrin, i.e., –OH and halide groups are attached to the adjacent carbon atoms, its interaction with DBU and CO_2 molecules could result in the formation of cyclic carbonate such as 3-chloro-1,2-propylenecarbonate. To confirm the formation of 3-chloro-1,2-propylenecarbonate, the reaction mixture obtained was analyzed using two dimensional (2D) HMBC (Heteronuclear Multiple Bond Correlation),

HSQC (Heteronuclear Single Quantum Coherence), and COSY (Correlated Spectroscopy) NMR analysis techniques and corresponding spectra are shown in Figure 3 and supporting information is shown in Figure S3a,b.

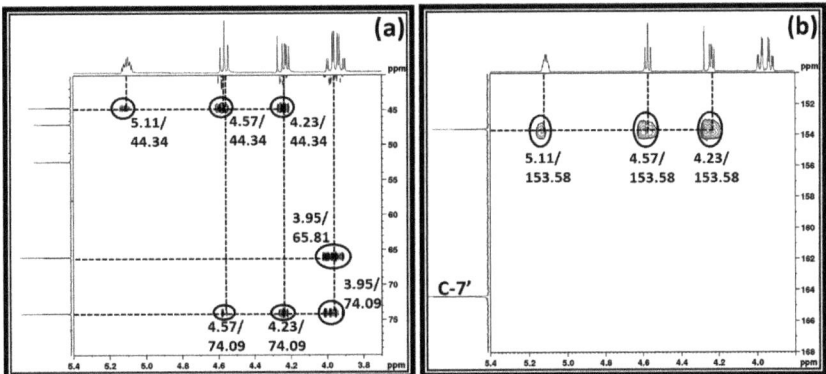

Figure 3. (**a**,**b**) ^{1}H-^{13}C HMBC NMR spectra of the reaction mixture after equivalent interaction of 1,3-dichloro-2-propanol, DBU and CO_2 in DMSO.

The HSQC NMR analysis shows that the protons in an unknown chemical species with chemical shifts 3.95, 4.23, 4.57, and 5.11 ppm, respectively, belong to the proton–carbon correlation signals with their corresponding carbon atoms (supporting information S3a). In the case of COSY NMR analysis, the proton with the chemical shift 5.11 ppm proton–proton correlated with all the remaining protons, whereas the protons at 4.23 and 4.57 ppm did not show any correlation with the proton resonating at 3.95 ppm (supporting information S3b). This suggests that the distribution of the protons in this unknown chemical species is identical to the 1,3-dichloro-2-propanol that was used in the synthesis. The HMBC NMR analysis showed that protons with chemical shifts 4.23, 4.57, and 5.11, respectively, are in correlation with the carbon atom resonating at 153.5 ppm. The signal for the carbon atom at 153.5 ppm was a new one and it usually belongs to the carbon atom in a carbonyl group. This observation suggests that the activation of the CO_2 molecule took place through the equivalent interaction between the reagents applied in the synthesis. Since the [DBUH]$^+$ cation forms in the reaction composition, the formation of this cation proceeds through the activation of CO_2 by the DBU superbase. In this case, as shown in Scheme 3a, DBU removed the proton from the –OH group in 1,3-dichloro-2-propanol and the resultant alkoxide species reacted with CO_2 and alkyl carbonate anion, whereupon the [DBUH]$^+$ cation formed in the reaction mixture. However, since the COSY and HMBC NMR analysis suggests that the protons with the chemical shifts 4.23, 4.57, and 5.11, respectively, are adjacent to each other and in long correlation with 153.5 ppm, further consecutive cyclization in alkyl carbonate anion took place, which further allowed the formation of 3-chloro-1,2-propylenecarbonate along with the release of [DBUH][Cl] salt (Scheme 3b). Hence, similar to the previously reported cyclic carbonate synthesis, the equivalent interaction between 1,3-dichloro-2-propanol, DBU, and CO_2 results in a 3-chloro-1,2-propylenecarbonate, which formed in the process [20]. This DBU mediated fixation of CO_2 in the form of 3-chloro-1,2-propylenecarbonate was carried out at room temperature and atmospheric pressure. Therefore, this method can be considered safer and sustainable compared to epichlorohydrin encompassed with high energy-consuming catalytic approaches. Even though both 1,3-dichloro-2-propanol and epichlorohydrin are derived from the hydro-chlorination of glycerol, the processing with epichlorohydrin in 3-chloro-1,2-propylenecarbonate synthesis is not safe considering its toxic and flammable nature.

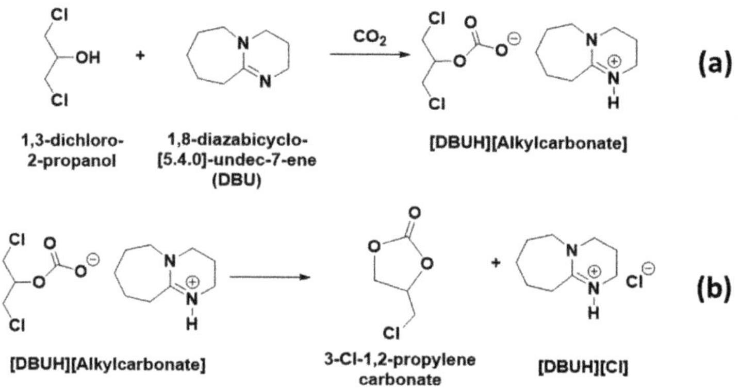

Scheme 3. (**a**) Activation of CO_2 by DBU through proton abstraction from 1,3-dichloro-2-propanol and (**b**) formation of 3-chloro-1,2-propylenecarbonate and [DBUH][Cl] salt.

After the synthesis of 3-chloro-1,2-propylenecarbonate, the reaction mixture was further explored in terms of its recovery via water and ethyl acetate-involved solvent extraction methods. In this case, after the separation of [DBUH][Cl] salt using ethyl acetate, water was further used to separate DMSO from 3-chloro-1,2-propylenecarbonate, which remained in the organic phase. After the removal of ethyl acetate, a 68% recovery of 3-chloro-1,2-propylenecarbonate was achieved. This represents that even though 3-chloro-1,2-propylenecarbonate is insoluble in water due to DMSO, a part of it remained in the aqueous phase. Further, a similar synthesis of 3-chloro-1,2-propylenecarbonate was carried out in 2-Me-THF solvent. During the bubbling of CO_2 and the simultaneous addition of DBU in the reaction mixture, containing 1,3-dichloro-2-propanol in a 2-Me-THF, it was observed that a white crystalline solid precipitate was forming and became separated in the reaction mixture. The ^{13}C NMR analysis of the reaction mixture along with the white precipitate was carried out and the obtained spectra are shown in Figure 4.

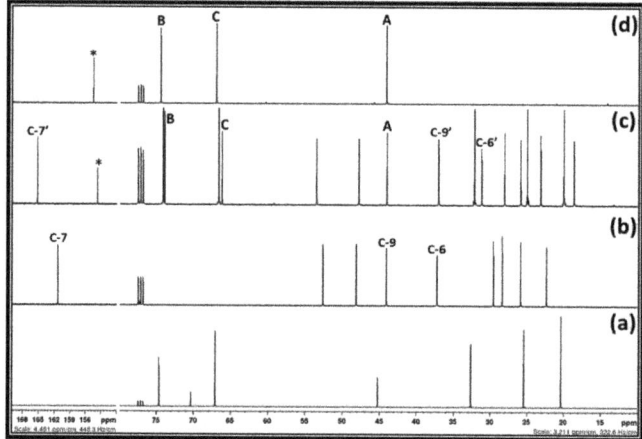

Figure 4. ^{13}C NMR spectra of the (**a**) 1,3-dichloro-2-propanol in 2-Me-THF, (**b**) DBU and, (**c**) Reaction mixture after equivalent interaction of 1,3-dichloro-2-propanol, DBU, and CO_2 in 2-Me-THF, and (**d**) 3-chloro-1,2-propylenecarbonate (NMR analysis in $CDCl_3$).

Figure 4 depicts that the 3-chloro-1,2-propylenecarbonate as well as [DBUH][Cl] salt were forming after an equivalent interaction between 1,3-dichloro-2-propanol, DBU, and CO_2 when 2-Me-THF was used as the solvent during the synthesis. Recently, Jupke et al.

reported that 2-Me-THF has a higher CO_2 solubility in the reaction system than water under identical experimental conditions [23]. Jessop and Matsuda et al. showed that 2-Me-THF has lower values of Kamlet-Taft parameters such as polarizability, π^* (0.5–1.1), which further allowed for the higher solubility of hydrophobic CO_2 in 2-Me-THF than water [24,25]. Matsuda et al. also further reported that CO_2 expanded 2-Me-THF, and other bio-based solvents turned out to be an excellent solvent media for biotransformation. The author explained that with an increase in the CO_2 pressure, the polarizability (π^*) value of the 2-Me-THF linearly decreased as a result of the higher solubility of CO_2, which further increased the transformation rate in this CO_2 expanded solvent system [26].

Therefore, similar to DMSO, 2-Me-THF can be used as a solvent in the synthesis of 3-chloro-1,2-propylenecarbonate. In this case, 2-Me-THF is preferred more considering its renewable nature and this solvent is already referred to as an alternative to THF and other organic solvents [27,28]. After the completion of the reaction, 2-Me-THF was further removed by evaporation and [DBUH][Cl] salt was separated from 3-chloro-1,2-propylenecarbonate using ethyl acetate solvent and a filtration technique. Further, 93% of the 3-chloro-1,2-propylenecarbonate was recovered when ethyl acetate was removed from the organic phase. Hence, the use of 2-Me-THF in the synthesis not only facilitated the separation of components from the reaction mixture but also further allowed a higher level of recovery of 3-chloro-1,2-propylenecarbonate.

To valorize 3-chloro-1,2-propylenecarbonate, it was further explored in the base-catalyzed transesterification in methanol where DBU was used as a base and the synthesis was carried out at different temperatures. The current catalytic approaches for the synthesis of 3-chloro-1,2-propylenecarbonate from ECH are considered as only an ideal example to demonstrate the valorization of CO_2 upon the synthesis of cyclic carbonates. Hence, 3-chloro-1,2-propylenecarbonate remains underutilized and needs to be upgraded to value-added chemical entities considering the value of the active form of CO_2 in the molecule. The concept of the transesterification of 3-chloro-1,2-propylenecarbonate under alkaline conditions was designed based on the previous reports where the cyclic carbonates such as ethylene carbonates, catechol carbonate, etc. were used to synthesize various aliphatic carbonates [13,29]. As shown in Scheme 4, the reaction of the alkaline transesterification involved the interaction of methanol with 3-chloro-1,2-propylenecarbonate and results in the formation of the DMC and 3-Chloro-1, 2-propanediol.

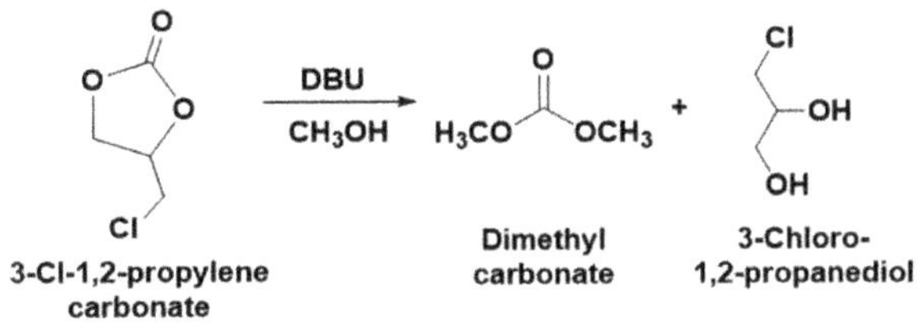

Scheme 4. Base catalyzed transesterification of the 3-chloro-1,2-propylenecarbonate to DMC and 3-Chloro-1, 2-propanediol.

After the interaction of the equivalent amounts of DBU and 3-chloro-1,2-propylenecarbonate in methanol for 30 min, it was observed that the expected products such as DMC and 3-chloro-1,2-propanediol formed in the reaction mixture (Figure 5b). However, besides the signal for these chemical species, new signals at 44.3, 52.4, and 62.1 ppm were also observed. The reaction mixture was stirred for various reaction times such as 2, 6 and 18 h and it was observed that these newly observed signals belong to unknown chemical species, the amounts of which gradually increased. Simultaneously, the signal belonging to 3-chloro-1, 2-propanediol steadily decreased as the reaction time increased. Besides that, it was also observed that the chemical shifts for the carbon atoms of DBU, especially at positions six and nine, respectively, were shielded, whereas the carbon atom at position seven became de-shielded under the given reaction period.

Considering the changes in the chemical shifts for the carbon atoms in the DBU molecule, it is evident that the [DBUH]$^+$ cation is gradually forming in the reaction mixture.

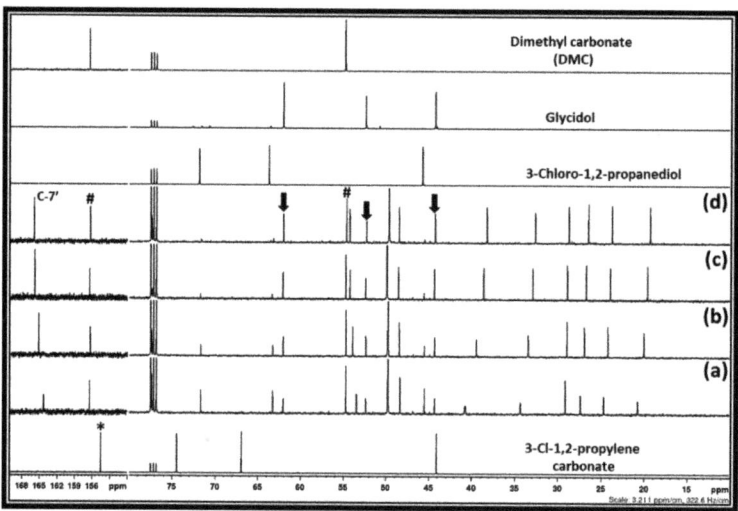

Figure 5. ^{13}C NMR spectra for the room temperature and DBU catalyzed transesterification of 3-chloro-1,2-propylenecarbonate in methanol, (**a**) 30 min, (**b**) 2 h, and (**c**) 6 h and, (**d**) 18 h (Downward arrows and # sign showed carbon atoms belongs to the glycidol and DMC, respectively).

It was previously reported that the synthesis of epoxides such as epichlorohydrin and glycidol from the hydro-chlorinated analogs of glycerol such as 1,3-dichloro-2-propanol and 3-Chloro-1, 2-propanediol, respectively, is a base-catalyzed reaction where an equivalent interaction of the base with either of these chlorinated species results in the formation of the corresponding epoxides [5,7]. In the present work, since the transesterification of 3-chloro-1,2-propylenecarbonate was carried out with an equivalent amount of DBU, the possibility is that the in situ-formed 3-chloro-1, 2-propanediol can transform further to an oxiranic function-comprising molecule, i.e., glycidol through the release of a Cl atom with a DBU base (Scheme 5). To confirm the glycidol formation, the NMR spectra of the reaction mixture after the 18 h and commercially available glycidol were compared and it was observed that identical signals related to glycidol (shown by downward arrow) were observed. Since the signal related to the [DBUH]$^+$ cation was also observed, this also confirmed that the formation of glycidol has occurred through the formation of [DBUH][Cl] salt in the reaction composition. Hence, the base-catalyzed transesterification of 3-chloro-1,2-propylenecarbonate in methanol results in the formation of DMC and glycidol along with [DBUH][Cl]. Overall, the dechlorination of the glycerol-derived 1,3-dichloro-2-propanol has occurred during the synthesis of DMC as well as glycidol, which not only facilitates the uptake of CO_2 but also allowed for the synthesis of industrially important value-added chemicals. Besides that, the DBU molecule not only assisted in the efficient CO_2 capture and served as a base catalyst in DMC synthesis but also performed as a reservoir for the chloride ion through the formation of its non-volatile and thermally stable chloride salt.

The synthesis of DMC and glycidol from 3-chloro-1,2-propylenecarbonate was further carried out at higher temperatures such as 35 and 50 °C, where the rate of formation of glycidol increased with the temperature and the complete conversion of the in situ-formed 3-Chloro-1,2-propanediol to glycidol took place in 2 h and 30 min, respectively (supporting information Figure S4a,b). Hence, in this synthesis, the applied temperatures significantly influenced glycidol synthesis levels, whereas the rate of DMC formation remained unaltered.

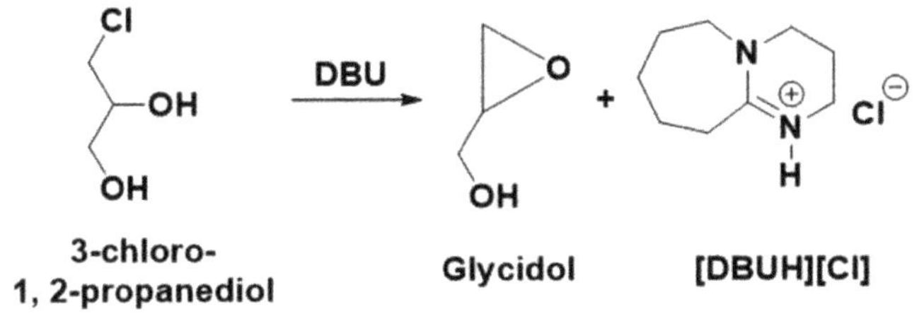

Scheme 5. Synthesis of glycidol and [DBUH][Cl] salt from 3-chloro-1,2-propanediol and DBU.

Using the distillation method, 92% of DMC was recovered along with methanol from the reaction mixture, whereas the remaining glycidol and [DBUH][Cl] salt were separated using solvent extraction. In the case of solvent extraction initially, 2-Me-THF was added in a mixture of [DBUH][Cl] salt and glycidol in order to remove glycidol selectively from the mixture. However, after the addition of 2-Me-THF, no solid [DBUH][Cl] salt precipitated out from the mixture. On the other hand, the turbid solution was obtained after 1 h and the transparent viscous liquid settled at the bottom of the flask. The separation of [DBUH][Cl] from the glycidol is not possible, probably as a result of the formation of a deep eutectic mixture through hydrogen bond acceptor (HBA) and hydrogen bond donor (HBD) interactions. In this case, considering previous reports regarding the compositions of various deep eutectic solvents (DES), the chloride anion containing ionic liquids such as choline chloride and a hydroxyl group comprised of molecules such as glycerol, ethylene glycol, etc., performed as a HBD and HBA, respectively to form a DES [30]. Sato et al. also showed that a similar hydrogen bonding interaction between the –OH group of glycidol and the Cl^- anion of the quaternary alkylammonium salt was established and resulted in the formation of a binding complex [31]. Therefore, the hydrogen bonding interaction possibly does not allow the precipitation of [DBUH][Cl] salt in 2-Me-THF. In order to trigger the separation of these two chemical species, the brine solution was added to their mixture, followed by the addition of 2-Me-THF to extract glycidol. In this case, NaCl in the brine solution possibly disturbed the hydrogen bonding between [DBUH][Cl] salt and glycidol and allowed the transfer of the latter to the 2-Me-THF phase. After the evaporation of 2-Me-THF from the organic phase, an 89% recovery of the pure form of the glycidol was achieved. The 13C NMR spectra of the recovered glycidol is shown in Figure 6. Water was removed from the aqueous phase and dry methanol was added to precipitate and separate NaCl from the [DBUH][Cl] salt. The alkaline alcoholic solution (NaOH in methanol) was further used to recover the molecular DBU from its chloride salt. In this context, 83% of the pure form of the DBU was obtained and the spectra of the recovered DBU is shown in supporting information Figure S5.

Hence, in this overall reaction approach, the hydro-chlorinated derivative of glycerol, i.e., 1,3-dichloro-2-propanol, was further used for the CO_2 capture and its further valorization was demonstrated to synthesize DMC along with the glycidol. The specialty of this work is that DBU superbase was applied for various tasks where it performed as an efficient, selective, and recoverable base catalyst along with CO_2 capturing as well as a dechlorinating agent in the synthesis. Apart from this, we introduce a new alternative and simultaneous synthetic approach to produce DMC and glycidol compared to the existing various individual catalytic approaches. In terms of the solvent system, 2-Me-THF emerged as a new and renewable solvent media for the CO_2 activation to value-added chemicals, which was previously limited to CO_2 capture in the form of expanded liquids.

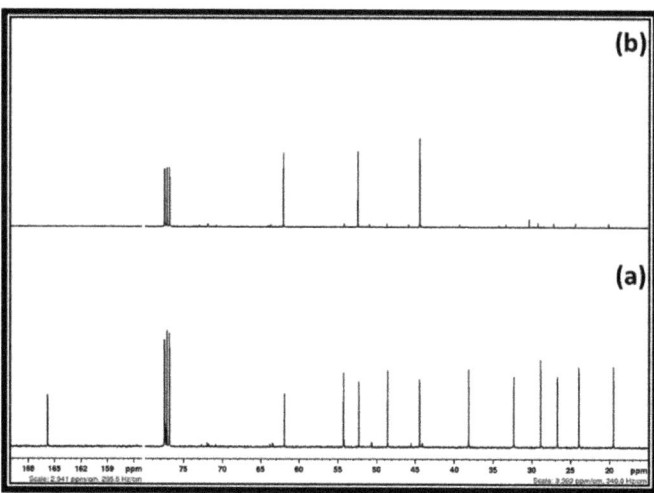

Figure 6. ^{13}C NMR spectra: (**a**) [DBUH][Cl] salt and glycidol after removal of DMC and methanol and, (**b**) recovered glycidol.

4. Conclusions

The glycerol-derived 1,3-dichloro-2-propanol as well as CO_2 was successfully valorized for the synthesis of an industrially important dimethyl carbonate and glycidol integrated process approach under mild reaction conditions. The synthesis of 3-chloro-1,2-propylenecarbonate from 1,3-dichloro-2-propanol was carried out in the DBU superbase-triggered CO_2 capture process followed by a cyclization approach at room temperature, whereupon the formation of cyclic carbonate was confirmed with both one- and two-dimensional NMR spectroscopy techniques. Amongst the applied solvents, upon the use of DMSO, 69% of 3-chloro-1,2-propylenecarbonate was recovered, on the other hand, 2-Me-THF emerged as a more efficient solvent system compared to DMSO, whereupon a 97% recovery was achieved without the use of solvent extraction. The synthesized 3-chloro-1,2-propylenecarbonate was further explored for the base-catalyzed transesterification to synthesize DMC in methanol, whereupon in situ-formed 3-chloro-1, 2-propanediol simultaneously converted to glycidol as a result of its equivalent interaction with DBU superbase. In this integrated process for the synthesis of DMC and glycidol, DBU superbase performed versatile tasks where it participated in the CO_2 activation and base-catalyzed alcoholysis process along with an efficient dechlorinating agent through the formation of a thermally stable [DBUH][Cl] salt. In the case of the alcoholysis of 3-chloro-1,2-propylenecarbonate to DMC and glycidol, the rate of the formation of DMC was not influenced by the applied temperature, whereas the rate of the formation of glycidol linearly increasing with the applied temperature. In this case, 3-chloro-1,2-propanediol was completely converted to glycidol in 30 min at 50 °C while at room temperature. Some of the 3-chloro-1,2-propanediol remained unreacted even after 18 h. The 93% of DMC along with methanol was recovered from the reaction mixture by evaporation, whereas 89% of glycidol was obtained from the mixture of [DBUH][Cl] salt using the brine solution and 2-Me-THF involved solvent extraction. In this case, the brine solution facilitated the separation of glycidol from the [DBUH][Cl] salt. The DBU superbase was also obtained with 83% recovery from the [DBUH][Cl] salt following a neutralization approach. Hence, in this process, the sustainable valorization of CO_2 along with a glycerol derivative such as 1,3-dichloro-2-propanol was demonstrated using recoverable DBU superbase and a renewable solvent, 2-Me-THF. This new reaction pathway can be further explored for the synthesis of other dialkyl carbonates along with glycidol, where, in this case, various types of alcohols can be utilized.

Supplementary Materials: The following are available online at https://www.mdpi.com/article/10.3390/cleantechnol3040041/s1, Figure S1: Calibration curve for the quantification of recovered dimethyl carbonate (Gas chromatography method). Figure S2: ^{1}H NMR spectra of the (a) 1,3-dichloro-2-propanol, (b) DBU, and (c) Reaction mixture after equivalent interaction of 1,3-dichloro-2-propanol, DBU, and CO_2 in DMSO (NMR analysis with capillary filled with D_2O). Figure S3: ^{1}H-^{13}C HSQC (a) and ^{1}H-^{1}H COSY (b) NMR spectra of the reaction mixture after equivalent interaction of 1,3-dichloro-2-propanol, DBU, and CO_2 in DMSO. Figure S4a: ^{13}C NMR spectra for the DBU catalyzed transesterification of 3-chloro-1,2-propylenecarbonate in methanol at 35 °C, (a) 30 min, (b) 1 h, and (c) 2 h. (Downward arrows and # sign showed carbon atoms belongs to the glycidol and DMC, respectively). Figure S4b: ^{13}C NMR spectra for the DBU catalyzed transesterification of 3-chloro-1,2-propylenecarbonate in methanol at 50 °C, (a) 15 min and (b) 30 min. (Downward arrows and # sign showed carbon atoms belongs to the glycidol and DMC, respectively). Figure S5: ^{13}C NMR spectra (a) [DBUH Cl] salt and (b) recovered DBU.

Author Contributions: Conceptualization, S.K.; methodology, S.K.; validation, S.K. and G.S.; resources, S.K.; data curation, S.K. and G.S.; writing—original draft preparation, S.K.; writing—review and editing, S.K., G.S. and J.-P.M.; supervision, J.-P.M.; project administration, S.K. and J.-P.M.; funding acquisition, J.-P.M. All authors have read and agreed to the published version of the manuscript.

Funding: This research received no external funding.

Institutional Review Board Statement: Not applicable.

Informed Consent Statement: Not applicable.

Acknowledgments: This work is part of activities of the Technical Chemistry, Department of Chemistry, Chemical-Biological Centre, Umeå University, Sweden as well as the Johan Gadolin Process Chemistry Centre at Åbo Akademi University in Finland. The Swedish Research Council (Drn: 2016-04090), Bio4Energy programme, Kempe Foundations and Wallenberg Wood Science Center under auspices of Alice and Knut Wallenberg Foundation are gratefully acknowledged for funding this project.

Conflicts of Interest: The authors declare no conflict of interest.

References

1. Azelee, N.I.W.; Ramli, A.N.M.; Manas, N.H.A.; Salamun, N.; Man, R.C.; Enshasy, H.E. Glycerol In Food, Cosmetics And Pharmaceutical Industries: Basics And New Applications. *Int. J. Sci. Technol. Res.* **2019**, *8*, 553–558.
2. Kaur, J.; Sarma, A.K.; Jha, M.K.; Gera, P. Valorisation of crude glycerol to value-added products: Perspectives of process technology, economics and environmental issues. *Biotech. Rep.* **2020**, *27*, e00487. [CrossRef]
3. Tan, H.W.; AbdulAziz, A.R.; Aroua, M.K. Glycerol production and its applications as a raw material: A review. *Renew. Sustain. Energy Rev.* **2013**, *27*, 118–127. [CrossRef]
4. Katryniok, B.; Kimura, H.; Skrzynska, E.; Girardon, J.S.; Fongarland, P.; Capron, M.; Ducoulombier, R.; Mimura, N.; Paula, S.; Dumeignil, F. Selective catalytic oxidation of glycerol: Perspectives for high value chemicals. *Green Chem.* **2011**, *13*, 1960–1979. [CrossRef]
5. Lari, G.M.; Pastore, G.; Mondelli, C.; Ramirez, J.P. Towards sustainable manufacture of epichlorohydrin from glycerol using hydrotalcite derived basic oxides. *Green Chem.* **2018**, *20*, 148–159. [CrossRef]
6. Morodo, R.; Gerardy, R.; Petit, G.; Monbaliu, J.M. Continuous flow upgrading of glycerol toward oxiranes and active pharmaceutical ingredients thereof. *Green Chem.* **2019**, *21*, 4422–4433. [CrossRef]
7. Cespi, D.; Cucciniello, R.; Ricciardi, M.; Capacchione, C.; Vassura, I.; Passarini, F.; Proto, A. A simplified early stage assessment of process intensification: Glycidol as a value-added product from epichlorohydrin industry wastes. *Green Chem.* **2016**, *18*, 4559–4570. [CrossRef]
8. Krafft, P.; Gilbeau, P. Process for Producing Epichlorohydrin. U.S. Patent 2009/0275726 A1, 4 November 2009.
9. Vitiello, R.; Russo, V.; Turco, R.; Tesser, R.; Di Serio, M.; Santacesaria, E. Glycerol chlorination in a gas-liquid semibatch reactor: New catalysts for chlorohydrin production. *Chin. J. Catal.* **2014**, *35*, 663–669. [CrossRef]
10. Harvey, L.; Kennedya, E.; Dlugogorski, B.Z.; Stockenhuber, M. Influence of impurities on the epoxidation of allyl alcohol to glycidol with hydrogen peroxide over titanium silicate TS-1. *Appl. Catal. A General.* **2015**, *489*, 241–246. [CrossRef]
11. Fiorani, G.; Perosa, A.; Selva, M. Dimethyl carbonate: A versatile reagent for a sustainable valorization of renewables. *Green Chem.* **2018**, *20*, 288–322. [CrossRef]
12. Labafzadeh, S.R.; Helminen, K.J.; Kilpelinen, I.; King, A.W.T. Synthesis of Cellulose Methylcarbonate in Ionic Liquids using Dimethylcarbonate. *ChemSusChem* **2015**, *8*, 77–81. [CrossRef]

13. Wang, J.Q.; Sun, J.; Shi, C.Y.; Cheng, W.G.; Zhang, X.P.; Zhang, S.J. Synthesis of dimethyl carbonate from CO_2 and ethylene oxide catalyzed by K_2CO_3-based binary salts in the presence of H_2O. *Green Chem.* **2011**, *13*, 3213–3217. [CrossRef]
14. Huang, H.; Can Samsun, R.; Peters, R.; Stolten, D. Greener production of dimethyl carbonate by the Power-to-Fuel concept: A comparative techno-economic analysis. *Green Chem.* **2021**, *23*, 1734–1747. [CrossRef]
15. Soyler, Z.; Meier, M.A.R. Sustainable functionalization of cellulose and starch with diallyl carbonate in ionic liquids. *Green Chem.* **2017**, *19*, 3899–3907. [CrossRef]
16. Anugwom, I.; Eta, V.; Virtanen, P.; Arvela, P.M.; Hedenstrcm, M.; Hummel, M.; Sixta, H.; Mikkola, J.P. Switchable Ionic Liquids as Delignification Solvents for Lignocellulosic Materials. *ChemSusChem* **2014**, *7*, 1170–1176. [CrossRef] [PubMed]
17. Jessop, P.G.; Mercer, S.M.; Heldebrant, D.J. CO_2-triggered switchable solvents, surfactants, and other materials. *Energy Environ. Sci.* **2012**, *5*, 7240–7253. [CrossRef]
18. Heldebrant, D.J.; Yonker, C.R.; Jessop, P.G.; Phan, L. Organic liquid CO_2 capture agents with high gravimetric CO_2 capacity. *Energy Environ. Sci.* **2008**, *1*, 487–493. [CrossRef]
19. Khokarale, S.G.; Bui, T.Q.; Mikkola, J.P. One-Pot, Metal-Free Synthesis of Dimethyl Carbonate from CO_2 at Room Temperature. *Sustain. Chem.* **2020**, *1*, 298–314. [CrossRef]
20. Khokarale, S.G.; Mikkola, J.P. Metal free synthesis of ethylene and propylene carbonate from alkylene halohydrin and CO_2 at room temperature. *RSC Adv.* **2019**, *9*, 34023–34031. [CrossRef]
21. Yadav, M.; Linehan, J.C.; Karkamkar, A.J.; Eide, E.V.; Heldebrant, D.J. Homogeneous Hydrogenation of CO_2 to Methyl Formate Utilizing Switchable Ionic Liquids. *Inorg. Chem.* **2014**, *53*, 9849–9854. [CrossRef]
22. Khokarale, S.G.; Mikkola, J.P. Efficient and catalyst free synthesis of acrylic plastic precursors: Methyl propionate and methyl methacrylate synthesis through reversible CO_2 capture. *Green Chem.* **2019**, *21*, 2138–2147. [CrossRef]
23. Aigner, M.; Echtermeyer, A.; Kaminski, S.; Viell, J.; Leonhard, K.; Mitsos, A.; Jupke, A. Ternary System CO_2/2-MTHF/Water-Experimental Study and Thermodynamic Modeling. *J. Chem. Eng. Data* **2020**, *65*, 993–1004. [CrossRef]
24. Hoang, H.N.; Fernandez, E.G.; Yamada, S.; Mori, S.; Kagechika, H.; Gonzalez, Y.M.; Matsuda, T. Modulating Biocatalytic Activity toward Sterically Bulky Substrates in CO_2-Expanded Biobased Liquids by Tuning the Physicochemical Properties. *ACS Sustain. Chem. Eng.* **2017**, *5*, 11051–11059. [CrossRef]
25. Jessop, P.G. Searching for green solvents. *Green Chem.* **2011**, *13*, 1391–1398. [CrossRef]
26. Nam Hoang, H.; Nagashima, Y.; Mori, S.; Kagechika, H.; Matsuda, T. CO_2-expanded bio-based liquids as novel solvents for enantioselective biocatalysis. *Tetrahedron* **2017**, *73*, 2984–2989. [CrossRef]
27. Pace, V.; Hoyos, P.; Castoldi, L.; Mara, P.D.; Alcantara, A.R. 2-Methyltetrahydrofuran (2-MeTHF): A Biomass-Derived Solvent with Broad Application in Organic Chemistry. *ChemSusChem* **2012**, *5*, 1369–1379. [CrossRef] [PubMed]
28. Alcantara, A.R.; Maria, P.D. Recent Advances on the Use of 2-methyltetrahydrofuran (2-Me-THF) in Biotransformations. *Curr. Green Chem.* **2018**, *5*, 86–103. [CrossRef]
29. Tabanelli, T.; Monti, E.; Cavani, F.; Selva, M. The design of efficient carbonate interchange reactions with catechol carbonate. *Green Chem.* **2017**, *19*, 1519–1528. [CrossRef]
30. Zhang, Q.; Vigier, K.O.; Royer, S.; Jerome, F. Deep eutectic solvents: Syntheses, properties and applications. *Chem. Soc. Rev.* **2012**, *41*, 7108–7146. [CrossRef]
31. Tanaka, S.; Nakashima, T.; Maeda, T.; Ratanasak, M.; Hasegawa, J.; Kon, Y.; Tamura, M.; Sato, K. Quaternary Alkyl Ammonium Salt-Catalyzed Transformation of Glycidol to Glycidyl Esters by Transesterification of Methyl Esters. *ACS Catal.* **2018**, *8*, 1097–1103. [CrossRef]

Article

Modeling of Vacuum Temperature Swing Adsorption for Direct Air Capture Using Aspen Adsorption

Thomas Deschamps [1,2], Mohamed Kanniche [1,*], Laurent Grandjean [1] and Olivier Authier [1]

[1] EDF R&D Lab Chatou, 78400 Chatou, France; thomas.deschamps@mines-paristech.fr (T.D.); laurent.grandjean@edf.fr (L.G.); olivier.authier@edf.fr (O.A.)
[2] Center of Energy Efficiency of Systems (CES), MINES ParisTech, PSL Research University, 75006 Paris, France
* Correspondence: mohamed.kanniche@edf.fr

Abstract: The paper evaluates the performance of an adsorption-based technology for CO_2 capture directly from the air at the industrial scale. The approach is based on detailed mass and energy balance dynamic modeling of the vacuum temperature swing adsorption (VTSA) process in Aspen Adsorption software. The first step of the approach aims to validate the modeling thanks to published experimental data for a lab-scale bed module in terms of mass transfer and energy performance on a packed bed using amine-functionalized material. A parametric study on the main operating conditions, i.e., air velocity, air relative moisture, air temperature, and CO_2 capture rate, is undertaken to assess the global performance and energy consumption. A method of up-scaling the lab-scale bed module to industrial module is exposed and mass transfer and energy performances of the industrial module are provided. The scale up from lab scale to the industrial size is conservative in terms of thermal energy consumption while the electrical consumption is very sensitive to the bed design. Further study related to the engineering solutions available to reach high global gas velocity are required. This could be offered by monolith-shape adsorbents.

Keywords: adsorption; CO_2 capture; modeling

1. Introduction

Climate change has become a critical issue during the last decades and is attributed to the increased levels of greenhouse gases (GHG) in the atmosphere. Today, carbon dioxide (CO_2) is present in the atmosphere at a concentration over 415 ppmv, i.e., about 0.04 volume percent, equivalent to an atmospheric reservoir of about 3200 $GtCO_2$ [1]. To address the increase in global CO_2 emissions to the atmosphere, GHG emission reduction targets and a wide range of greenhouse gas mitigation technologies are being considered, such as CO_2 capture from flue gases at power plants and other industrial sites followed by CO_2 transport and long-term geological storage [2,3]. The direct capture of CO_2 in the ambient air (DAC: Direct Air Capture) through a contactor is an alternative pathway among the negative emissions technologies to capture CO_2 directly from the atmosphere and should be deployed to achieve emission trajectories in line with the carbon neutrality objective and climate change mitigation [4,5]. Indeed, CO_2 removal is expected to play a key role in the transition to a net-zero system, in particular, to offset the emissions of industrial sectors that are difficult to decarbonize. In this regard, CO_2 removal technologies and changes in land-use sinks account for 450–1000 $GtCO_2$ of negative emissions to keep the global warming below 1.5 °C by 2100 [6]. The CO_2 can be permanently stored in deep geological formations, resulting in negative emissions, or used in the production of building materials, chemical intermediates, or synthetic fuels to replace conventional fossil fuels. The DAC technique seems attractive given its potential to decarbonize the atmosphere: it can address distributed emissions such as aviation and transport and can be installed close to suitable storage sites and to low- or zero-carbon energy sources

which are needed to run the plant, with little degradation performance in the case of low-pollutant air. However, it still presents multiple technical and economic uncertainties and barriers [7]. Furthermore, the relatively high dilution of CO_2 in the atmosphere leads to higher energy needs (approximately three times more energy) and costs for DAC relative to other conventional CO_2 capture technologies and applications. Thus, DAC is energetically and economically challenging to be deployed at large scale.

DAC technologies have attracted new interest for several years [8,9] but the first developments of CO_2 capture in ambient air date back to the 1930s, with first applications in gas separation upstream of cryogenic air separation ($N_2/O_2/Ar$) in order to avoid CO_2 solidification, then for the control of air composition in confined systems (submarine, spacecraft) to keep the air breathable with no possibility of renewal [10]. There are currently a number of small pilot and demonstration DAC plants operating worldwide [11], mainly in Europe, the United States, and Canada, capturing around 10,000 tCO_2/year, with large-scale facilities (1 $MtCO_2$/year) in advanced development in the United Kingdom and in the United States, e.g., for use in enhanced oil recovery. After production is scaled up, the costs are expected to become competitive and fall to USD 200 per ton of CO_2 [12]. However, the DAC sector is still in an early stage of commercial development, and research and development are still needed to overcome some challenges, e.g., to redesign and optimize the materials and processes to achieve low-cost and low-carbon performances [13]. Several approaches to DAC are technically feasible but the development efforts are mainly focused on two reversible techniques for CO_2 capture in air: chemical absorption [14], which relies on the property of a liquid basic solution to solubilize CO_2, such as sodium and potassium hydroxide, and adsorption, where CO_2 is fixed on the surface of a porous solid sorbent [15,16]. Both techniques require roughly 80% thermal energy (e.g., sorbent regeneration) and 20% electricity (e.g., contactor fans, vacuum pumps) for operation [11], and have to be fueled by low-carbon energy sources [17], in particular close to industrial sites where low-carbon heat can be recovered, to reduce lifecycle emissions [18,19].

The DAC technique based on low-temperature adsorption is developed by several companies among which are Climeworks in Switzerland and Global Thermostat in the United States. Such technologies have limited land and water footprints [7]. However, the co-adsorption of H_2O in wet air treatment increases the thermal energy consumption for regeneration and the sorbent regeneration requires large amounts of low temperature heat [15]. Compared to absorption that works continuously with solvent looping, the adsorption process works generally in batches with several beds filled in parallel. Processes have been developed to allow efficient contact of air with adsorbent and efficient regeneration of the material. The adsorption phenomena are schematically divided into two large families according to the nature of the bonds between the adsorbate and the solid: physisorption (weak interaction), e.g., on activated carbons, activated aluminas, silica gels or zeolites, and chemisorption (strong interaction), e.g., on chemical adsorbents based on amines immobilized on a solid support or alkali carbonates [20,21]. An optimal adsorbent would combine the following qualities: high adsorption capacity, high selectivity, easily regenerable, fast kinetics, high resistance (mechanical, chemical, and thermal) and lifetime, high availability, low pressure drop, low toxicity, and low cost. In practice, a major limitation of physical adsorbents is their low adsorption capacity at low CO_2 partial pressure, which leads to a preferable use of chemical adsorbents for CO_2 capture from air. On the other hand, while chemical adsorbents have a higher adsorption capacity compared to physical adsorbents, the energy consumption associated with their regeneration is higher for breaking the chemical bonds between the adsorbate and the adsorbent. The use of an amine bonded to a porous solid support, e.g., such as honeycomb monoliths, pellets, or other granular shapes, is therefore suitable [15,16,22].

In cyclic VTSA (vacuum temperature swing adsorption), the air ventilated from the atmosphere may flow through a fixed bed of solid adsorbent, where the CO_2 (and H_2O) is adsorbed on a porous material at ambient temperature and pressure. The flow to be treated is depleted as it progresses through the solid bed, which gradually becomes saturated. After operation and when the adsorbent approaches saturation, the adsorbent is regenerated in situ by heating (TSA around 80–130 °C) and rough vacuum (VSA below 300 mbar$_{abs}$). The CO_2 and H_2O are then released in the gas phase, which allows, after gas drying, to recover a high-purity CO_2 stream that can be compressed for transport and storage. The adsorbent is finally cooled to room temperature for reuse in a new cycle (adsorption, purge, regeneration, repressurization). The VTSA proves to be a promising process from analysis of productivity and energy consumption [23].

Despite their apparent simplicity, the design and development of adsorption processes is time consuming and requires cost reductions to be competitive at full scale. Process modularization is a strong feature in order to reduce investment costs through series and learning effects [24,25]. Compared to an all-in-one architecture, the modular design allows to realize and test a large number of modules in a first development phase, before completely setting the optimal module architecture which can then be standardized and produced in large series in order to lower the costs by learning rate. Modularity allows scale-up by increasing the number of components operating in parallel.

There have been a few studies on the modeling of a DAC fixed bed process [16,26–28]. Simulations are most commonly performed at the laboratory scale, i.e., with sorbent material masses in the range of grams to kilograms using amine-functionalized nano-fibrillated cellulose [16], metal organic framework [26,28], or polystyrene [27]. In this paper, we propose to model in Aspen Adsorption software a DAC modular process that employs VTSA on a packed bed using amine-functionalized material. The one-dimensional model accounts for adsorption isotherms, mass, energy, and momentum balances to simulate temperature and concentration dynamics along the bed. The simulations are performed for several adsorption–desorption cycles at two scales: first at the laboratory scale (2 kgCO_2/year), then at the larger scale of a pilot module (50 tCO_2/year). A parametric study on the main operating conditions, i.e., air velocity (superficial gas velocity lower to minimum fluidization velocity), air relative moisture (0–80%), air temperature (5–35 °C), and CO_2 capture rate (5–98%), is undertaken to assess the global performances and energy consumption. The results obtained in this work can guide future research on the design of a DAC modular process.

2. Process and Model Description

Aspen Adsorption software (V12) was used for modeling as it offers a framework of numerical and physical methods facilitating the dynamic mass and energy balance evaluation of the process' adsorption/desorption full cycle. This software is a comprehensive flowsheet simulator for the design, simulation, and analysis of dynamic adsorption processes.

2.1. Process Overview

The simplified scheme of the DAC adsorption process is given in Figure 1: air is drawn into the fixed adsorbent bed using fans. The CO_2 of the air is physically and chemically bound to the solid sorbent material. CO_2-free air is released back into the atmosphere. Once the sorbent is saturated with CO_2, it is heated using indirect internal heating and/or steam stripping. The internal heating and steam stripping could use free-carbon electricity and/or low-grade heat source thanks to vacuum conditions ensured by vacuum pumps during the stripping step. The CO_2 is then released from the sorbent and collected as concentrated gas thanks to a condenser. This continuous cycle is then ready to start again. The sorbent is reusable and lasts for several thousand cycles (approx. 2–3 years).

The dynamic bed model in Aspen Adsorption is based on the fundamental equations of an adsorbent bed composed of multiple layers: isotherms (thermodynamics), mass and energy balances of gas and solid, including mass and heat transfers, and pressure drop. The following equations are used to model continuous cycles of CO_2 and H_2O adsorption/desorption through a packed bed of amine-functionalized sorbent material.

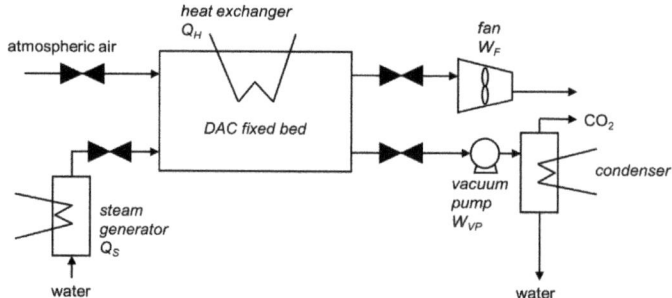

Figure 1. Simplified process flowsheet for adsorption-based DAC technology.

2.2. Adsorption Isotherms of CO_2 and H_2O

An adsorption isotherm is the relation between the amount of adsorbate (loading) and the gas phase composition (partial pressure) at thermodynamic equilibrium for a given temperature. The adsorption capacity of the solid increases with CO_2 partial pressure and decreases with temperature. A simple theoretical model for single component adsorption is the Langmuir model well-suited for chemisorption. This model uses the assumptions that all adsorption sites on the homogeneous solid surface are energetically equivalent, and that the adsorbate forms a monolayer without interactions between the adsorbate particles. The Toth model [29] used for pure CO_2 adsorption on amine-functionalized nano-fibrillated cellulose material [30] differs from the Langmuir isotherm by considering the heterogeneity of the adsorbent surface characterized by the Toth parameter $t(T)$:

$$q_{CO_2}(T, p_{CO_2}) = n_s(T) \cdot \frac{b(T) \cdot p_{CO_2}}{\left(1 + (b(T) \cdot p_{CO_2})^{t(T)}\right)^{\frac{1}{t(T)}}} \quad (1)$$

where b, t, and n_s are temperature-dependent parameters:

$$b(T) = b_0 \cdot e^{-\frac{\Delta h_{ads,CO_2,0}}{R \cdot T_0} \cdot \left(\frac{T_0}{T} - 1\right)} \quad (2)$$

$$t(T) = t_0 + \alpha \cdot \left(1 - \frac{T_0}{T}\right) \quad (3)$$

$$n_s(T) = n_{s,0} \cdot e^{\chi \cdot \left(\frac{T_0}{T} - 1\right)} \quad (4)$$

The H_2O adsorption isotherm on the same material [30] is modeled by the Guggenheim–Anderson–de-Boer (GAB) isotherm [31] defined by the following equation.

$$q_{H_2O}(T, p_{H_2O}) = C_m(T) \cdot \frac{C_G(T) \cdot K_{ads}(T) \cdot \frac{p_{H_2O}}{p_{vap}(T)}}{\left(1 - K_{ads}(T) \cdot \frac{p_{H_2O}}{p_{vap}(T)}\right) \cdot \left(1 + (C_G(T) - 1) \cdot K_{ads}(T) \cdot \frac{p_{H_2O}}{p_{vap}(T)}\right)} \quad (5)$$

where $\frac{p_{H_2O}}{p_{vap}(T)} = \phi$ is the relative humidity.

The following coefficient provides the water content of a monolayer at the surface of the adsorbent:

$$C_m(T) = C_{m,0} \cdot e^{\frac{\beta}{T}} \quad (6)$$

The following coefficients are following Arrhenius-type and temperature-dependent equations:

$$C_G(T) = C_{G,0} \cdot e^{\frac{\Delta H_C}{RT}} \quad (7)$$

$$K_{ads}(T) = K_{ads,0} \cdot e^{\frac{\Delta H_K}{RT}} \quad (8)$$

In addition, the binary CO_2 and H_2O adsorption isotherms consider the competitive adsorption of the two substances. The CO_2 has very little effect on the H_2O adsorption capacity, so the H_2O adsorption isotherm remains largely unaffected by the presence of CO_2 [32]. On the contrary, the CO_2 adsorption capacity is generally enhanced in the presence of H_2O [22]. An empirical function applicable in a small pressure range is used at first approximation to describe the CO_2 adsorption under humid conditions by multiplying the Toth isotherm by an enhancing parameter $f_{RH}(p_{CO_2}, \phi)$ which is function of the relative humidity and the CO_2 partial pressure:

$$q_{CO_2}^{binary}(T, p_{CO_2}, \phi) = f_{RH}(p_{CO_2}, \phi) \cdot q_{CO_2}(T, p_{CO_2}) \tag{9}$$

$$f_{RH}(p_{CO_2}, \phi) = 1 + \phi \cdot \left(0.6 - \frac{p_{CO_2}}{59 \; mbar} \cdot 0.47\right) \tag{10}$$

where ϕ is the relative humidity of air.

The coefficients obtained on the same material from [30,32] are summarized in Table 1.

Table 1. Material properties (isotherms).

CO₂—Toth Isotherms			H₂O—GAB Isotherms		
b_0	1/bar	22,500	$C_{m,0}$	kmol/kg	0.0000208
T_0	K	296	β	K	1540
t_0	-	0.422	$C_{G,0}$	-	6.86
α	-	0.949	ΔH_C	kJ/kmol	−4120
$n_{s,0}$	kmol/kg	0.00197	$k_{ads,0}$	-	2.27
χ	-	2.37	ΔH_K	kJ/kmol	−2530
$\Delta h_{ads,CO_2,0}$	kJ/mol	60	$\Delta h_{ads,H_2O,0}$	kJ/mol	49

2.3. Main Assumptions

The bed model makes the main following assumptions:

1. A one-dimensional model using axial dispersion is considered. The effect of axial dispersion along the gas flow direction is studied according to the axial Péclet dimensionless number. The Péclet number is the product of the Reynolds number and the Schmidt number and is used to compare the mass transport by dispersion and diffusion:

$$Pe_z = \frac{v_g H_b}{E_z} \tag{11}$$

where v_g is the gas velocity in m/s, H_b the height of the bed in m, and E_z the dispersion coefficient in m²/s.

As a gas flows through a packed bed, axial dispersion occurs by molecular diffusion, turbulent mixing arising from the splitting and recombining of flows around the solid particles, and wall effects due to non-uniformity of packing. In general, wall effects can be controlled by a large ratio of bed-to-particle diameters. The molecular diffusion and turbulent mixing are additive and can be lumped into an effective dispersion coefficient [33]. Mass transport by convection takes control, i.e., the bed operates near ideal plug-flow conditions, when the mass Péclet number is significantly larger than 1. Therefore, the dispersion term is included in the material balance described later. Moreover, to limit the radial dispersion, three conditions are required: the ratio of the bed diameter to the particle diameter must be high (typically larger than 15 [34]), the gas must be uniformly mixed before entering the bed, and the pressure drop must not be too low to allow the gas to be equidistributed. A high value of the radial Péclet number can be considered according to the radial dispersion coefficient [35]:

$$E_r = \frac{v_g r_p}{4} \tag{12}$$

Thus, radial dispersion is negligible, and the bed model is one-dimensional.

2. CO_2 and H_2O are the only adsorbed substances. Adsorption of the other substances of air, such as N_2, O_2, and Ar, are neglected given the dominant role of chemisorption. In addition, there are no parasitic reactions between the gas and the adsorbent.
3. The substances in ideal gas phase flow first from the bulk gas to the macropore, and then from the macropores to the solid surface via the micropores with uniform pore structure (isotropic material). The microporous diffusion term is assumed to be negligible in comparison to the external film resistance term and the macropore diffusion term.
4. The bed is non-isothermal and is considered adiabatic with uniform heating from the walls. The clogging resistance is neglected.

2.4. Mass Balance

The overall mass balance for a multi-component gas phase considers the convection, axial dispersion, and mass transfer from the gas to the solid phase. Each substance in the gas phase is governed by the equation below:

$$-\varepsilon_i E_{zk} \frac{\partial^2 c_k}{\partial z^2} + \frac{\partial(v_g c_k)}{\partial z} + \varepsilon_B \frac{\partial c_k}{\partial t} + J_k = 0 \quad (13)$$

where ε_i is the interparticle voidage, ε_B the total bed voidage, and E_{zk} the dispersion coefficient given by the correlation below [33]:

$$E_{zk} = 0.73 \, D_{mk} + \frac{v_g r_p}{\varepsilon_i \left(1 + 9.49 \frac{\varepsilon_i D_{mk}}{2 v_g r_p}\right)} \quad (14)$$

where D_{mk} is the molecular diffusivity in m^2/s (estimated by Aspen properties database) and r_p the particle radius in m.

The rate of flux to the solid surface per unit volume is:

$$J_k = -\rho_s \frac{\partial w_k}{\partial t} \quad (15)$$

where ρ_s is the particle bulk density in kg/m^3 (i.e., the mass of solid per unit volume of column) and the rate of adsorption is expressed as:

$$\frac{\partial w_k}{\partial t} = k_k (w^*_k - w_k) = k_k K_{Kk} (c^*_k - c_k) \quad (16)$$

and where Henry's coefficient K_{Kk} is derived from the isotherms:

$$K_{Kk} = \frac{\partial w^*_k}{\partial c_k} = RT \frac{\partial w^*_k}{\partial P_k} \quad (17)$$

To get to the adsorption surface, the substances must diffuse from bulk gas phase into the pores of solid particles, then diffuse from pore phase into the surface of solid particles. The overall mass transfer coefficient between the gas and solid phases is given as a lumped term comprising the external film resistance term and the macropore diffusion term:

$$\frac{1}{k_k} = \frac{r_p}{3 \, k_{fk}} + \frac{r_p^2}{15 \, \varepsilon_p \, K_{pk}} \quad (18)$$

k_{fk} is the film resistance coefficient defined below:

$$k_{fk} = Sh_k \frac{D_{mk}}{2 r_p} \quad (19)$$

The Sherwood number Sh_k is obtained by [36] for a Reynolds number Re in the range 3–10,000:

$$Sh_k = 2.0 + 1.1 \, Sc_k^{1/3} Re^{0.6} \quad (20)$$

where the Schmidt number and the Reynolds number are:

$$Sc_k = \frac{\mu}{D_{mk}\rho_g} \quad (21)$$

$$Re = \frac{v_g \, 2r_p \rho_g}{\mu} \quad (22)$$

where μ is the dynamic gas viscosity in Pa·s.

The macropore diffusion coefficient K_{pk} is obtained from:

$$\frac{1}{K_{pk}} = \tau \left(\frac{1}{D_{kk}} + \frac{1}{D_{mk}} \right) \quad (23)$$

where τ is the particle tortuosity factor and the Knudsen coefficient D_{kk} is obtained from the following correlation:

$$D_{kk} = 97 r_{pore} \left(\frac{T}{M_k} \right)^{0.5} \quad (24)$$

where r_{pore} is the macropore radius in m and M_k is the molecular weight of the substance in kg/mol.

2.5. Energy Balance

The energy balance is separated into two expressions, the gas energy balance and the solid energy balance.

2.5.1. Gas-Phase Energy Balance

The energy balance of the gas phase in the bed includes the following terms, i.e., the gas conductivity, the convection, and the transfers with the adsorbent bed and with the internal wall:

$$-\varepsilon_i k_{gz} \frac{\partial^2 T_g}{\partial z^2} + C_{vg} v_g \rho_g \frac{\partial T_g}{\partial z} + \varepsilon_B C_{vg} \rho_g \frac{\partial T_g}{\partial t} + P \frac{\partial v_g}{\partial z} + HTC a_p (T_g - T_s) \\ + a_{Hx} Q_{Hx} = 0 \quad (25)$$

The heat transfer between the gas and solid is modeled with the film resistance model according to:

$$HTC = j C_{pg} v_g \rho_g Pr^{-\frac{2}{3}} \quad (26)$$

where the Prandtl number and the j-factor are:

$$Pr = \frac{\mu C_{vg}}{k_{gz}} \quad (27)$$

$$j = 1.66 \, Re^{-0.51} \text{ if } Re < 190 \quad (28)$$

$$j = 0.983 \, Re^{-0.41} \text{ otherwise} \quad (29)$$

The specific particle surface per unit volume bed is:

$$a_p = (1 - \varepsilon_i) \frac{3}{r_p} \quad (30)$$

A jacket heat exchanger is used to heat the bed and for cooling according to:

$$Q_{Hx} = U_{Hx} (T_g - T_{Hx}) \quad (31)$$

where U_{Hx} is the heat transfer coefficient of the heat exchanger in W/(m^2·K) and T_{Hx} the heating source temperature in K at the axial position of the bed.

2.5.2. Solid Energy Balance

The energy balance on the adsorbent considers the heat transfer by convection between the gas flow and the adsorbent but also the heat released during the adsorption:

$$-k_{sz}\frac{\partial^2 T_s}{\partial z^2} + \rho_s C_{ps}\frac{\partial T_s}{\partial t} + \sum_k H_k + \rho_s \sum_k \left(\Delta H_k \frac{\partial w_k}{\partial t}\right) - HTCa_p(T_g - T_s) = 0 \quad (32)$$

The heat of the adsorbed phase for each component is given by the following equation:

$$H_k = \rho_s C_{pak} w_k \frac{\partial T_s}{\partial t} \quad (33)$$

where C_{pak} is the heat capacity of the adsorbed phase component in J/(kg·K).

2.6. Energy Requirements

The main electric utilities through the proposed VTSA model are the electrically driven fan used to compensate for the pressure drop inside the bed (W_F) and the vacuum pump (W_{VP}).

The Ergun equation expresses the pressure drop (in Pa/m) in a fixed bed by the Karman–Kozeny equation for laminar flow and the Burke–Plummer equation for turbulent flow:

$$\frac{\partial P}{\partial z} = -\left(\frac{150(1-\varepsilon_i)^2}{(2r_p\psi)^2 \varepsilon_i^3}\mu v_g + 1.75\, M\rho_g \frac{(1-\varepsilon_i)}{2r_p\psi\varepsilon_i^3}v_g^2\right). \quad (34)$$

where ψ is the particle shape factor and M refers to the molecular weight of the gas in kg/mol.

The fan work (in GJ/tCO$_2$) considering the fan isentropic efficiency η_f (typically 0.7) and the overall pressure drop from the Ergun equation is assessed using Aspen Plus:

$$W_F = \frac{P_F t_{ads}}{m_{CO_2,ads}} \quad (35)$$

where P_F is the net work required by the fan in GW, t_{ads} the adsorption duration in s, and $m_{CO_2,ads}$ the mass of adsorbed CO$_2$ in t.

The work of the vacuum pump (in GJ/tCO$_2$) considering the pump efficiency (typically 0.7) is also assessed using Aspen Plus:

$$W_{VP} = \frac{P_{VP} t_V}{m_{CO_2,ads}} \quad (36)$$

where P_{VP} is the net work required by the vacuum pump in GW and t_V the vacuum duration in s.

The thermal energy requirement (Q_H) is provided by indirect heating through a jacket heat exchanger or by direct heating using steam during the purge step. The heating need (in GJ/tCO$_2$) is assessed using Aspen Adsorption from the heat exchanger energy:

$$Q_H = \frac{Q_{HX}}{m_{CO_2,ads}} \quad (37)$$

where Q_{HX} is the heat exchanger energy (in GJ) during heating.

Similarly, the heat (Q_S) provided by steam (in GJ/tCO$_2$) is given from:

$$Q_S = \frac{\int_{t_S} \dot{m}_{H_2O} C_{pH_2O} \Delta T\, dt}{10^6 n_{CO_2,ads} M_{CO_2}}. \quad (38)$$

where $\dot{m}_{H_2O}$ is the steam mass flow rate in kg/s, C_{pH_2O} the steam heat capacity in J/(kg·K), ΔT the temperature difference from the inlet and outlet in K, $n_{CO_2,ads}$ the quantity of adsorbed CO_2 in mol, M_{CO_2} the molecular weight of CO_2 in kg/mol, and t_S the steam step duration in s.

2.7. Thermodynamic Comparison

The energy requirements can be compared to the thermodynamic minimum separation work. For a substance to be separated from an ideal mixture, e.g., CO_2 in a mixture of CO_2 and N_2, the minimum separation work can be calculated as a function of the CO_2 capture rate between 0 and 100%, as follows according to the thermodynamic principles:

$$W_{min}^S = -\frac{RT}{3600M}\left(ln(x) + \frac{(1-x)}{x}ln(1-xy) + (1-y)ln\left(\frac{1-xy}{x(1-y)}\right)\right) \quad (39)$$

where W_{min}^S is the minimum separation work (kWh/tCO_2), R the perfect gas constant (8.314 J/(mol·K)), T the mixture temperature (K), x the molar fraction of CO_2 in the mixture, M_{CO_2} is the molecular weight of CO_2 in kg/mol, and y the CO_2 capture rate.

The minimum separation work at 415 ppmv and 293.15 K varies between 6 kWh/tCO_2 for a 5% CO_2 capture rate, 119 kWh/tCO_2 for a 90% CO_2 capture rate, and 132 kWh/tCO_2 for a 98% CO_2 capture rate.

2.8. Parameters and Cycle Description

The typical air conditions considered in this study are the following: pressure of 1.013 bar, temperature of 283.15 K, relative humidity of 40%, molar composition with N_2 (78.70%), O_2 (20.50%), H_2O (0.76%), and CO_2 (0.04%). The model parameters are given in Table 2.

Table 2. Adsorbent properties and model parameters.

Parameter	Symbol	Value	
Interparticle voidage	ε_i	0.446	
Particle radius	r_p	0.0036	m
Macropore radius	r_{pore}	10^{-6}	m
Particle shape factor	ψ	0.83	
Particle tortuosity	τ	3.36	
Molecular diffusivity	D_m	2.02×10^{-5}	m^2/s
Particle heat capacity	C_{ps}	2.07	kJ/(kg·K)
Adsorbed CO_2 heat capacity	C_{paCO_2}	88	kJ/(kmol·K)
Adsorbed H_2O heat capacity	C_{paH_2O}	75	kJ/(kmol·K)
Steam heat capacity	C_{pH_2O}	1.86	kJ/(kg·K)
Particle thermal conductivity	k_{sz}	0.0445	W/(m·K)
Gas dynamic viscosity	μ	1.8×10^{-5}	Pa·s
Gas density	ρ_g	1.2	kg/m^3
Heating source temperature	$T_{Hx,0}$	383.15	K
Heat transfer coefficient	U_{Hx}	60	W/(m^2·K)

The cycle is composed of eight different steps (Table 3) described below:

1. Adsorption. Air from the atmosphere is sent into the bed where the CO_2 and H_2O are adsorbed. The step must end before the bed reaches saturation and cannot adsorb CO_2 anymore. Therefore, the overall recovery of CO_2 is maximized. Adsorption is stopped when the concentration of CO_2 at the bed outlet reaches 10% of the inlet concentration.
2. Evacuation. The inlet air flow is stopped, and the vacuum pump is turned on until a pressure of 300 mbar$_{abs}$ is reached at the end of the bed. The vacuum pump continues to work to maintain this low pressure during the regeneration.
3. Pre-heating. To eliminate N_2 and O_2 residues in the bed and meet the purity target, the sorbent is heated up to 60 °C via indirect heating, to start the desorption of H_2O and evacuate the gas in the atmosphere.

4. Pre-purge. The purge is finished by a short steam step that completely removes the remaining N_2 and O_2 from the bed. The duration of the step is designed so that the purity of the desorbed CO_2 is higher than 95% and must be short to avoid desorbing too much CO_2 in the atmosphere.
5. VTSA. To recover CO_2, the bed is heated indirectly to 110 °C to desorb H_2O and CO_2. The outlet gas composed of CO_2 and H_2O is recovered.
6. Steam-stripping under vacuum. The desorption is finished by injection of overheated steam in direct contact with the adsorbent to desorb and recover the maximum amount of CO_2. The duration of steps 5 and 6 have a strong impact on the cyclic capacity and on the energy penalty.
7. Cooling. To avoid the bed poisoning, e.g., by urea formation in the presence of O_2 at high temperature because of amine oxidation, the bed is cooled at least below 60 °C via the heat exchanger before sending the air in the adsorber for a new adsorption cycle.
8. Repressurization. Air is progressively injected to reach atmospheric pressure to be ready for adsorption.

Table 3. Cycle parameters.

	Step	Duration	Temperature (°C)	Pressure (bar)
1	Adsorption	Until C_{out}/C_0 = 10%	10	1.013
2	Evacuation	10 s	10	<0.150
3	Pre-heating	10 min	60	<0.300
4	Pre-purge	3 min	-	<0.300
5	VTSA	1 h 20 min	110	<0.300
6	Steam purge under vacuum	20 min	-	<0.300
7	Cooling	24 min	<60	<0.300
8	Repressurization	Until p = 1.013 bar	10	1.013

The detailed simulations are performed with the commercial software Aspen Adsorption (V12) by applying the "single bed approach" and a cycle organizer (Figure 2) to control the duration of the different sequential steps, the manipulated variables, and the setting-up of the boundary conditions (flowrate, composition, temperature, pressure). The evolution of concentrations, pressure, and velocity profiles is followed in real time throughout the simulation of successive entire cycle steps.

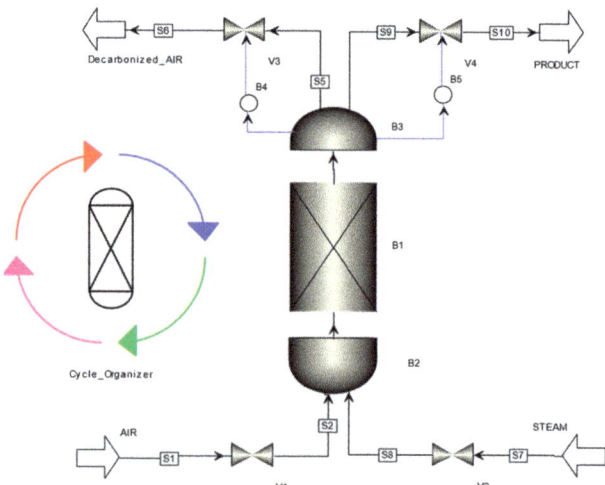

Figure 2. Process flowsheet for DAC in Aspen Adsorption.

3. Results

The modeling approach was at first validated within experimental data of [32] performed on a lab-scale pilot, around 2.8 kgCO$_2$/year. A parametric study was then performed on the lab-scale pilot before the latter was scaled-up to the module of around 50 tCO$_2$/year. The module's bed was equipped with a jacket indirect heat exchanger. Low-grade overheated steam was provided co-currently via a separate inlet. Two separate outlets were used, one rejected the bed gas into the atmosphere during adsorption and the purge steps, while the second one transported the desorbed gas during the desorption step. Both outlet streams were evacuated by dedicated vacuum pumps.

The CO$_2$ was recovered from the gas exiting the bed during the desorption steps and then experiences a condensation and phase separation step to fully eliminate the steam. The remaining gas was CO$_2$ with a purity superior to 95%.

3.1. Lab-Scale Configuration

The lab-scale characteristics are given in Table 4. Depending on initialization values provided to the cycle organizer block (Figure 2), steady state cycles would not be reached within the first cycle but only starting from the second and successive following cycles. Figure 3 shows established regime of the temperature during cycles.

Table 4. Lab-scale bed characteristics.

		Lab Scale	
Bed height	H_b	0.299	m
Bed internal diameter	D	0.081	m
Intraparticle voidage	ε_p	0.937	
Particle bulk density	ρ_s	55.4	kg/m^3
Air flow rate	Q_{air}	14	NL/min

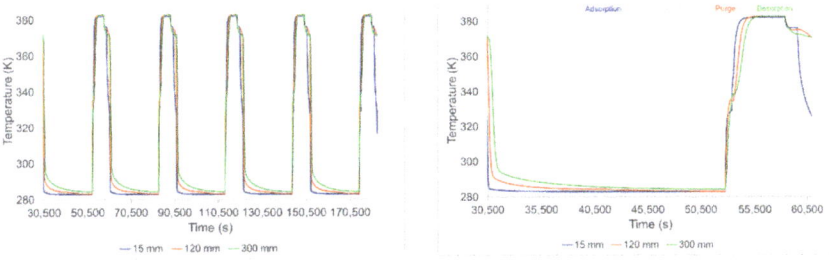

Figure 3. Temperature profiles of successive cycles (**left**) and a zoom on a single cycle (**right**).

The largest spatial temperature gradient was about 30 K (Figure 3) meaning that the bed was globally homogeneous in temperature. A non-obvious behavior occurred at the end of the desorption step when the temperature decreased before the cooling step. This was linked to the endothermic nature of desorption during the steam-stripping step. During this latter step, a higher amount of CO$_2$ was desorbed in a relatively short time, therefore the temperature of the bed slightly decreased.

Figure 4 represents the adsorbed quantities of CO$_2$ and H$_2$O over time. The evolution during the adsorption phase was relatively different for the H$_2$O and CO$_2$. In fact, the adsorbed quantity of H$_2$O was more than four times larger than for CO$_2$. This is linked to the partial pressures of the substances and to the fact that the two substances have different isotherms with different mathematical expressions derived from the experimental measurement on the lab-scale pilot test facility. Moreover, H$_2$O adsorption seems to saturate over adsorption phase time while CO$_2$ adsorption continues to increase linearly. The H$_2$O adsorption peak, around t = 52,500 s (Figure 4), is linked to the pre-purge step using steam because there is a steep increase in the vapor partial pressure resulting in a quick increase in the adsorbed quantities. Finally, during the vapor stripping step, the vapor partial pressure increases compared to the previous step. Therefore, there is adsorption of H_2O and desorption of CO_2 because of high temperature and low CO_2 partial pressure.

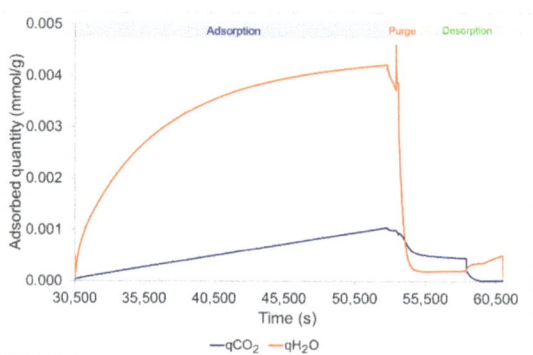

Figure 4. Adsorbed quantity of CO_2 (blue curve) and H_2O (red curve) over time during a full cycle.

The adsorption uptake was calculated via the results of the CO_2 loading variable of adsorbent material over time during the adsorption step. The cycle's CO_2 cyclic capacity was calculated by time integration of the CO_2 quantity at the outlet of the bed during desorption and was equal to around 0.83 mmol/g, i.e., 0.037 g/g. The capture rate, i.e., the quantity of CO_2 recovered at the end of a whole cycle divided by the quantity of CO_2 treated during the adsorption step, was around 77% at a purity larger than 99%.

The specific energy requirements are summarized in Table 5 below.

Table 5. Energy performances of the lab-scale module.

Fan work (kWh/tCO_2)	2.78
Vacuum work (kWh/tCO_2)	97.22
Total electrical energy need (kWh/tCO_2)	100
Steam heating need (GJ/tCO_2)	0.26
Heating need (GJ/tCO_2)	12.88
Active cooling need (GJ/tCO_2)	0.64
Total thermal energy need (GJ/tCO_2)	13.14

Finally, the specific energy need given by the model was consistent with the energy need evaluated by Wurzbacher et al. [15]. In fact, the data provided values around 9.3 GJ/tCO_2 (i.e., 2600 kWh/tCO_2) for the total thermal energy need and 100 kWh/tCO_2 for the electrical energy need. Furthermore, the significant difference from the minimum separation work, around 100 kWh/tCO_2 (25 times smaller factor in order of magnitude) must be pointed out, which shows the energetic improvement potential to improve this capture technique.

3.2. Parametric Study of the Lab-Scale Process

The objective of this section is to investigate the impact of ambient air parameters on the performance of the DAC process. This way, it is possible to determine the bed dimensions for the most stringent operating point with the objective to minimize energy requirements, while maintaining the highest possible cyclic capacity. Otherwise, it is necessary to increase the bed volume to achieve similar annual capture values.

The parameters of the study and the associated variations are summarized in Table 6 and the reference points used in the different studies are presented in Table 7.

Table 6. Studied parameters and range of variation.

Parameters of the Analysis	Variation Range
Air flow velocity (m/s)	[0.001–0.6]
Relative humidity (RH) (%)	[0–80]
Air temperature (K)	[268.15–293.15] at 0%RH [278.15–308.15] at 40% RH
Breakthrough curve position C_{out}/C_{in} (%)	[5–98]

Table 7. Reference points for the different parametric studies.

	Variation Parameter	Temperature	Humidity Rate	Breakthrough Curve Position	Air Flow Speed
	Temperature	/	293.15 K	293.15 K	293.15 K
Characteristics of the reference point	Humidity rate	40%	/	40%	40%
	Breakthrough curve position	10%	10%	/	10%
	Air flow speed	0.044 m/s	0.044 m/s	0.044 m/s	/
	CO_2 concentration	400 ppm	400 ppm	400 ppm	400 ppm

The results are presented in Figure 5. The evolutions of the CO_2 cyclic capacity (Figure 5, upper left) are shown to decrease with increasing ambient air temperature. This is directly linked to the isotherm equations presented before that indicate a better adsorption at lower temperature. Surprisingly, as shown in upper-right of Figure 5, the increase in ambient air relative humidity is favorable for increasing CO_2 cyclic capacity. This is also explained by the expression of the humidity enhancement factor in the binary CO_2 isotherm (Equations (9) and (10)). Indeed, there is competition between increasing the relative humidity and temperature; the first trend increases the CO_2 capacity linearly while the latter is decreased with increasing temperature. These trends are coherent with VTSA desorption technology characteristics observed in laboratory-scale beds [32].

Figure 5 (bottom-left and right) shows the evolution of the CO_2 cyclic capacity and the mass transfer coefficient (MTC) as functions of the air velocity in the bed. The condition stopping the adsorption remains the same for all simulations, i.e., when the CO_2 breakthrough reaches 10% of the inlet CO_2 concentration. Thus, increasing the air flow leads to an adsorption duration that is too short because of two competitive phenomena: the first one is that adsorption is kinetically faster because there is more CO_2 entering the bed in the same time period leading to higher CO_2 bed-volumetric mole concentration, and the second one is that by increasing air velocity, the residence time of the CO_2 inside the bed becomes closer and closer to the characteristic time of adsorption ($\sim \frac{1}{MTC}$) (see bottom-right of Figure 5). Therefore, the thermodynamics of adsorption become less and less predominant over the kinetics. Some CO_2 cannot be adsorbed and goes directly to the outlet. The bed is consequently less charged than in the lower air velocity cases when applied to the CO_2 concentration, at the bed exit, a fixed breakthrough of 10% of the inlet CO_2 concentration.

Figure 5 (bottom-right) shows the sensitivity analysis of the mass transfer coefficients of CO_2 and H_2O on air velocity which was performed for air velocity variation. It shows the interest in working with high air velocity, because the higher the air velocity the lower the predominant diffusion and, consequently, the adsorption step of the cycle is quicker. However, if the increased air velocity is obtained via forced ventilation, then the energy penalty of the fan will increase with the flow velocity. Therefore, there is a tradeoff between the adsorption kinetics, the CO_2 cyclic capacity, and the energy need of the process.

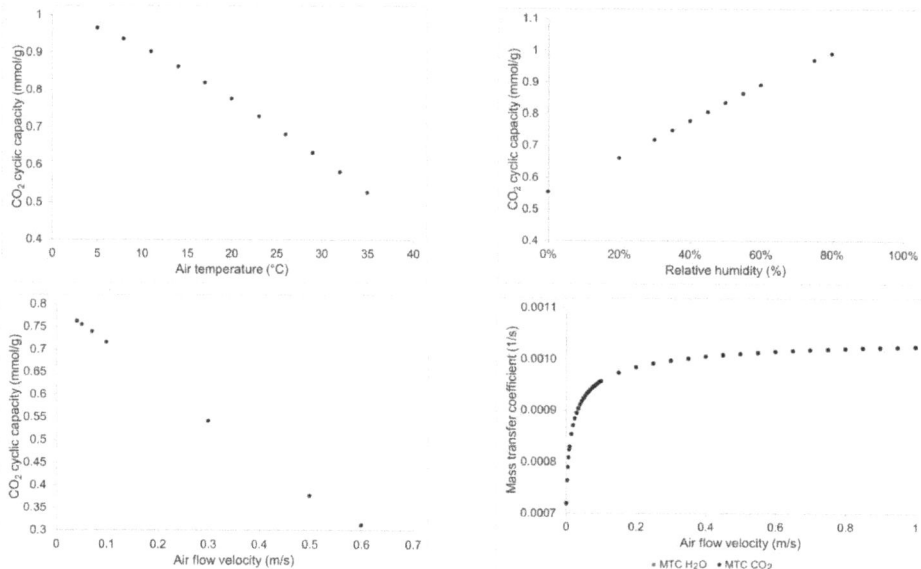

Figure 5. Results of the different parametric studies on the CO_2 cyclic capacity of the process at the lab-scale.

Figure 6 summarizes the specific heating energy results of the parametric study. The results can be explained for ambient temperature and velocity impacts by the same arguments as above: CO_2 cyclic capacity decreases by increasing these parameters, but the cycle's conditions are unchanged particularly for a fixed breakthrough cutoff level, this leads to a higher specific heating need because all the solid sorbent is heated during the stripping step, while it is not sufficiently loaded with CO_2.

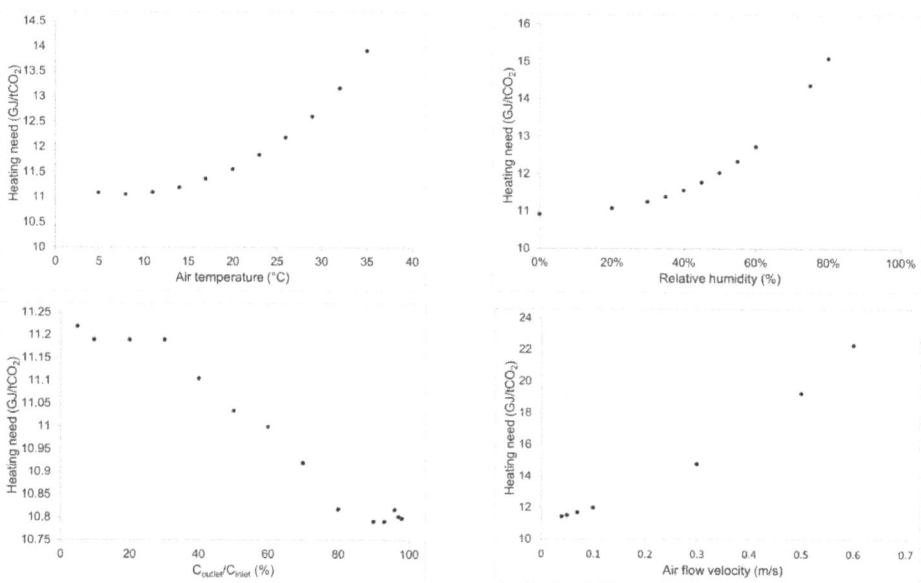

Figure 6. Results of the parametric studies on the specific heating need of the process at the lab scale.

When increasing the breakthrough level (Figure 6, bottom-left), the cyclic capacity increases, which means that the solid sorbent is loaded with more CO_2 and the heating need decreases slightly. Finally, increasing the humidity rate also increases the partial pressure of H_2O during adsorption, therefore, there is more H_2O being adsorbed and more H_2O to desorb. As desorption is endothermic, an increased amount of H_2O in the air increases the heating need; indeed, increasing the cyclic capacity is not enough to maintain at least a constant specific heating need.

3.3. Modular-Scale Configuration

The lab-scale bed design is very similar to the design presented in [32], apart from the shape of the bed. The scale-up methodology aims to define the size of the adsorbent bed and the flow rates so that the yearly capture of 50 tCO_2 could be reached.

Given that a large bed diameter allows to increase the volumetric rate of inlet air without increasing the superficial velocity, however, the bed diameter was limited to 1.8 m in order to avoid radial dispersion. In addition, the bed length is also dependent on the material characteristics such as the density and the amine content along with the superficial velocity. A bed with a length of 2 m and a density of 55.4 kg/m³ was considered at first. Increasing the density by a factor of four to reach 221.6 kg/m³, but conserving the adsorbent mass and the same adsorption duration implies having a shorter length of the bed of 0.50 m. Finally, starting from the latter configuration, increasing the amine content of the adsorbent material by 14% implies reducing the length of the bed more, to 0.44 m, to keep the adsorption time constant. Therefore, when increasing both the volume of the bed and the sorbent density or amine content, it allows to increase the superficial velocity while keeping the adsorption time constant.

The chosen design has a bed density of 221.6 kg/m³, a length of 1.78 m, to reach a total weight of bed of 1 ton of adsorbent material. The bulk air velocity during adsorption is about 1.97 m/s.

3.3.1. Up-Scaling Approach

At the lab scale, the global Reynolds number, used to determine whether the fluid flow is laminar or turbulent, reaches a value of 21 (i.e., 1 < Re < 100). It is in a transitory regime where both the inertia and viscous terms have an impact. This configuration allows to limit the pressure drop and still have high values for the mass transfer. For the chosen modular-scale design, the value of the Reynolds number is much higher, around 985. In this configuration, the fluid flow is much more turbulent; therefore, the viscous term becomes negligible in comparison with inertia forces and, consequently, the pressure drop is around 1500 times higher than the lab-scale bed case.

For the selected design (Table 8), the Sherwood number is around 64, which is roughly in the same order of magnitude as the lab-scale dimension, around 8. It has little impact on the MTC because the latter is only increased by 8% during the scale-up operation. In both configurations, the axial Péclet numbers are superior to 10^4. Those high values imply that the dispersion is weak.

Table 8. Industrial module scale.

		Module Scale	
Bed height	H_b	1.78	m
Bed internal diameter	D	1.80	m
Intraparticle voidage	ε_p	0.811	
Particle bulk density	ρ_s	221.6	kg/m³
Air flow rate	Q_{air}	1.8	Nm³/s

3.3.2. Performances

The goal of the scaling-up is to design an adsorption bed that could achieve the capture of 50 tCO_2 yearly. The chosen design with high bed weight and short cycle duration (around

6 h 30 min) allows the module to perform more than three cycles per day. The cyclic capacity of this configuration was evaluated as 0.958 mmol of CO_2/g of sorbent, i.e., 0.042 g/g. In this configuration, the module can capture up to 56 tons of CO_2 yearly with a purity superior to 95%. The specific energy requirements are summarized in Table 9 below.

Table 9. Energy performance of the industrial module scale.

Fan work (kWh/tCO_2)	3419.44
Vacuum work (kWh/tCO_2)	30.56
Total electrical energy need (kWh/tCO_2)	3450
Steam heating need (GJ/tCO_2)	0.12
Heating need (GJ/tCO_2)	12.60
Active cooling need (GJ/tCO_2)	1.16
Total thermal energy need (GJ/tCO_2)	13.87

The strong increase in the fan work (by 345%) is linked to the pressure loss which is much higher in the industrial bed comparatively to the lab-scale bed: -1395 Pa compared to -0.95 Pa for the lab-scale pilot.

4. Conclusions

Performance of an adsorption-based technology for CO_2 capture directly from the air was evaluated numerically at the industrial scale. Detailed mass and heat balance dynamic modeling of the vacuum temperature swing adsorption process was developed in adsorption-dedicated commercial software. The first step of the proposed work validated the modeling thanks to published experimental data of a lab-scale bed module in terms of mass transfer and energy performance. Modeling results of energy performance of the lab-scale bed were found close to experimental data. A method of up-scaling the lab-scale bed to an industrial module was also exposed and mass transfer and energy performances of industrial module were provided.

The modeling design of 50 tCO_2/year industrial unit was achieved while conserving almost the same technology which was experimented at lab scale, but for the solid sorbent which is supposed to be enhanced for industrial modules. The scale up from a lab scale to the industrial size is conservative in terms of thermal energy consumption while the electrical consumption is very sensitive to the bed design. Because of the very low concentrations of CO_2 in the air, the specific electrical energy consumption of fans to overcome the bed pressure drop is very high per captured CO_2 unit. The compression work should be minimized by minimizing the bed pressure drop, otherwise, electric consumption of air fans could be prohibitive. A parametric study over the characteristics of the air in terms of temperature, relative humidity, and gas velocity in the bed could also have a large impact on mass transfer and energy performance.

Further research related to the engineering solutions available to reach high global gas velocity without fluidization of the bed is required. This could be offered by monolith-shape adsorbents. The impact on the mass transfer and the pressure drop must be considered. Good energy integration of the available heat and electrical consumption minimization are paramount to achieve the specific energetical need targeted by this type of technology.

Author Contributions: Conceptualization, M.K. and O.A.; methodology, M.K. and O.A.; software, T.D. and M.K.; validation, M.K. and O.A.; formal analysis, T.D., M.K. and O.A.; investigation, T.D.; resources, L.G.; data curation, M.K. and O.A.; writing—original draft preparation, T.D., M.K. and O.A.; writing—review and editing, M.K. and O.A.; visualization, T.D.; supervision, M.K. and O.A.; project administration, L.G.; funding acquisition, L.G. All authors have read and agreed to the published version of the manuscript.

Funding: This research received no external funding.

Institutional Review Board Statement: Not applicable.

Informed Consent Statement: Not applicable.

Conflicts of Interest: The authors declare no conflict of interest.

References

1. Hepburn, C.; Adlen, E.; Beddington, J.; Carter, E.A.; Fuss, S.; Mac Dowell, N.; Minx, J.C.; Smith, P.; Williams, C.K. The Technological and Economic Prospects for CO_2 Utilization and Removal. *Nature* **2019**, *575*, 87–97. [CrossRef] [PubMed]
2. Kanniche, M.; Le Moullec, Y.; Authier, O.; Hagi, H.; Bontemps, D.; Neveux, T.; Louis-Louisy, M. Up-to-Date CO_2 Capture in Thermal Power Plants. *Energy Procedia* **2017**, *114*, 95–103. [CrossRef]
3. Bui, M.; Adjiman, C.S.; Bardow, A.; Anthony, E.J.; Boston, A.; Brown, S.; Fennell, P.S.; Fuss, S.; Galindo, A.; Hackett, L.A.; et al. Carbon Capture and Storage (CCS): The Way Forward. *Energy Environ. Sci.* **2018**, *11*, 1062–1176. [CrossRef]
4. Fuss, S.; Lamb, W.F.; Callaghan, M.W.; Hilaire, J.; Creutzig, F.; Amann, T.; Beringer, T.; de Oliveira Garcia, W.; Hartmann, J.; Khanna, T.; et al. Negative Emissions—Part 2: Costs, Potentials and Side Effects. *Environ. Res. Lett.* **2018**, *13*, 063002. [CrossRef]
5. Sustainable Development Scenario—World Energy Model—Analysis. Available online: https://www.iea.org/reports/world-energy-model/sustainable-development-scenario (accessed on 1 October 2021).
6. Rogelj, J.; Luderer, G.; Pietzcker, R.C.; Kriegler, E.; Schaeffer, M.; Krey, V.; Riahi, K. Energy System Transformations for Limiting End-of-Century Warming to below 1.5 °C. *Nat. Clim. Chang.* **2015**, *5*, 519–527. [CrossRef]
7. Viebahn, P.; Scholz, A.; Zelt, O. The Potential Role of Direct Air Capture in the German Energy Research Program—Results of a Multi-Dimensional Analysis. *Energies* **2019**, *12*, 3443. [CrossRef]
8. Fasihi, M.; Efimova, O.; Breyer, C. Techno-Economic Assessment of CO_2 Direct Air Capture Plants. *J. Clean. Prod.* **2019**, *224*, 957–980. [CrossRef]
9. Zolfaghari, Z.; Aslani, A.; Moshari, A.; Malekli, M. Direct Air Capture from Demonstration to Commercialization Stage: A Bibliometric Analysis. *Int. J. Energy Res.* **2022**, *46*, 383–396. [CrossRef]
10. House, K.Z.; Baclig, A.C.; Ranjan, M.; van Nierop, E.A.; Wilcox, J.; Herzog, H.J. Economic and Energetic Analysis of Capturing CO_2 from Ambient Air. *Proc. Natl. Acad. Sci. USA* **2011**, *108*, 20428–20433. [CrossRef]
11. McQueen, N.; Gomes, K.V.; McCormick, C.; Blumanthal, K.; Pisciotta, M.; Wilcox, J. A Review of Direct Air Capture (DAC): Scaling up Commercial Technologies and Innovating for the Future. *Prog. Energy* **2021**, *3*, 032001. [CrossRef]
12. Bourzac, K. We Have the Technology. *Nature* **2017**, *550*, S66–S69. [CrossRef] [PubMed]
13. Sanz-Pérez, E.S.; Murdock, C.R.; Didas, S.A.; Jones, C.W. Direct Capture of CO_2 from Ambient Air. *Chem. Rev.* **2016**, *116*, 11840–11876. [CrossRef] [PubMed]
14. Keith, D.W.; Holmes, G.; Angelo, D.S.; Heidel, K. A Process for Capturing CO_2 from the Atmosphere. *Joule* **2018**, *2*, 1573–1594. [CrossRef]
15. Wurzbacher, J.A.; Gebald, C.; Piatkowski, N.; Steinfeld, A. Concurrent Separation of CO_2 and H_2O from Air by a Temperature-Vacuum Swing Adsorption/Desorption Cycle. *Environ. Sci. Technol.* **2012**, *46*, 9191–9198. [CrossRef] [PubMed]
16. Wurzbacher, J.A.; Gebald, C.; Brunner, S.; Steinfeld, A. Heat and Mass Transfer of Temperature–Vacuum Swing Desorption for CO_2 Capture from Air. *Chem. Eng. J.* **2016**, *283*, 1329–1338. [CrossRef]
17. Fahr, S.; Powell, J.; Favero, A.; Giarrusso, A.J.; Lively, R.P.; Realff, M.J. Assessing the Physical Potential Capacity of Direct Air Capture with Integrated Supply of Low-carbon Energy Sources. *Greenh. Gases* **2022**, *12*, 170–188. [CrossRef]
18. Terlouw, T.; Treyer, K.; Bauer, C.; Mazzotti, M. Life Cycle Assessment of Direct Air Carbon Capture and Storage with Low-Carbon Energy Sources. *Environ. Sci. Technol.* **2021**, *55*, 11397–11411. [CrossRef]
19. Deutz, S.; Bardow, A. Life-Cycle Assessment of an Industrial Direct Air Capture Process Based on Temperature–Vacuum Swing Adsorption. *Nat. Energy* **2021**, *6*, 203–213. [CrossRef]
20. Shi, X.; Xiao, H.; Azarabadi, H.; Song, J.; Wu, X.; Chen, X.; Lackner, K.S. Sorbents for the Direct Capture of CO_2 from Ambient Air. *Angew. Chem. Int. Ed.* **2020**, *59*, 6984–7006. [CrossRef]
21. Singh, G.; Lee, J.; Karakoti, A.; Bahadur, R.; Yi, J.; Zhao, D.; AlBahily, K.; Vinu, A. Emerging Trends in Porous Materials for CO_2 Capture and Conversion. *Chem. Soc. Rev.* **2020**, *49*, 4360–4404. [CrossRef]
22. Wijesiri, R.P.; Knowles, G.P.; Yeasmin, H.; Hoadley, A.F.A.; Chaffee, A.L. Desorption Process for Capturing CO_2 from Air with Supported Amine Sorbent. *Ind. Eng. Chem. Res.* **2019**, *58*, 15606–15618. [CrossRef]
23. Sabatino, F.; Grimm, A.; Gallucci, F.; van Sint Annaland, M.; Kramer, G.J.; Gazzani, M. A Comparative Energy and Costs Assessment and Optimization for Direct Air Capture Technologies. *Joule* **2021**, *5*, 2047–2076. [CrossRef]
24. Dahlgren, E.; Göçmen, C.; Lackner, K.; van Ryzin, G. Small Modular Infrastructure. *Eng. Econ.* **2013**, *58*, 231–264. [CrossRef]
25. Wilson, C.; Grubler, A.; Bento, N.; Healey, S.; De Stercke, S.; Zimm, C. Granular Technologies to Accelerate Decarbonization. *Science* **2020**, *368*, 36–39. [CrossRef]
26. Sinha, A.; Darunte, L.A.; Jones, C.W.; Realff, M.J.; Kawajiri, Y. Systems Design and Economic Analysis of Direct Air Capture of CO_2 through Temperature Vacuum Swing Adsorption Using MIL-101(Cr)-PEI-800 and Mmen-Mg_2(Dobpdc) MOF Adsorbents. *Ind. Eng. Chem. Res.* **2017**, *56*, 750–764. [CrossRef]
27. Yu, Q.; Brilman, D.W.F. Design Strategy for CO_2 Adsorption from Ambient Air Using a Supported Amine Based Sorbent in a Fixed Bed Reactor. *Energy Procedia* **2017**, *114*, 6102–6114. [CrossRef]

28. Zhu, X.; Ge, T.; Yang, F.; Wang, R. Design of Steam-Assisted Temperature Vacuum-Swing Adsorption Processes for Efficient CO_2 Capture from Ambient Air. *Renew. Sustain. Energy Rev.* **2021**, *137*, 110651. [CrossRef]
29. Do, D.D. *Adsorption Analysis: Equilibria and Kinetics (with Cd Containing Computer Matlab Programs)*; World Scientific: Singapore, 1998; ISBN 978-1-78326-224-3.
30. Gebald, C. Development of Amine-Functionalized Adsorbent for Carbon Dioxide Capture from Atmospheric Air. Ph.D. Thesis, ETH Zurich, Zurich, Switzerland, 2014.
31. Portugal, I.; Dias, V.M.; Duarte, R.F.; Evtuguin, D.V. Hydration of Cellulose/Silica Hybrids Assessed by Sorption Isotherms. *J. Phys. Chem. B* **2010**, *114*, 4047–4055. [CrossRef]
32. Wurzbacher, J.A. Development of a Temperature-Vacuum Swing Process for CO_2 Capture from Ambient Air. Ph.D. Thesis, ETH Zurich, Zurich, Switzerland, 2015.
33. Kast, W. *Adsorption aus der Gasphase: Ingenieurwissenschaftliche Grundlagen und Technische Verfahren*; VCH: Vancouver, BC, Canada, 1988; ISBN 978-3-527-26719-4.
34. Delgado, J.M.P.Q. A Critical Review of Dispersion in Packed Beds. *Heat Mass Transf.* **2006**, *42*, 279–310. [CrossRef]
35. Carberry, J.J. *Chemical and Catalytic Reaction Engineering*; McGraw-Hill: New York, NY, USA, 1976; ISBN 978-0-07-009790-2.
36. Wakao, N.; Funazkri, T. Effect of Fluid Dispersion Coefficients on Particle-to-Fluid Mass Transfer Coefficients in Packed Beds: Correlation of Sherwood Numbers. *Chem. Eng. Sci.* **1978**, *33*, 1375–1384. [CrossRef]

Article

A Numerical Analysis of the Effects of Supercritical CO_2 Injection on CO_2 Storage Capacities of Geological Formations

Kamal Jawher Khudaida * and Diganta Bhusan Das *

Department of Chemical Engineering, Loughborough University, Loughborough LE11 3TU, UK
* Correspondence: kjkamal2014@gmail.com (K.J.K.); D.B.Das@lboro.ac.uk (D.B.D.)

Received: 31 May 2020; Accepted: 20 August 2020; Published: 1 September 2020

Abstract: One of the most promising means of reducing carbon content in the atmosphere, which is aimed at tackling the threats of global warming, is injecting carbon dioxide (CO_2) into deep saline aquifers (DSAs). Keeping this in mind, this research aims to investigate the effects of various injection schemes/scenarios and aquifer characteristics with a particular view to enhance the current understanding of the key permanent sequestration mechanisms, namely, residual and solubility trapping of CO_2. The paper also aims to study the influence of different injection scenarios and flow conditions on the CO_2 storage capacity and efficiency of DSAs. Furthermore, a specific term of the permanent capacity and efficiency factor of CO_2 immobilization in sedimentary formations is introduced to help facilitate the above analysis. Analyses for the effects of various injection schemes/scenarios and aquifer characteristics on enhancing the key permanent sequestration mechanisms is examined through a series of numerical simulations employed on 3D homogeneous and heterogeneous aquifers based on the geological settings for Sleipner Vest Field, which is located in the Norwegian part of the North Sea. The simulation results highlight the effects of heterogeneity, permeability isotropy, injection orientation and methodology, and domain-grid refinement on the capillary pressure–saturation relationships and the amounts of integrated CO_2 throughout the timeline of the simulation via different trapping mechanisms (solubility, residual and structural) and accordingly affect the efficiency of CO_2 sequestration. The results have shown that heterogeneity increases the residual trapping of CO_2, while homogeneous formations promote more CO_2 dissolution because fluid flows faster in homogeneous porous media, inducing more contact with fresh brine, leading to higher dissolution rates of CO_2 compared to those in heterogeneous porous medium, which limits fluid seepage. Cyclic injection has been shown to have more influence on heterogenous domains as it increases the capillary pressure, which forces more CO_2 into smaller-sized pores to be trapped and exposed to dissolution in the brine at later stages of storage. Storage efficiency increases proportionally with the vertical-to-horizontal permeability ratio of geological formations because higher ratios facilitate the further extent of the gas plume and increases the solubility trapping of the integrated gas. The developed methodology and the presented results are expected to play key roles in providing further insights for assessing the feasibility of various geological formations for CO_2 storage.

Keywords: geological sequestration; CO_2 storage capacity; CO_2 storage efficiency; CO_2 sequestration; deep saline aquifers

1. Introduction

Injecting CO_2 into deep saline aquifers (DSAs) has been proposed as one of the most viable means of tackling global warming [1]. This is because the technology has developed sufficiently due to the experience gained from oil and gas exploration and waste disposal methodologies. Moreover, the DSAs offer more extensive storage potential than other geological formations, such as oil and gas fields or coal seams [1]. Consequently, many studies have been conducted to assess their storage capacity and efficiency to safely sequester the injected gas [2–8].

As discussed earlier in our previous works [9,10], CO_2 storage methodology in saline aquifers can be categorised into hydrodynamic and chemical mechanisms. The first one includes the structural and residual trapping of CO_2 within the aquifer pore space, while the second one comprises of the solubility and mineral trapping of CO_2.

Two important factors that should be considered while assessing the suitability of an aquifer for sequestering CO_2 are its capacity and injectivity. They should allow for the safer and cost-effective storage of large amounts of the disposed gas [11]. Additionally, hydrostatic conditions play a crucial role in increasing the storage of saline aquifers because the higher pressure in deeper formations induces gas compression, resulting in more storage of CO_2 in a specific volume of the aquifer [12–14]. In this regard, the integrity of the caprock with low permeability is an essential consideration because any existing faults or cracks in the aquifer rock will result in the injected gas escaping to the surface. Porosity and permeability of the formation have significant influence on the selection of the appropriate site for carbon storage because the higher permeability of a medium allows fluids to migrate easily through the better-connected pores away from the injection well, which subsequently magnifies the capacity and efficiency of the aquifer to store CO_2 [15–18].

Theoretically, the storage capacity of an aquifer is the substantial limit of CO_2 that can be admitted into it [19,20]. However, this limit is not practically achievable due to various geological factors and engineering barriers (e.g., pore connectivity, lack of geological data, economic feasibility, legal regulations and infrastructure benchmarks). Therefore, a term called effective storage capacity has been coined [21,22], which has been a subject of a number of studies using different calculation methods. These methods involve the use of volumetric and compressibility methods [23–25], mathematical models [26], dimensional analyses [27], analytical investigations [28] and numerical modelling [29,30] to assess the efficiency of geological formations to sequester CO_2. Most of these studies are theoretical or analytical in nature based on 2D models that seem to lack sufficient interests for practical employment. Detailed comparison studies have been conducted to evaluate the impact of a variety of approaches and methodologies on estimating CO_2 sequestration in geological formations [20,31,32].

One basic estimation method, which is widely adopted, is the U.S. Department of Energy (US-DOE) method. As explained in detail by Goodman et al. [33], the method assumes an infinitive boundary and defines the efficiency of an aquifer to store CO_2 by the pore volume that is available to be occupied by the injected gas. It determines the CO_2 mass storage capacity and efficiency for an aquifer as:

$$G_{CO_2} = A_t h_g \varnothing_t \rho_{CO_2} E_{aq} \tag{1}$$

where A_t is the total cross-sectional area of the domain, h_g defines the gross thickness of the formation, $\varnothing_t$ is the total porosity of the rock, ρ_{CO_2} and E_{aq} represent the density of the injected CO_2 and the storage efficiency of the aquifer, respectively.

An earlier approach, proposed by Zhou et al. [23] predicts the pressure build-up history and the impact on the actual storage efficiency in response to the CO_2 injection process. The authors define the storage efficiency factor as the volumetric fraction of the sequestered CO_2 per unit volume of pores in the potential domain. In spite of achieving good agreement between the analytical results and the numerically predicted values, the authors state that this method is not suitable for geological formations of low permeability that lead to lower injectivity, and creates more non-uniformity in the pressure build-up within the simulated domain. This is due to many simplifications and assumptions in the analytical solutions in their research work.

Another method, developed by Szulczewski [34], considers both the residual and solubility trapping mechanisms in addition to the CO_2 migration capacity. The method is applicable to both open-boundary and pressure-limited systems. Additionally, this method counts the net thickness of the aquifer to calculate the pore volume instead of the gross thickness in heterogeneous domains. This is because most of the injected CO_2 targets the high-porosity layers, such as sandstone or carbonate rocks, rather than any intermingled layers of shale or clay that store negligible amounts of the injected gas.

For open-boundary systems, the total mass of CO_2 (C_t) stored in an aquifer can be determined by [34]:

$$C_t = \rho_g L_t WH \varnothing (1 - S_{wc}) \frac{2}{\varepsilon_T} \qquad (2)$$

where ρ_g is the density of CO_2 at prevailing temperature and pressure, L_t is the total length of the aquifer, W is the width of the well, H is the net thickness of the aquifer $\varnothing$ is the porosity of the rock, S_{wc} is the connate water saturation and $\frac{2}{\varepsilon_T}$ is the storage efficiency factor.

If the aquifer is classified as a pressure-limited system, the CO_2 mass is calculated as follows [34]:

$$C_p = \rho_g HW \sqrt{\frac{k_{aq}cT}{\mu_w}} \frac{P_{fraq} - (p_o + \overline{\rho_w} gD)}{4\widetilde{p}_{max}} \qquad (3)$$

where k_{aq} is the permeability of the aquifer, c is the compressibility, T is the temperature, μ_w is the brine viscosity, P_{fraq} is the fracture pressure of the rock, p_o is the hydrostatic pressure, $\overline{\rho_w}$ is the average density of brine, g is the gravitational acceleration, D is the depth to the top of the aquifer and $\widetilde{p}_{max}$ represents the maximum dimensionless pressure in the system which needs to be determined numerically, solving a different set of partial differential equations (PDEs) for the pressure-limited flow system [34].

Several techniques can be used to increase the capacity and efficiency of CO_2 sequestration in saline aquifers that will consequently support the efforts by the Intergovernmental Panel on Climate Change (IPCC) to incite policy makers with the importance of deploying carbon capture and sequestration (CCS) as one of the cost effective technologies for confronting climate change and global warming concerns.

Geological formation capacity can be increased by improving the injectivity through increasing the injection mass flow rate or pressure to compensate the loss of permeability due to salt precipitation in the well vicinity. Furthermore, injecting into adjacent layers with high permeability helps with attenuating pressure build-up and, consequently, higher injection rates can be employed [35,36].

Using horizontal injection wells instead of vertical ones is one of the methods implemented to increase the injectivity and capacity of aquifers because it helps to diminish the pressure-build-up peaks around the injection well and spread pressure uniformly within the domain. Deploying this technique requires the determination of the minimum length of the horizontal well that is dependent on the effective radius of pressure disturbance around the vertical injection well [37–40].

It has been evidenced that the solubility of CO_2 into brine can be accelerated by injecting slugs of fresh brine on top of the storage formation during and after CO_2 injection. This can increase CO_2 dissolution by more than 40% within a period of 200 years, which reduces the risk of CO_2 leakage in long-term sequestration, according to the study by Hassanzadeh et al. [41]. The study also investigated further factors that have significant impact on increasing the storage efficiency in saline aquifers, including optimizing the rate of the injected brine and transporting the injected and existing fluids within the reservoir in addition to the effect of aquifer properties, such as thickness, vertical anisotropy and layers of heterogeneity included within the media.

In their work, De Silva and Ranjith [39] concluded that, while using horizontal injection wells in the absence of chase brine injection improves the storage of aquifers, vertical injection wells with chase brine injection performs better storage efficiency. However, the authors suggest that the injection process should be carried out over the whole thickness of the aquifer to maximize the storage capacity. In contrast, Khudaida and Das [9] observed that injecting CO_2 into the lower section of a reservoir enhances the solubility trapping mechanism and subsequently increases the storage efficiency.

Introducing hydraulic fractures in formation rock can improve the injectivity by increasing the effective permeability of the aquifer, which facilitates migration and, consequently, preserves more contact between the injected CO_2 and the existing brine, in addition to preventing any pressure build-up within the aquifer. However, this technique needs a detailed characterization of the formation and has to be implemented with extra care to avoid causing any gas leakage [38].

Keeping the above discussions in mind, this work aims to provide further understanding on how to assess the feasibility of a potential storage site by investigating the behaviour and migration of CO_2-brine as a two-phase flow system in porous geological formations under various injection conditions and scenarios. It also demonstrates the effects of various site characteristics, such as heterogeneity and anisotropy, on the injectivity and safe storage of the injected CO_2. To aid the above, the capillary pressure relationships for CO_2-brine as a two-phase flow is also studied. It is envisaged that the results will address the applicability of different injection techniques in terms of orientation and continuity to enhance the capacity and efficiency of sequestering CO_2 in geological formations.

2. Modelling Setup

To assess the storage capacity and efficiency of an unconfined aquifer (i.e., migration-limited domain), a hypothetical cylindrical computational domain, extending from 0.3 m (the radius of the injection-well case) to 6000 m laterally and 96 m vertically, was simulated with two types of numerical grid resolutions, namely, coarse and fine grids. For the coarse-grid, the domain was horizontally discretized into 88 grid-blocks with a finer mesh in the vicinity of the injection well, which became gradually coarser further away. Vertically, the domain was discretized into 24 blocks of 4 m high. This mesh refinement has made 2112 elements as shown in Figure 1. For the fine resolution, the grid spacing was increased by 100% in both directions, producing 8448 cells. Supercritical CO_2 (scrCO_2) was injected into the centre of the domain at a constant rate of 32.0 kg/s (about 1 MMT/year), which represents a typical benchmark value [38], via a number of cells either at the bottom section or through the whole thickness of the reservoir.

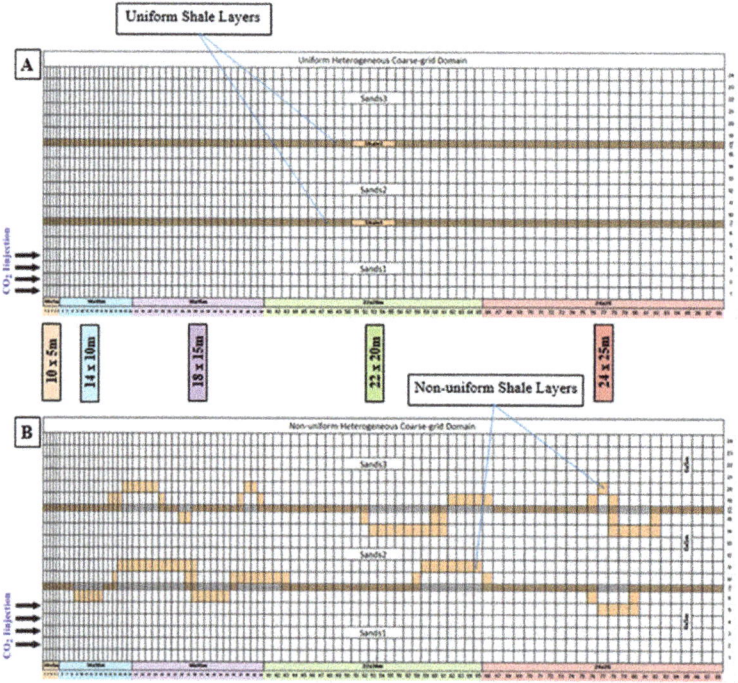

Figure 1. A schematic 2D diagram of the modelled heterogeneous domains: (**A**) uniform heterogeneity; (**B**) non-uniform heterogeneity.

Heterogeneity is defined as the variability of porosity/permeability within the simulated domain and is usually quantified using various geostatistical techniques, including the Lorenz coefficient (Lc) and the coefficient of variation (Cv) methods that are commonly used in establishing porosity and permeability models in exploration. For this study, this variability was not calculated because the simulation parameters were taken from geological settings for Sleipner Vest Field (Tables 1 and 2).

Table 1. Lithostratigraphic division and petrophysical parameters from Sleipner Vest Field, after [42]. These are model parameters used in this study; however, the developed methodology is generic and may be applied in other geological formations.

Layer	Units	Sand 1	Sand 2	Sand 3	Shale 1	Shale 2
Thickness	(m)	30	30	30	3	3
Porosity	-		0.35			0.1025
Horizontal permeability	(md) (m^2)		304 3.0×10^{-13}			10.13 0.1×10^{-13}
Vertical permeability	(md) (m^2)		304 3.0×10^{-13}			10.13 0.1×10^{-13}
Density	(kg/m^3)			2650		
Pore Compressibility	(Pa^{-1})			4.5×10^{-10}		
Aquifer pressure	(MPa)			11.2		
Pressure gradient	(KPa/m)			10.012		
Aquifer temperature	(°C)			37		
Salinity (mass fraction)	-			0.032		
Aquifer depth	(m)			800–1100		
Water depth	(m)			110		

Table 2. Capillary pressure–saturation; permeability functions parameters of the simulated aquifer, after [42,43].

Description	Symbol	Value	Units
Irreducible aqueous saturation [42]	S_{lr}	0.2	-
Irreducible gas saturation [42]	S_{gr}	0.05	-
Saturation function parameters for (Sand) [42]	α	2.735	M^{-1}
Saturation function parameters for (Shale) [42]	α	0.158	M^{-1}
Saturation function parameters [43]	m	0.4	-
	n	1.667	-
Pore index parameter $\lambda = \frac{m}{1-m}\left(1 - 0.5^{\frac{1}{m}}\right) = n - 1$	λ	0.667	-
Maximum residual gas saturation for aquifer [43]	S_{grm}	0.208	-
Maximum residual gas saturation for aquitard [43]	S_{grm}	0.448	-

2.1. Parameters and Calculations

STOMP-CO$_2$ simulation code [42] was used to carry out the simulation runs in this research work and conduct the P_c-S_w calculations. This has been discussed earlier [9,42] and therefore it is not discussed in detail in this paper and, additionally, the simulation code results were validated through a reasonable mapping with a lab-scale setup which was described in detail in a dedicated section of previous research by the authors [44]. The simulation parameters used in this work are based on the geological settings for the Sleipner Vest Field, which is located in the Norwegian part of the North Sea at an approximate depth of 1100 m. It is identified to be one of the typical CO$_2$ disposal sites offering anticipated hydrostatic conditions to keep the injected CO$_2$ in supercritical conditions. Moreover, this depth is far enough away from the fresh water sources, which are usually located at around a 500 m depth. All petrophysical parameters and formulations factors are listed in Tables 1 and 2. CO$_2$ properties adopted in the simulation were arranged in a data table developed from the equation of state (EOS) by Span and Wagner [45], which is widely considered to be an accurate reference EOS for

CO_2 for its ability to provide accurate results in the most technically relevant pressures up to 30 MPa and temperatures up to 523 K, the conditions that are common in the geological sequestration of CO_2.

The Span and Wagner equation is based on an extensive range of fitted experimental thermal properties in the single-phase region, the liquid–vapour saturation curve, the speed of sound, the specific heat capacities, the specific internal energy and the Joule–Thomson coefficient [46]. This equation is expressed in the form of the Helmholtz energy (ϕ) as follows:

$$\phi(\delta, \mathcal{T}) = \phi(\delta, \mathcal{T}) + \phi^r(\delta, \mathcal{T}) \qquad (4)$$

where $\phi = \rho/\rho_c$, $\mathcal{T} = T_c/T$, ρ_c and T_c are the critical density and critical temperature of CO_2, respectively. ϕ is the ideal gas part of the Helmholtz energy and ϕ^r is the residual part of the Helmholtz energy. The two parts of the Helmholtz energy (the basic and phase diagram elements) of this equation of state are explained in detail in the literature published by Span and Wagner [45] and are not repeated in this study. The Span and Wagner EOS has been employed in this research to calculate the density of CO_2 at different simulation conditions under the following assumptions:

1. It is based on a wide range of experimental data with uncertainty values of +0.03 to +0.05% in the density values.
2. It is valid for a wide range of pressure and temperature values, even beyond the triple (critical) point in the phase-diagram of CO_2.
3. The EOS can be extrapolated up to the limits of the chemical stability of CO_2.

The only limitation of this EOS is that it is time-expensive to evaluate in dynamic numerical simulations because it consists of a large number of algorithms and exponentials.

The phase equilibria calculations in STOMP-CO_2 code [42] are conducted via a couple of formulations by Spycher et al. [47] and Spycher and Preuess [48] that are based on Redlich–Kwong equation of state with fitted experimental data for water–CO_2 flow systems. The mole fraction of water in the gas phase ($X_g^{H_2o}$) and mole fraction of the CO_2 in the aqueous phase ($X_l^{CO_2}$) are calculated by the following equations:

$$X_g^{H_2o} = \frac{(1-B)}{\left(\frac{1}{A}-B\right)} \qquad (5)$$

$$X_l^{CO_2} = B\left(1 - X_g^{H_2o}\right) \qquad (6)$$

where:

$$A = \frac{K_{H_2o}^0}{\varnothing_{H_2o} P} exp\left[\frac{(P-P)\overline{V}_{H_2o}}{RT(K)}\right] \qquad (7)$$

$$B = \frac{\varnothing_{CO_2} P}{(10^3/M^{H_2o})K_{CO_2}^0} exp\left[-\frac{(P-P^0)\overline{V}_{CO_2}}{RT(K)}\right] \qquad (8)$$

In Equations (7) and (8), K^0 is the thermodynamic equilibrium constant for water or CO_2 at temperature T in Kelvin (K), and reference pressure $P^0 = 1$ bar, P is the total pressure, $\overline{V}$ represents the average partial molar volume of each pure condensed phase, ϕ is the fugacity coefficient of each component in the CO_2-rich phase and R is the gas constant [42].

The aqueous saturation (S_w) is calculated by Van Genuchten [49] formulation that correlates to the capillary pressure (P_c) to the effective saturation (S_e):

$$S_e = \frac{S_w - S_{wr}}{1 - S_{wr}} = \left[1 + (\alpha \cdot P_c)^n\right]^m; \text{ for } P_c > 0 \qquad (9)$$

where S_{wr} is the water residual saturation and α, n and m are the Van Gunechten parameters that describe the characteristics of the porous media.

During the injection period (drainage process), there is no gas entrapment because it only occurs during the imbibition process when the displaced water invades the domain back as soon as CO_2 injection stops leaving some traces of it trapped behind in some small-sized pores. As a result, the injected CO_2 can either exist as free or trapped gas.

The effective trapped gas is computed using a model developed by Kaluarachchi and Parker [50]:

$$\hat{S}_{gt}^{potential} = \left[\frac{1 - \hat{S}_l^{min}}{1 - R(\hat{S}_l^{min})}\right] \quad (10)$$

where $\hat{S}_l^{min}$ is the minimum aqueous saturation (irreducible water saturation) and R is the Land's parameter [51] which relates to the maximum trapped gas saturation:

$$R = \frac{1}{\hat{S}_{gt}^{max}} - 1 \quad (11)$$

$\hat{S}_{gt}^{max}$ is the maximum trapped gas saturation that can be achieved during the drainage process. Maximum trapped gas and minimum aqueous (irreducible water) saturation are calculated by the following correlations by Holts [43]:

$$S_{gr}^{max} = 0.5473 - 0.969\,\phi \quad (12)$$

$$S_{wirr} = 5.6709\left[\log\frac{(k)}{\phi}\right]^{-1.6349} \quad (13)$$

where ϕ and k are the porosity and intrinsic permeability of the medium, respectively. The aqueous and gas relative permeabilities are computed by Mualem [52] correlation in combination with Van Genuchten [49] formulations, according to the following Equations (14) and (15), respectively:

$$k_{rl} = \left(\overline{S}_l\right)^{\frac{1}{2}}\left[1 - \left(1 - \left(\overline{S}_l\right)^{\frac{1}{m}}\right)^m\right]^2 \quad (14)$$

$$k_{rg} = \left(1 - S_l\right)^{\frac{1}{2}}\left[1 - \left(1 - \left(S_l\right)^{\frac{1}{m}}\right)^m\right]^2 \quad (15)$$

where m is the pore distribution index, $\overline{S}_l$ is the effective aqueous saturation, which is calculated from Equation (4) and S_l represents the apparent aqueous saturation which is defined as the sum of the effective aqueous and entrapped CO_2 saturations [53].

2.2. Initial and Boundary Conditions

Three types of simulated domains—namely, homogeneous, uniform heterogeneous and non-uniform heterogeneous—were modelled in this study. They were assumed to be isotropic for most simulation runs and isothermal under a hydrostatic pressure gradient of 10.012 KPa/m with an open boundary condition, leading to scattered pressure build-up. The models were presumed to have no heterogeneity in the azimuthal direction, but different vertical-to-horizontal permeability ratios were studied, in some specific cases, to investigate the effect of anisotropy on the storage capacity and efficiency. The system was modelled as a 3D cylindrical domain and the results were compared to those when the system was considered as a two-dimensional radial flow to save computational time and requirements. The gravity and inertial effects were neglected.

Prior to injecting ScrCO$_2$ into the centre of the domain, it was considered to be fully saturated with brine with initial conditions, as illustrated in Tables 1 and 2. ScrCO$_2$ was injected through four grid-cells at the bottom layer of the grid for 30 years, followed by a lockup period of 4970 years. No flux boundary condition was considered for the aqueous wetting phase (brine) at the injection well case as a west boundary, whilst the east boundary was set to be infinite with zero flux for CO_2 as a non-wetting gas phase. Zero flux was also considered at the top and bottom confining layers, forcing the injected CO_2 to swell crossways.

As an open storage system, the pressure build-up was not considered to be a limiting factor; however, the value of the maximum bottom-hole pressure at the injection well and hydrological effect on shallow groundwater sources had to be taken into consideration [23,54]. The injection rate for this simulation system was set according to the rock fracture pressure (P_{frac}) using the simplified model adopted by Szulczewski et al. [55], which calculates the pressure-limited storage capacity by:

$$M_p = 2\rho_{CO_2} HW \sqrt{\frac{kC}{\mu_b T} \frac{P_{frac}}{\hat{p}_{max}}} \qquad (16)$$

where ρ_{CO_2} is the density of the injected gas, H and W are the height and width of the domain, respectively, k represents the intrinsic permeability, and C is the compressibility of the formation; μ_b is the bulk viscosity and P_{frac} is the fracture pressure.

For infinite aquifer, the value of the maximum dimensionless pressure ($\hat{p}_{max}$) in Equation (16) was ~0.87, according to Szulczewski et al. [55].

All parameters in Equation (16) were known, except for the fracture pressure of the rock, which can be defined as the effective vertical stress for deep aquifers and is determined by the following equation, given by Szulczewski et al. [56]:

$$\sigma'_v = (\rho_b - \rho_w)Z \qquad (17)$$

where ρ_b, ρ_w represent bulk and brine densities, respectively, and Z is the depth at which the aquifer is located.

$$\rho_b = \rho_r \phi \qquad (18)$$

where ρ_r is the rock density and $\emptyset$ is the formation porosity.

From Equations (16)–(18), the value of the injection rate for the model is set at 32 kg/s, which, according to Equation (16), results in a pressure build-up value of less than 1.5 magnitudes of the hydrostatic pressure. This value is far away from the average default values of the sustainable pressure (181% of the hydrostatic pressure gradient) reported for Dundee Limestone in the Michigan Basin in the USA, which is located at a 1200 m depth [23].

2.3. Storage Efficiency Calculation

Theoretically, the CO_2 sequestration efficiency in saline aquifers can be assessed by calculating the efficiency storage factor, which refers to the volume fraction of the pores occupied by the injected CO_2:

$$E_{aq} = \frac{V_{CO_2}}{V_\emptyset} \qquad (19)$$

V_{CO_2} is the volume of injected CO_2, which can be calculated from the known mass rate of the injected gas under the hydrostatic conditions of the geological formation. $V_\emptyset$ is the volume of the pores in the domain:

$$V_\emptyset = V_t \phi_t \qquad (20)$$

where V_t and $\emptyset_t$ are the total volume and total porosity of the domain, respectively.

To calculate the storage capacity in this research work, the modern equation, developed by Szulczewiski [34], was employed:

$$E_{aq} = \frac{G_{CO_2}}{\rho_g L_t WH \emptyset (1 - S_{wc})} \qquad (21)$$

where G_{CO_2} is the total mass of the integrated CO_2 (dissolved and residually trapped), ρ_g is the density of CO_2 at hydrostatic conditions and W, L_t and H represent the width, total length and net thickness of the aquifer, respectively. $\varnothing$ defines the porosity of the rock and S_{wc} defines the connate (irreducible) water saturation.

This methodology has been implemented because it accounts for the net thickness of the aquifer rather than the whole thickness. This can be justified by the fact that only the higher permeability layers are targeted by the injected gas [34].

This study aims to investigate the impact of heterogeneity, permeability, grid resolution and injection methodology on CO_2-water system mobility and the behaviour of the injected scrCO$_2$ at different time steps on the CO_2 storage capacity and efficiency at a field-scaled domain. An archetype of actual field heterogeneity has been developed in a domain that consists of three stratums of sands intermingled with two layers of low permeability shales, as illustrated in Figure 1. All petrophysical and simulation parameters are shown in Tables 1 and 2, respectively.

A series of simulation cases (presented in Table 3) were setup to demonstrate different models of a computational domain, including homogeneous, uniform and non-uniform heterogeneous models with coarse and fine grid resolutions. The simulation runs comprised two different employed schemes of injection (continuous and cyclic). The continuous injection scheme involved 30 years of continuous injection at a constant rate of 32 kg/s (about 1 million metric tons (MMT) per year) while in the second scenario, the injection period was implicated in three cycles of 10 years, separated by two stopping periods of 5 years in between in order to ensure that the structural trapping mechanism ended and other trapping mechanism took their role before injecting a new cycle. Furthermore, three cases with different values of vertical-to-horizontal permeability ratio (k_v/k_h) were developed along with other models to assess the influence of injection scope and orientation of the injection well on the flow behaviour and CO_2 sequestration efficiency. In all 14 cases, the total simulation time was 5000 years, including injection and pausing times. This value was set up after many trial simulations to detect the steady state time scales. Before 1000 years, most of the injected gas would be in a free gas phase, which is subject to escape through any existing cracks or faults in the caprock. As the permanent sequestration of the injected CO_2 is the focus of this work, a new term of permanent sequestration factor of the aquifer (E_{aq}^{perm}) was introduced. This factor was calculated from the numerical simulation results by STOMP-CO_2 code [42] and compared for different cases under different conditions through various time scales.

Due to the density difference between the injected supercritical CO_2 (scCO$_2$) (about 280 kg/m^3) and the existing brine (about 1100 kg/m^3) (i.e., gravity driving forces), initially the former fluid percolates upwards to be physically trapped under the upper impervious layer (caprock). During this time, part of the gas dissolves in the existing brine to form an aqueous phase rich in CO_2, which is heavier than the ambient liquid and hence sinks down to settle at the bottom of the aquifer. As soon as the injection stops, the replaced brine invades the domain to reinstate the CO_2, leaving some traces of it behind in some small-sized pores in a process called residual or capillary trapping. These amounts of CO_2 are determined by the simulation code for different cases and utilized to calculate the capacity and efficiency of the simulated aquifer. The latter values are used to calculate the sequestration efficiency by:

$$\text{Total integrated } CO_2 = \text{Integrated aqueous } CO_2 \left(CO_2^{aq}\right) + \text{Integrated gas } CO_2 \left(CO_2^{g}\right) \quad (22)$$

$$\text{Integrated gas } CO_2^{g} = \text{Trapped gas } CO_2 \left(CO_2^{gt}\right) + \text{Free gas } CO_2 \left(CO_2^{gf}\right) \quad (23)$$

$$E_{aq}^{perm} = \frac{CO_2^{aq} + (CO_2^{gt})}{\rho_g L_t W H \varnothing (1 - S_{wc})} \quad (24)$$

where all parameters are explained in Equation (21).

Table 3. Simulation cases and conditions.

Case No.	Domain	Heterogeneity and Layers Thickness	Grid Resolution	Nodes Distribution (r, θ, z)	Permeability Ratio (k_v/k_h)	Injection Scheme (30 years)
Base-3D	Homogeneous	N/A 1 × 96 m	Coarse	(75, 4, 24)	1	Vertical continuous into lower section
Base-2D	Homogeneous	N/A 1 × 96 m	Coarse	(75, 1, 24)	1	Vertical continuous into lower section
1	Homogeneous	N/A 1 × 96 m	Coarse	(88, 1, 24)	1	Vertical continuous into lower section
2	Heterogeneous	Uniform 3 × 30 m, 2 × 3 m	Coarse	(88, 1, 24)	1	Vertical continuous into lower section
3	Heterogeneous	Non-uniform 3 × 30 m 2× variable	Coarse	(88, 1, 24)	1	Vertical continuous into lower section
4	Heterogeneous	Non-uniform 3 × 30 m, 2× var.	Coarse	(88, 1, 24)	1	Vertical batch * (10-5-10-5-10) into lower section
5	Homogeneous	N/A 3 × 30 m 2× variable	Coarse	(88, 1, 24)	1	Vertical batch * (10-5-10-5-10) into lower section
6	Homogeneous	N/A 3 × 30 m, 2× var.	Fine (+100%)	(176, 1, 48)	1	Vertical continuous into lower section
7	Homogeneous	N/A 3 × 30 m 2× variable	Coarse	(88, 1, 24)	1	Vertical continuous into lower section
8	Homogeneous	N/A 3 × 30 m 2× variable	Coarse	(88, 1, 24)	1	Vertical continuous Whole thickness
9	Homogeneous	N/A 3 × 30 m 2× variable	Coarse	(88, 1, 24)	0.1	Vertical continuous into the whole thickness
10	Homogeneous	N/A 3 × 30 m 2× variable	Coarse	(88, 1, 24)	0.01	Vertical continuous whole thickness 96 m **
11	Homogeneous	N/A 3 × 30 m 2× variable	Coarse	(88, 1, 24)	0.01	Horizontal continuous 96 m **
12	Homogeneous	N/A 3 × 30 m 2× variable	Coarse	(88, 1, 24)	0.01	Horizontal continuous 192 m **

* Batch injection schemes refer to the years of (injection–stop–injection–stop–injection). ** Width of the injection well maintaining constant injection rate.

Because the system was assumed to be boundless (open boundary conditions) with no pressure build-up, most of the integrated free gas was subject to migration away from the injection well along the overlapping layer and a small amount of it may sweep out of the domain through any existing fractures or faults in the overlaying caprock. Therefore, in this work the focus was on the storage efficiency of the aquifer in terms of the permanent sequestration of the injected CO_2, which occurs mainly through solubility and residual trapping mechanisms due to the insignificant influence of the mineral trapping mechanism for a few thousand years, according to De Silva et al. [39].

3. Results and Discussion

Injecting $scCO_2$ into a brine-saturated porous formation produces spatial distribution maps of both fluids. Figure 2 illustrates the integrated gas saturation maps and spatial distribution of the aqueous CO_2 mass fraction within the 3D cylindrical model of the simulated domain (case Base-3D in Table 3) at different time scales. It is shown that, soon after injection, the gas bounces upwards due to the density difference between the two fluids and simultaneously migrates crossways due to the pressure gradient between the injected $ScrCO_2$ and the in situ hydrostatic pressure. During this drift, some of the injected gas disperses into the existed brine, producing a CO_2-saturated aqueous phase that is heavier than the pure brine and, consequently, tends to sink down towards the bottom of the domain, forming a fingered structure, as displayed in Figure 2 (right).

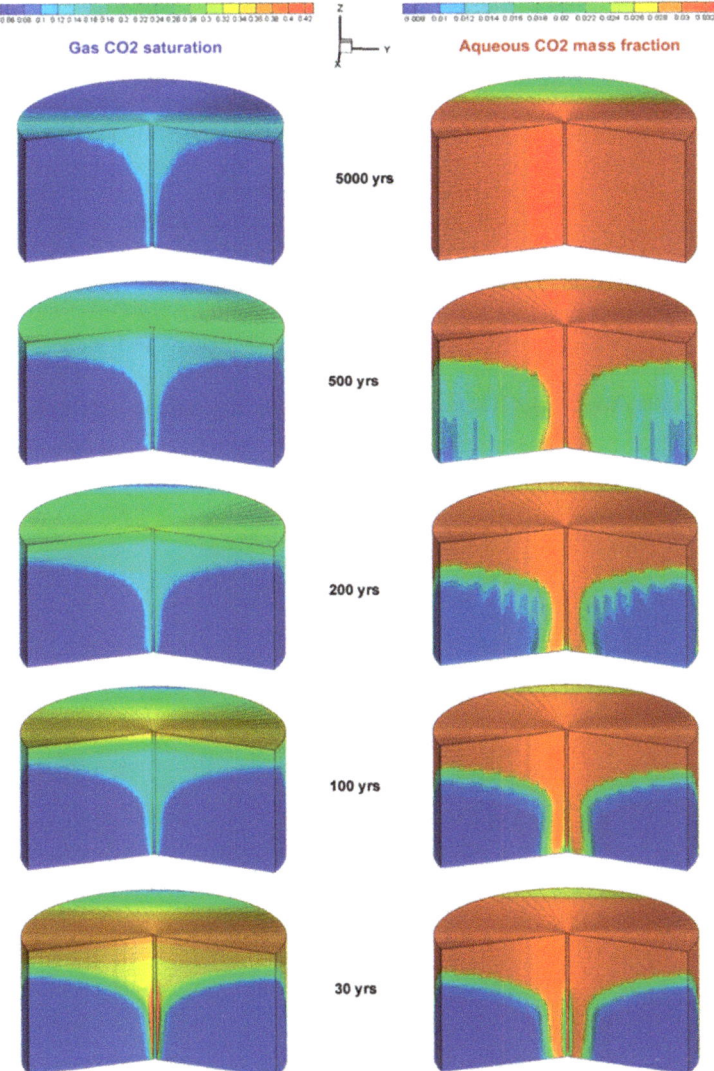

Figure 2. Free gas saturation and aqueous (dissolved) CO_2 mass fraction contours for a 3D homogeneous model at different time steps (case Base-3D in Table 1).

3.1. CO_2 Mobility and Behaviour

Due to the density difference between the injected gas and the hosted brine, the buoyancy forces initially dominate the water–CO_2 flow system. The injected scrCO_2 displaces the existing brine soon after the injection starts, and the gas moves upwards to be physically trapped under the overlaying impermeable layer (caprock). The flow system involves interfacial contact between the two fluids that results in considerable amounts of the free CO_2 gas to dissolve in the accommodated brine representing the solubility trapping mechanism. This is in addition to the amount of the gas that is trapped because of the capillarity and interfacial forces between the pore surface and the percolating gas.

Simultaneously, limited traces of CO_2 trapped in the locale (space) even during the injection lifetime due to the injection pressure that forces some drops of the gas into some small-sized pores.

However, these amounts are insignificant and further subject to be snapped off by the invading brine during imbibition. The actual capillary trapping is noticeable only after the gas injection stops. As soon as the injection ceases after 30 years, the residual trapping mechanism dominates when the replaced brine invades back the domain to sweep the integrated gas out of the pores. During this process, traces of CO_2 get detached from the trailing part of the gas plume and pierce into the small-sized pores due to the capillary forces.

The displayed trends in Figure 3A exhibit that, after 100 years, ~74% of the injected scrCO_2 was structurally trapped as a free integrated gas in the homogeneous domain. ~17.5% of the injected CO_2 was dissolved in brine and the remaining 8% was residually trapped in small-sized pores due to the capillarity. For the heterogeneous model in Figure 3B, on the other hand, it was observed that ~70% of the injected gas was trapped as a free gas, 20.49% was dissolved in the brine, while 9.4% was disconnected from the plume trailing edge and adhered to the rock surface inside some small-sized pores, due to the surface tension forces.

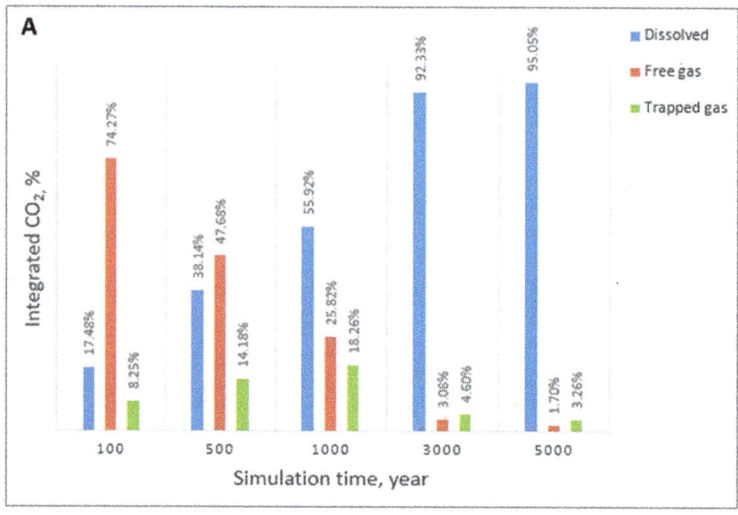

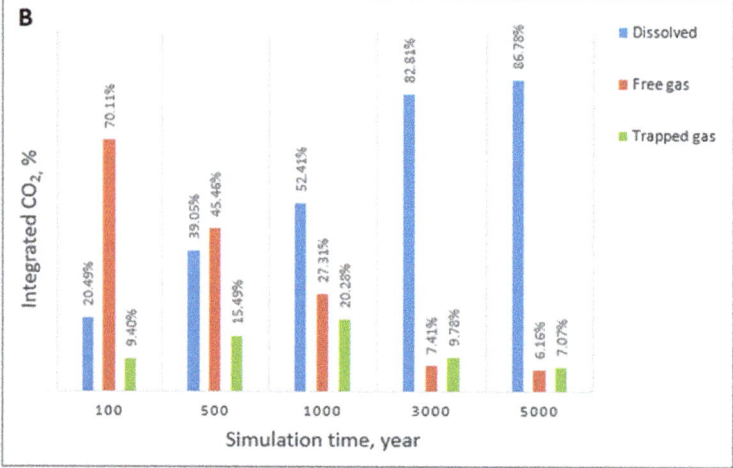

Figure 3. Various phases of integrated CO_2 trends in: (**A**) homogeneous porous domain (case 1); (**B**) uniform heterogeneous domain (case 2) in Table 3.

The findings from this study have shown that, under similar hydrostatic conditions and petrophysical characteristics, homogeneous formations promote more CO_2 dissolution, owing to the fact that, under the same hydraulic gradient, fluid flows faster in homogeneous porous media compared to that in the heterogeneous porous medium, which limits fluid seepage and is consistent with some previously published studies [57,58].

This fast migration induces more contact with fresh brine, leading to the higher dissolution rates of CO_2. This can be seen in Figure 3A,B, which shows that only 1.7% of the injected CO_2 was left as a free gas in the homogeneous model at the end of the simulation compared to the heterogeneous case, in which more than 6% of free gas was recorded.

The timing maps of CO_2 sequestration by each trapping mechanism are depicted in Figure 3 for homogeneous and heterogeneous formations. It can be seen from the figure that, during the first few hundreds of years, the structural trapping mechanism dominates (i.e., more free CO_2 gas) while, after thousands of years, the solubility trapping becomes the dominant mechanism (i.e., more CO_2 dissolves in the brine). The maximum amount of CO_2 is residually trapped at about 1000 years and declines later because some of it dissolves into the surrounding brine to form weak carbonic acid that reacts with the rock material and precipitates as solid carbonates after a few thousand years.

3.2. Impact of Heterogeneity

To investigate the impact of different types of heterogeneity (uniform and non-uniform) on the propagation of CO_2 profiles, P_c-S_w relationships and storage efficiency, three numerical cases—namely 1, 2 and 3—with their employed conditions, illustrated in Table 3, have been implemented in this study. It is a fact that the permeability of geological formations is strongly dependent on their porosity and heterogeneity, and it plays a key role in understanding water–CO_2 flow in the subsurface. This influence is clearly exposed in Figure 4, which demonstrates different maps of gas distribution in the modelled domains.

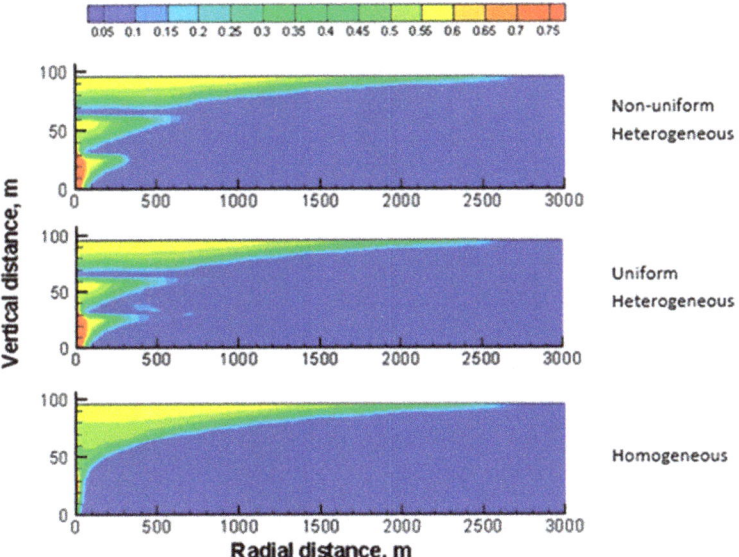

Figure 4. Spatial distribution of CO_2 after 30 years of simulation (end of injection) for homogeneous media (case 1) and heterogeneous media (cases 2 and 3) in Table 3.

The achieved maps depict that the homogeneous domain has produced sharp-edged contours, while the heterogeneous media resulted in irregular edges exposing more contact surface area between the CO_2 and the local brine, which enhances the storage efficiency. The irregular frontages in the heterogeneous media are due to the intermingled layers of shale that restrict the injected gas from moving across different layers of the domain that results in less contact with the ambient brine and less subjectivity to entrapment in more small-sized pores. Heterogeneity is found to have a substantial impact on the amounts of trapped and dissolved gases, as a result of the influence on the capillary pressure–saturation relationship, which is imitated by an increase in the amount of the residually trapped CO_2.

Figure 5A shows that, soon after gas injection stops (i.e., imbibition process starts), the amount of the trapped gas sharply increases when the replaced brine invades back into the domain and isolates some blobs of CO_2 from the trailing edge of the mobile CO_2 plume. After 200 years, this progress slightly retards because part of the trapped gas tends to dissolve in the brine. This increase continues until 1400 years of simulation, when the trapped gas profiles steeply decline before tending to settle after 3000 years. The figure further demonstrates more residually trapped gas in the heterogeneous models compared to the homogeneous ones at the end of the simulation.

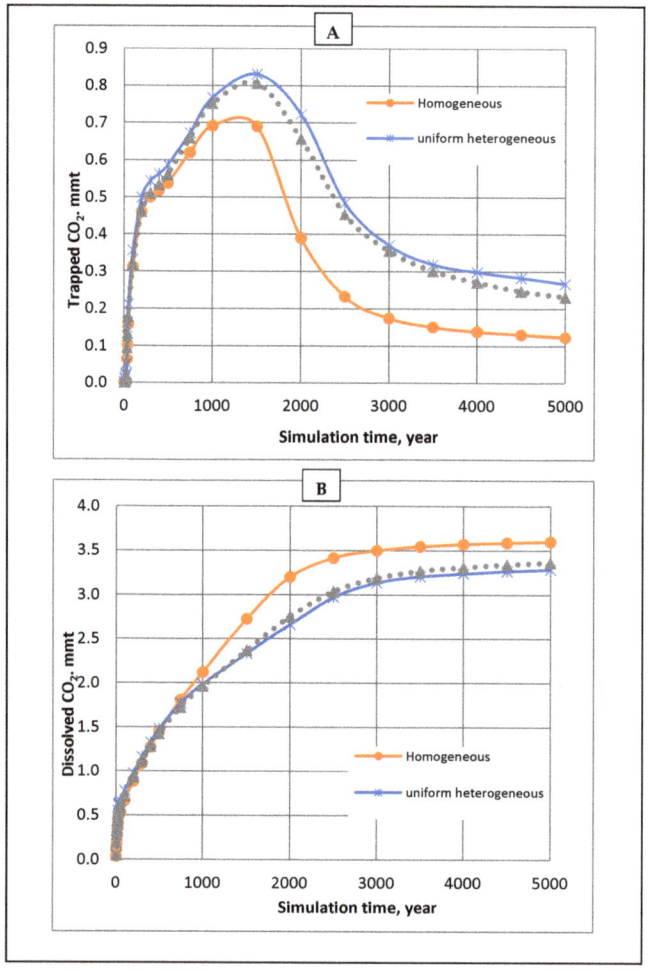

Figure 5. Effect of heterogeneity for cases 1, 2 and 3 on (**A**) CO_2 residual trapping and (**B**) dissolved CO_2.

In Figure 5B, no effect of heterogeneity on CO_2 solubility was detected before 800 years of simulation because the system was totally dominated by buoyancy and hydrostatic forces. Afterwards, it was observed that more CO_2 got dissolved in the homogeneous domain compared to both types of heterogeneous ones by about 17% after 2000 years. However, this influence approximately declined after 4000 years to 9%. It is apparent from the results, displayed in Figure 5B, that both types of heterogeneity provide almost identical but lower amounts of dissolved CO_2 compared to the homogeneous media throughout the simulation time. This suggests that gas migration is more straightforward through homogeneous media, owing to the lesser resistance to flow.

In contrast to the results by Chasset et al. [35], the increase in CO_2 dissolution can be justified by the presence of intermingled layers of shale that play a role as internal barriers to retard the vertical migration and the promote lateral flow of the injected CO_2. However, this horizontal movement retards after the injection period due to the limited hydraulic gradient, which limits gas contact with more fresh brine, leading to a reduction in gas assimilation and dissolution. The values of trapped and dissolved CO_2 surely affect the storage capacity of the site; however, this impact is applicable to a very limited extent in agreement with the results from a recent study by Zhao et al. [58], which revealed that strong heterogeneity in geological formations reduces the storage capacity because it limits gas seepage.

In spite of the two contrary trends, Figure 6, shows that heterogeneous domains are more efficient in storing the disposed gas by a factor of about 15% compared to the homogeneous ones under similar conditions. This does not comply with the numerical results depicted in Figure 3, that show higher values of free-gas CO_2 left off by the end of simulation in the homogenous domain compared to the heterogeneous one. This controversy is due to the net thickness parameter suggested by Szulczewski [34] for Equation (21) instead of the total thickness of the modelled domain. To implement this in our calculations, the thicknesses of the shale layers were excluded and this resulted in less values of the net thickness in cases 2 and 3 (see Table 3), leading to smaller pore volume available to store the injected CO_2 and, consequently, higher values of storage efficiency were achieved for the heterogeneous domains using Equation (21). This is an important point that needs further investigation to assess the effectiveness of this method to more accurately determine the storage efficiency in open-boundary domains.

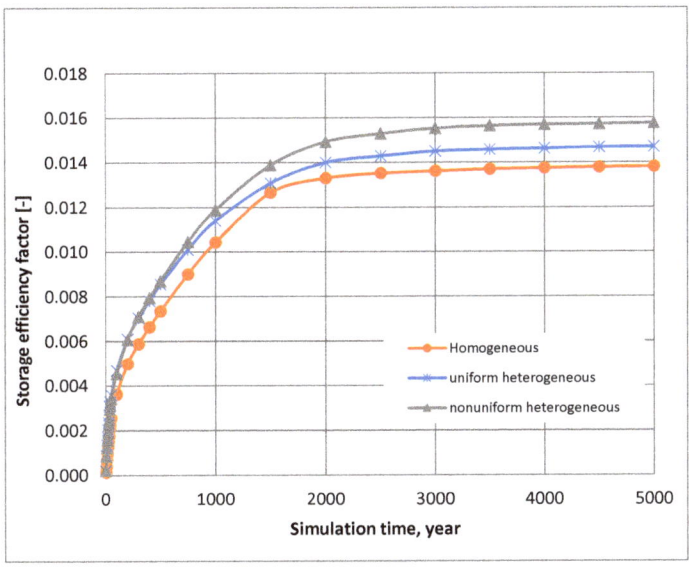

Figure 6. Influence of heterogeneity on storage efficiency factor for Cases 1, 2 and 3 (Table 3).

3.3. Effects of Cyclic Injection on CO_2 Mobility and Sequestration

Saturation (S_w)—relative permeability (k_r) relationship is a key feature that describes the CO_2–water flow system because it has a huge influence on the behaviour and fate of the injected gas in the subsurface. This study has investigated the impact of the injection methodology on the S_w-k_r relationships and eventually on the effectiveness of the disposed gas storage. Purposely, an observation point was setup at 200 m radially away from the injection well (to avoid the effect of the high pressure difference forces close to the wellbore) and 15 m from the bottom of the aquifer, which represents the midpoint of the lower segment of the domain into which the gas injection takes place.

The achieved results from implementing cyclic injection techniques are demonstrated in Figure 7A, that manifests the development of gas relative permeability profiles for continuous and cyclic injection methods (see cases 3 and 4 in Table 3). The influence of the cyclic injection is obvious from the fluctuating profiles from which it can be observed that, for the continuous injection method, the relative permeability of the CO_2 curve declined from a peak value of (0.43) after 10 years to zero by the end of the injection period (30 years). The figure further displays the three cycles of injection impact on the permeability curves with highest peak values of 0.43, 0.66 and 0.7. This impact has been directly imitated on the gas saturation trends in Figure 7B, which evidences the favourite of cyclic injection method because higher amounts of injected CO_2 were found to be safely trapped after the cease of injection.

This is comparable to the continuous injection case, which depicts higher values of gas saturation after the end of injection; however, these values decline soon after that to reach a value of 0.01 after 2000 years of simulation (this is not shown in Figure 7B, which is magnified to show more details about the drainage period).

This variation can be justified by the two additional cycles of imbibition process that lead to more blobs of CO_2 getting disconnected from the trailing edge of the ascending gas plume.

For the homogeneous models (see cases 1 and 5 in Table 3), cyclic injection confirmed no effect on CO_2 dissolution and almost equal amounts of free gas were left off in the domain by the end of the simulation runs, as shown in Figure 8. However, for the heterogeneous domains (cases 3 and 4), continuous injection produced slightly greater profiles of CO_2 dissolution.

In contrast, the residual trapping of CO_2 in heterogeneous media was found to be more sensitive to the cyclic injection because the simulation results revealed that more CO_2 was trapped using the cyclic injection method in the heterogeneous modelled domain compared to the continuous one after 5000 years of simulation, as illustrated in Figure 8.

Table 4 concludes that cyclic injection into homogeneous domains increases the amount of trapped CO_2 gas to some extent (compare cases 1 and 5). However, continuous injection into heterogeneous formations enhances the storage efficiency factor (determined by Equation (21)) by about 0.0003, that represents 1.7% (compare cases 3 and 4), because it can be seen from the table that by applying continuous injection (case 3), 0.773 MMT (approximately 0.17%) more of the injected gas was permanently sequestered either by residual or solubility sequestration mechanism using continuous injection techniques.

In agreement with the results by Juanes et al. [59], this can be justified by the increase in capillary pressure which forces more CO_2 into smaller-sized pores to be trapped and exposed to dissolution in the brine at later stages of storage. In contrary for the cyclic injection, releasing pressure after 10 years encourages the gas plume to percolate upwards through larger pores to accumulate at the top of the domain as a free gas.

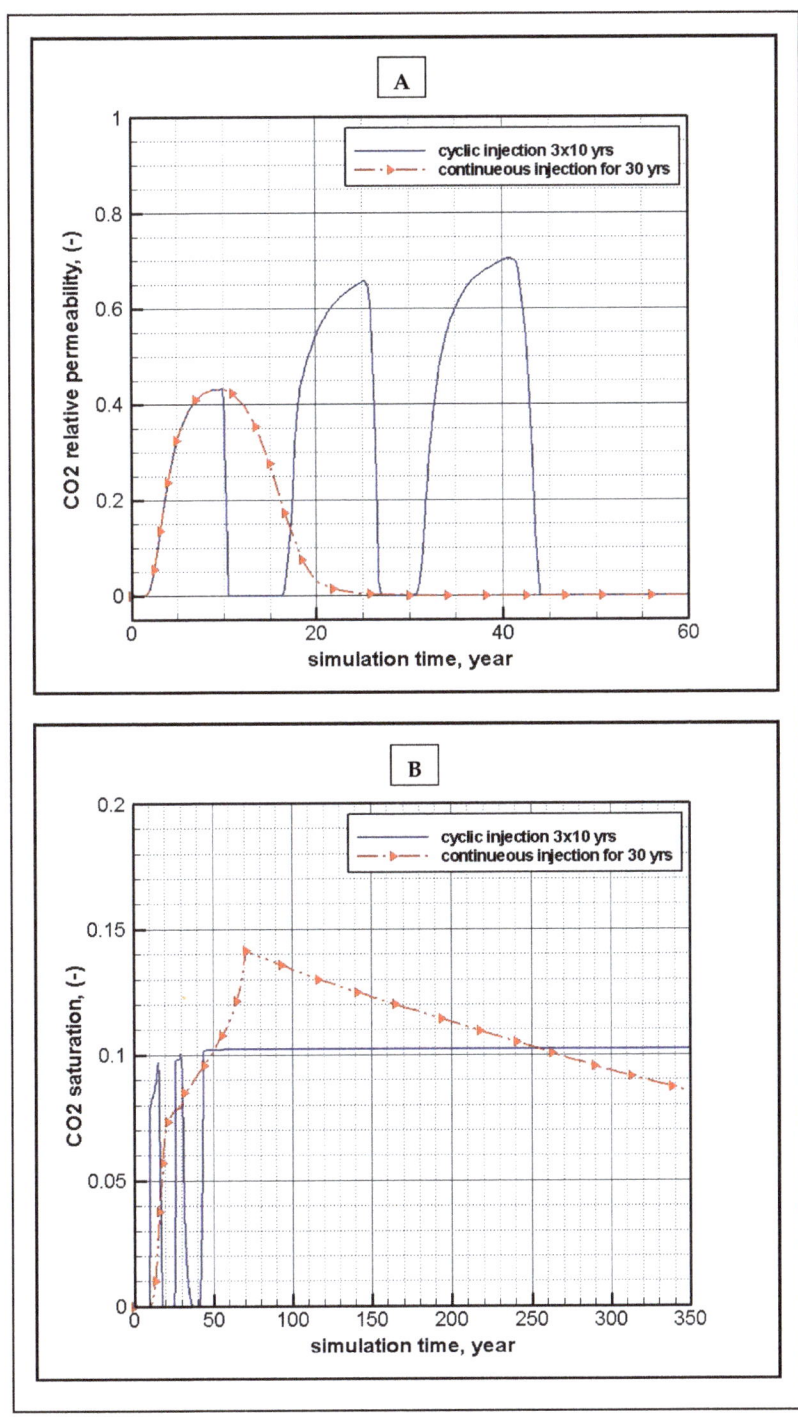

Figure 7. Impact of cyclic injection for cases 3 and 4 on: (**A**) CO_2 relative permeability; (**B**) CO_2 saturation.

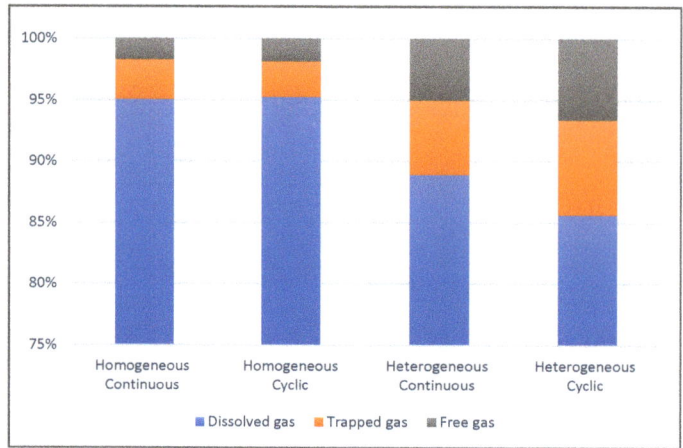

Figure 8. Impact of cyclic injection on different integrated CO_2 phases in DSAs for cases 1, 5, 3 and 4 in Table 3, respectively.

Table 4. Simulation results and calculated efficiency factor for all modelled cases.

Case No.	Description	Dissolved Gas MMT **	Trapped Gas MMT **	Free Gas MMT **	Storage Efficiency Factor [–]
Base	Homogeneous Isotropic Coarse grid, Cont. Vert. Inj. Into lower section	16.095	0.758	0.495	0.0372
1	Homogeneous Isotropic Coarse grid, Cyclic Vert. inj. Into lower section	28.800	0.986	0.513	0.01379
2	Homogeneous Anisotropic Coarse grid, Cont. Vert. inj. Into whole thickness	26.290	2.141	1.864	0.0417
3	Irregular Heterogeneous Isotropic Coarse grid Cont. Vert. inj., lower section	26.929	1.854	1.511	0.0158
4	Irregular Heterogeneous Isotropic Coarse grid Cyclic Vert. inj., lower section	25.941	2.355	1.999	0.0155
5	Homogeneous Isotropic Coarse grid, Cont. Vert. inj. Into lower section	28.858	0.886	0.551	0.0138
6	Homogeneous Isotropic Fine grid, Cont. Vert. Inj. Into lower section	29.161	0.637	0.497	0.0138
7	Homogeneous Isotropic Coarse grid, Cont. Vert. inj. Into lower section	28.147	1.282	0.866	0.0136
8	Homogeneous Isotropic Coarse grid, Cont. Vert. inj. Into whole thickness	28.241	1.217	0.838	0.0136
9	Regular Heterogeneous Isotropic Coarse grid Cont. Vert. inj., lower section	26.190	2.360	1.745	0.0132
10	Homogeneous Anisotropic Coarse grid, Cont. inj. Vertical inj. well 96 m	21.367	4.699	4.229	0.0121
11	Homogeneous Anisotropic Coarse grid, Cont. inj. Horizontal inj. well, 96 m *	18.787	6.977	4.532	0.0119
12	Homogeneous Anisotropic Coarse grid, Cont. inj. Horizontal inj. well 192 m *	19.885	5.167	5.244	0.0116

* Horizontal injection well at the bottom of the domain starting from the centre point. ** MMT = million metric ton (10^9 kg).

3.4. Effect of Vertical Injection

This section extends a previous work [9] to further optimize the vertical injection method with the aim to investigate the storage efficiency enhancement. To carry this out, two simulation cases were conducted, implementing two different scopes of vertical injection (see cases 7 and 8 in Table 3). In case 7, scrCO$_2$ was injected into the lower segment of the aquifer through four vertical grid cells of 4 m, while for case 8, the injection was executed over 24 blocks (i.e., over the whole thickness of the aquifer that extends to 96 m).

In terms of the dissolution of the injected scrCO$_2$, unexpectedly no significant effect was observed at all-time scales for case 7. On the other hand, slightly more trapped gas concentrations within the vicinity of the injection well were demonstrated for case 8, as shown in Figure 9. This is because, in this case, the whole amount of the gas was injected through the lower segment of the domain, most of which was influenced by the reversing brine tendency to disconnect more blobs of CO$_2$ from the rambling edge of the gas plume. In contrast, when the injection applied into the whole thickness of the domain (case 8), only a sixth of the scrCO$_2$ mass rate was injected into the lower section of the model and most of this amount had bounced upwards before the imbibition process started. This means that only a significantly small part of the injected gas was affected by the raiding brine, leading to a reduced amount of residually trapped gas. It is evidenced from the achieved results that injecting CO$_2$ through the whole thickness of the domain slightly reduced the amount of the free gas left within the domain in medium terms of storage by 0.028 MMT (about 3.2%) as depicted in Table 4 for cases 7 and 8. However, the injection scope has shown no sensible influence on the storage efficiency because both cases returned almost identical values of storage efficiency factor, as presented in Table 4.

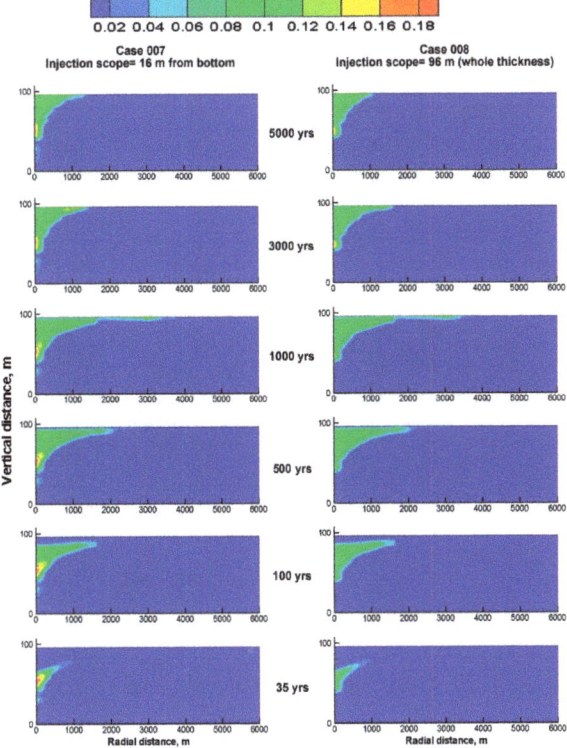

Figure 9. Impact of injection scope on trapped CO$_2$ mass distribution.

3.5. Impact of Directional Permeability Ratio

In the aim of assessing the effect of heterogeneity anisotropy in geological formations on the efficiency of storage, three models of a hypothetical aquifer with values of vertical-to-horizontal permeability ratios (k_v/k_h) equivalent to 1.0, 0.1 and 0.01 were developed and modelled (see cases 8, 9 and 10 in Table 3). This array was set up as a realistic figure for most sandstone rocks according to a relatively new study by Widarsono et al. [60].

The obtained results depicted in Figure 10, show the deceptive influence of the permeability ratio on the CO_2 plume shape and spatial distribution maps. While the plume tends to horizontally extend further along the overlaying layer at higher permeability ratios, more CO_2 shows the tendency to migrate laterally within the two lower layers of the domain for lower values of permeability ratio. This owes to the fact that lower permeability in the vertical direction restrains the upward movement of CO_2, forcing the injected gas to migrate horizontally across the domain, proposing more gas into small-sized pores where it is more likely to be permanently entrapped when the brine invades back into the domain after the injection stops, which is referred to as residual trapping.

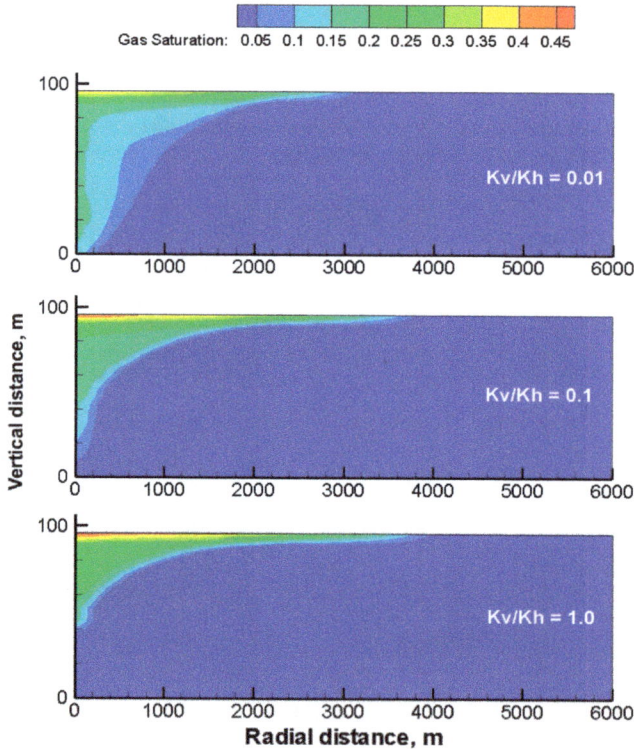

Figure 10. The impact of the vertical/horizontal permeability ratio on CO_2 distribution after 500 years in a homogeneous domain. Cases 8, 9 and 10 in Table 3.

The latter impact is evidenced in Figure 11A, which shows significantly greater amounts of trapped gas in cases of lower permeability ratios at all post injection time steps.

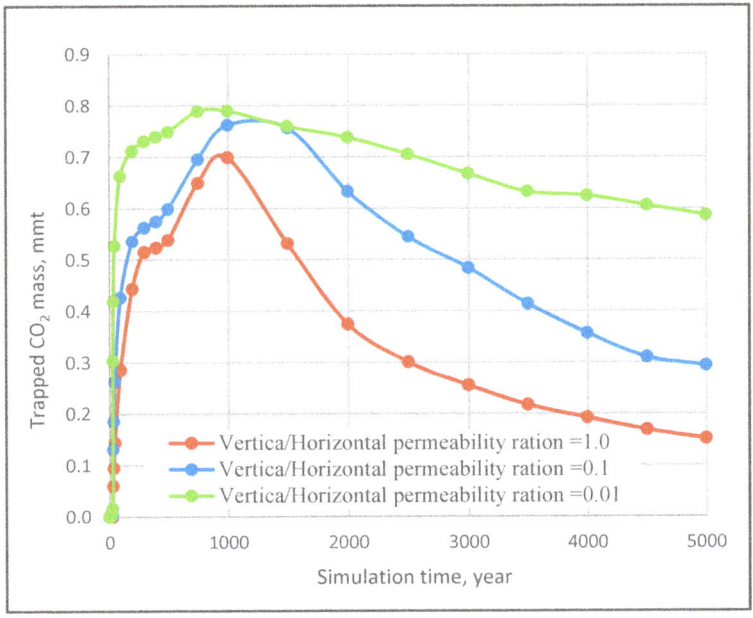

(**A**)

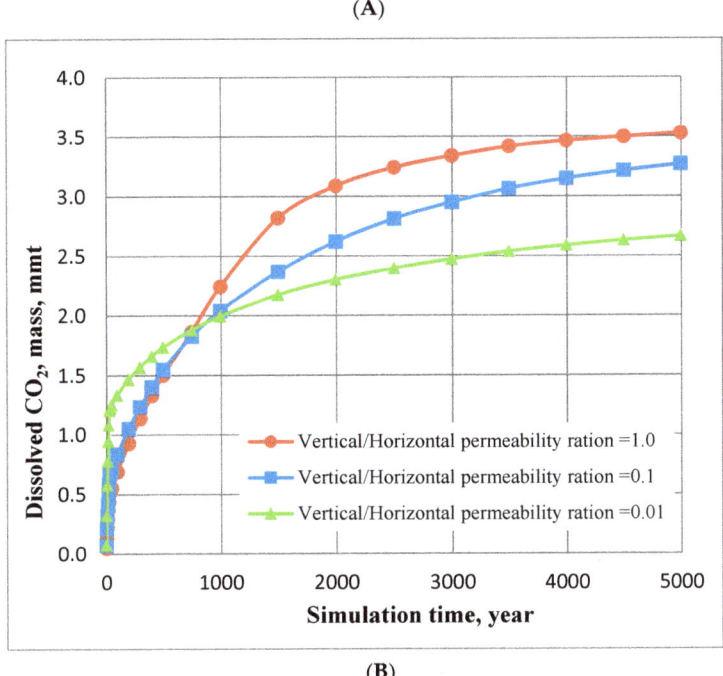

(**B**)

Figure 11. (**A**). Impact of vertical-to-horizontal permeability ratio on the integrated free CO_2 for cases 8, 9 and 10. (**B**). Impact of vertical-to-horizontal permeability ratio on dissolved CO_2 gas in a homogeneous domain for cases 8, 9 and 10 in Table 3.

One of the stupendous returns from this study is the inflection point in the gas dissolution trends after around 800 years of the simulation lifetime, as noticeable in Figure 11B. This deviation occurred because at early stages, the flow system was entirely dominated by the structural trapping mechanism, in which most of the injected gas remained in free phase. This mechanism is mainly dependent on the upwards movement of the free gas that is more effective at higher vertical permeability values (i.e., larger values of (k_v/k_h)), as explained previously in this section. The horizontal movement of the gas, due to the pressure gradient and low vertical permeability, promotes more contact between the two fluids, leading to more dissolution of CO_2 in the formation brine at early stages (solubility trapping). However, this migration has no significant impact compared to the large buoyancy forces at later stages.

This clarifies the larger amounts of dissolved CO_2 at lower values of k_v/k_h before 800 years in Figure 11B. By approaching 1000 years of simulation, the solubility trapping mechanism takes control because the density and pressure gradient driving forces decline when most of the integrated gas has either settled at the top of the domain or within the vicinity of the injection well, as illustrated in Figure 10 (see case 8 in Table 3). Consequently, the domain becomes dominated by the solubility trapping, which is based on the contact interfacial area between the two fluids and the hydrostatic conditions that influence the CO_2 dissolution rate in the surrounding brine. CO_2 dissolution into the formation brine creates a denser aqueous phase that tends to sink downwards when the vertical movement becomes important to maintain more gas contact with the fresh brine, leading to more dissolution of the gas into the brine. This convectional movement is easier at higher permeability ratios that result in larger amounts of gas dissolution.

Despite all this evidence, the determined storage efficiency for different permeability ratios using Equation (21) by Szulczewski [34] for open boundary domains has shown better storage efficiency at lower permeability ratios. This is significantly controversial and requires more investigation and discussion because our repeated numerical experiments have revealed contrasting results (see Figure 12A,B). This can be referred to as the length parameter used in Equation (21) to calculate the effective volume of the domain and, consequently, the CO_2 mass that can be stored. The author suggested using the maximum extent of the gas plume to calculate the effective volume; however, our simulation results revealed a huge difference in the obtained plume lengths for cases 8, 9 and 10 in Table 3. They were found to be 3965, 3802 and 3135 m for permeability ratios of 1.0, 0.1 and 0.01, respectively (see Figure 10).

This significant difference has returned unrealistic values of the storage efficiency when implemented in Equation (21), because looking at Figure 12A, it can be evidently noticed that higher permeability ratios produced greater amounts of dissolved and trapped CO_2, and less amounts of free-gas (i.e., enhanced solubility and residual trapping of the injected gas). It is apparent from the figure that, for k_v/k_h values of 1.0, 0.1 and 0.01, the achieved permanent trapping rates were 29.457 MMT (97%), 28.55 MMT (87%) and 26.066 MMT (71%), respectively.

Depending on the findings from this study, it is suggested that the plume length parameter in Equation (21) is reviewed and presented by a more realistic value to make the equation further applicable to various CO_2 injection scenarios into geological formations.

In this work, the total length of the domain, which extends far enough that the gas plume does not reach it, was used in order to avoid any boundary effects on the in situ pressure build-up or gas seepage from the computational domain for all simulation runs (i.e., to employ an open boundary condition away from the injection well). This means that the whole aquifer volume was used to calculate the storage efficiency factor, which explains the small-obtained values of the efficiency factors in Table 4. Using this total length in Equation (21) to determine the sequestration efficiency has returned more realistic values of the storage efficiency in terms of the directional permeability effect, as shown in Figure 12B, which indicates that the storage efficiency factor of any aquifer increases proportionally with the vertical-to-horizontal permeability ratio (see cases 8, 9 and 10 in Table 4).

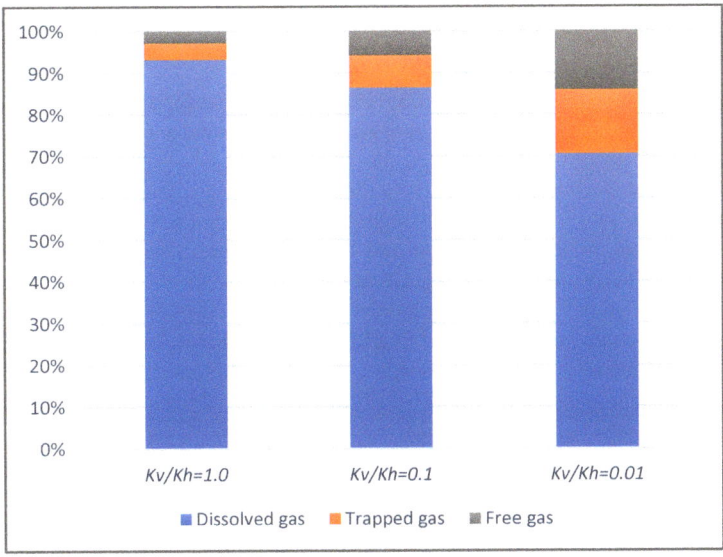

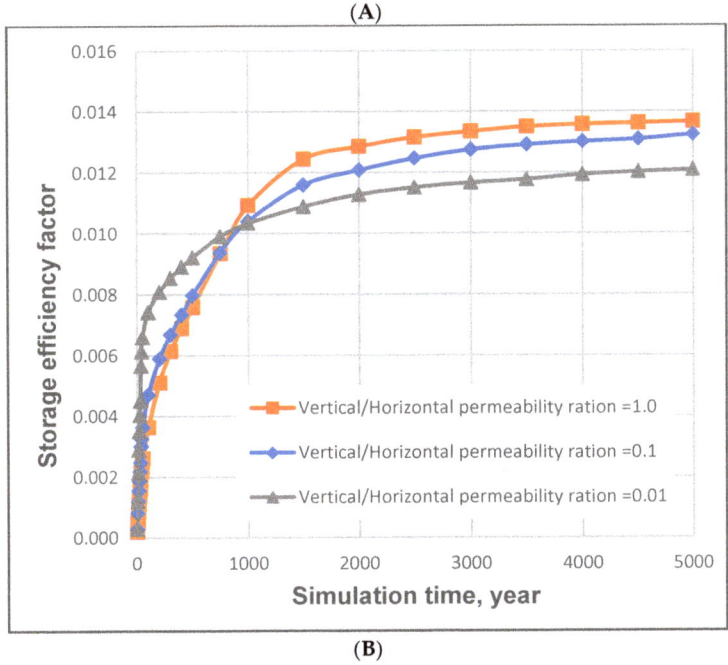

Figure 12. (**A**). CO_2 storage efficiency at different permeability ratios based on the numerical results plume length (see cases 8, 9 and 10 in Table 4). (**B**). CO_2 storage efficiency at different permeability ratios based on maximum plume length used in Equation (21) (see cases 8, 9 and 10 in Table 4).

3.6. Influence of Injection Orientation

Injecting CO_2 into sedimentary formations through horizontal wells has been a subject of many research works and review studies, most of which are based on applying horizontal injection into confined geological formations, considering the induced pressure build-up. While some authors conclude that injecting CO_2 via horizontal injection wells improves the trapping efficiency [39], others find that such a methodology influences the mechanical stability of the overlaid caprock and does not improve the storage efficiency in the long term [40,61].

In this study, we investigated the influence of the injection orientation on the hydrodynamic behaviour of CO_2 and storage efficiency for an open boundary model and compared the results with those for the conventional vertical injection methodology. Purposely three simulation models were developed (see cases 10, 11 and 12 in Table 3) to identify the impact of the injection orientation on the storage efficiency and P_c-S_w relationship in geological formations. Both vertical and horizontal injection wells were set up for 96 m for cases 10 and 11, while the horizontal well length was extended to 192 m for case 12. All other simulation conditions were similar for the three cases and a constant injection rate was maintained to run the simulations over different well lengths (96 and 192 m).

The achieved numerical results revealed that injection orientation has a significant influence on the gas migration and behaviour in unconfined geological formations, as depicted in Figure 13, that highlights the disparity between the gas distribution contours achieved through using horizontal injection wells (cases 11 and 12 in Table 3) and those obtained from the vertical injection methodology (case 10).

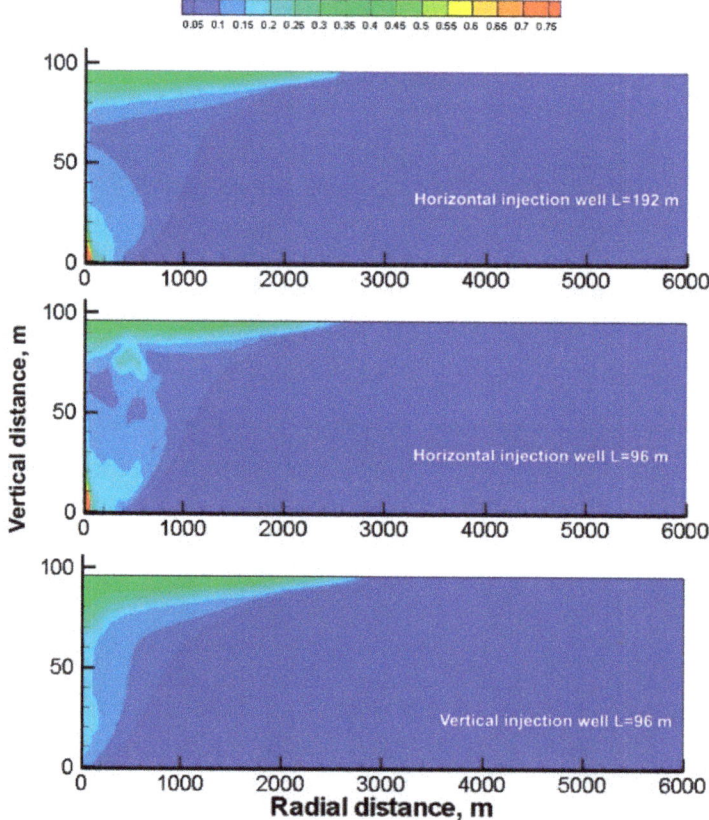

Figure 13. Integrated CO_2 distribution maps for different injection methods after 50 years in a homogeneous domain. Cases 10, 11 and 12 in Table 3.

The figure exhibits that a noticeable portion of the injected gas was trapped by the displacing brine in the case of the aquifer-thickness equivalent horizontal injection well (96 m) compared to the longer horizontal well (192 m). This can be attributed to the large injection mass flow rate per unit area (i.e., limited number of gridlocks) which delayed the upwards propagation of the buoyant CO_2, leading to only part of it reaching the top of the domain to create a thin tongue-like shape that migrated crossways. The other portion of the injected gas was exposed to be encountered by the invading brine that physically isolated blobs of it within the local pores network to be dissolved at later stages.

Using horizontal injection techniques was found to increase the quantities of the trapped gas, as presented in Table 4, due to the magnified values of average capillary pressure. Unexpectedly, the amount of the trapped gas was found to be significantly less than that which was achieved by the shorter horizontal well. This can be explained by the smaller injection rate per unit area in the first case, which promotes more percolation of the free-gas towards the top of the aquifer, as shown in Table 4. Despite the relative increase in the gas dissolution depicted in Figure 14A, for the longer horizontal well, the amount of the free-gas left off by the end of the simulation was higher, as shown in Figure 14B, leading to lower storage efficiency, as evidenced in Figure 15.

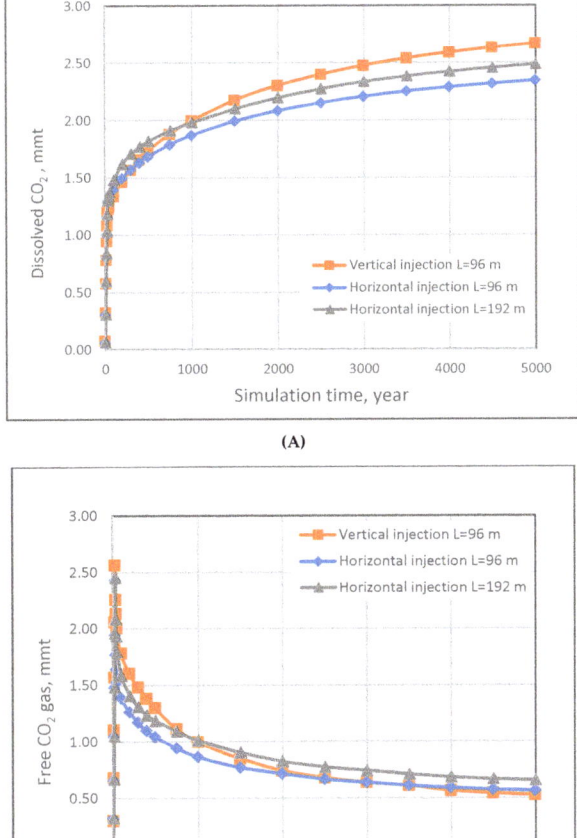

Figure 14. (**A**). Effect of injection orientation for cases 10, 11 and 12 (Table 3) on CO_2 solubility in a homogeneous domain. (**B**). Effect of injection orientation for cases 10, 11 and 12 (Table 3) on free-gas CO_2 in a homogeneous domain.

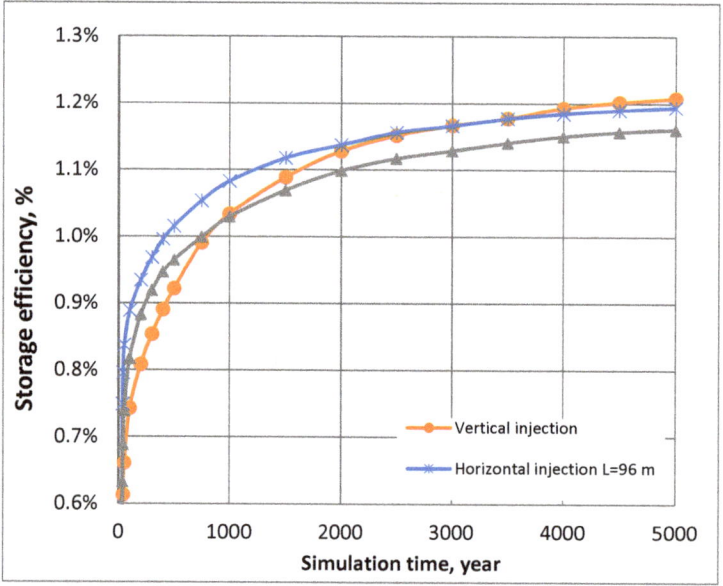

Figure 15. CO_2 storage efficiency for different injection orientation scenarios—cases 10, 11 and 12 in Table 3.

Additionally, the results reveal that horizontal injection into migration-controlled domains (i.e., open-boundary domains), returns slightly higher storage efficiency in short terms of simulation; however, after 2000 years, vertical injection methodology was found to be more efficient, as evidenced in Figure 15.

This is consistent with the findings by Okwen et al. [61], who suggest that using horizontal wells is preferable for pressure-limited domains and for sequestering large amounts of CO_2 in a short time frame. Accordingly, implementing longer horizontal injection wells does not significantly enhance the storage capacity and the economical factor has to be taken into consideration, should they need to be used for injecting large amounts of gas within limited periods of time.

3.7. Sensitivity to Domain Grid-Resolution

The grid discretization of any simulated domain is an important factor used to accurately capture the occurrence of different flow dynamics and assess the sensitivity of modelling results to the spatial gridding schemes. As mentioned earlier, in this study, two levels of grid refinement, a coarse grid ($88 \times 1 \times 24$) and a fine grid ($176 \times 1 \times 48$) were used to record the simulation code outputs (see cases 1 and 6 in Table 3).

The influence of the grid resolution is illustrated in Figure 16, where more detailed fingering maps of CO_2 dissolution can be observed in the fine-grid domain, compared to those for the coarse grid. Moreover, longer gas plumes were detected in the finer grid, which means that more accurate records of different forms of integrated gas were netted.

This is further evidenced in Figure 17A,B, in which it can be observed that, after 1500 years, a about 20% larger amount of dissolved gas and 54% lesser amount of free-gas were logged by the simulation code when a finer grid was implemented. These figures declined to about 1.3% and 3.2%, respectively, by the end of simulation. In Figure 17A, less impact of the grid resolution on CO_2 dissolution in the hosted brine was noted for both grids up to around 300 years of simulation. Then, after, an obvious increase in the dissolved gas trends for the finer grid, specifically between

800–2000 years, was noted. This deviancy diminishes after 2000 years, which agrees with the findings by Gonzalez-Nicolas et al. [62] and Bielinski [63].

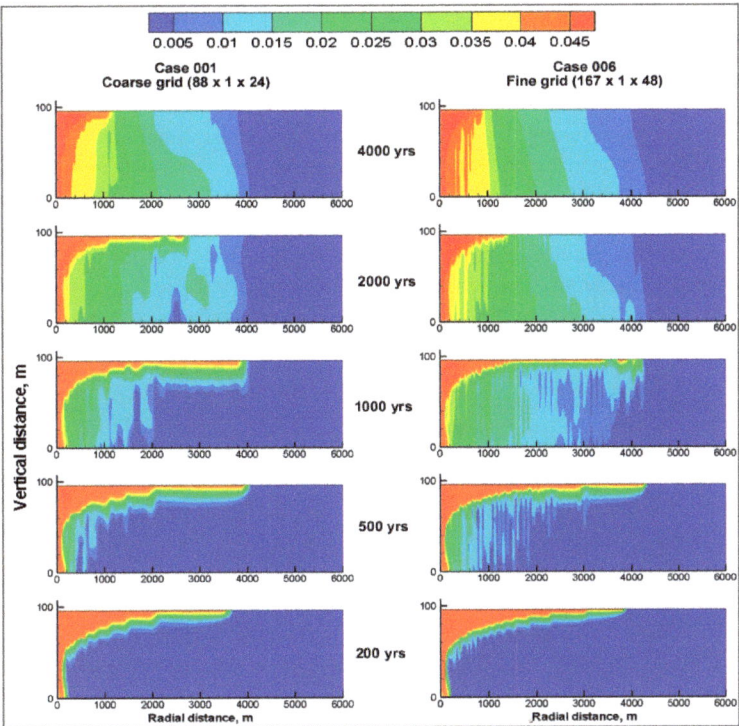

Figure 16. Aqueous CO_2 distribution in coarse and fine-grid homogeneous domains at different time scales.

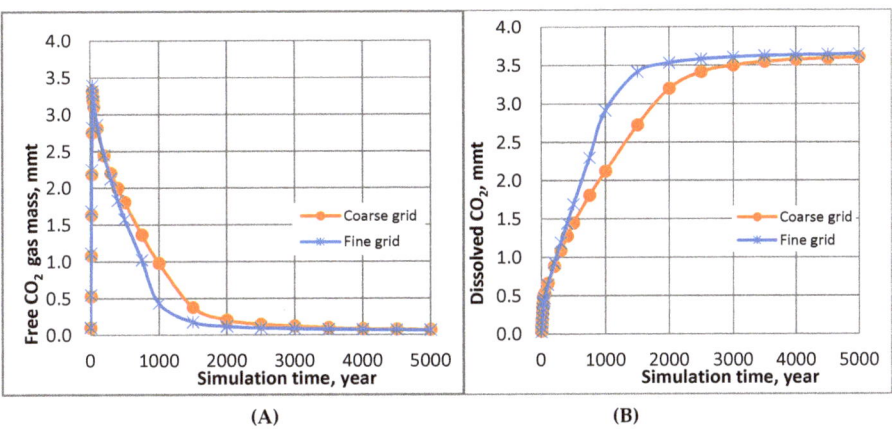

Figure 17. Effect of grid resolution for cases 5 and 6 in a homogeneous domain on: (**A**) dissolved CO_2; (**B**) free-gas CO_2.

This can be justified by the findings from this work (see Figure 15), which demonstrate the influence of grid resolution on the plume shape and number of formed fingers in the simulated domain due to the convection forces and gravity instability. Consequently, more accurate results were logged using finer grids which justifies the relatively higher efficiency achieved in the case of fine-resolution grid, as illustrated in Figure 17.

Unexpectedly, the results revealed higher gas entrapment in the coarse grid (case 1) than that for the fine grid (case 6), as shown in Table 4. This can be due to the fact that, by using larger blocks in the computational domain, part of the dissolved CO_2 might have been logged either as a free or trapped gas, which can be explained from the relatively larger amounts of the latter two forms of the content gas in the case of coarse refinement. In spite of this significant overestimation of the netted values of the trapped gas in the coarse grid, the amount of the free gas was found to be less in the finer grid by about 3.2%, as displayed in Table 4. However, the increase in the storage efficiency factor was only about 0.1% using a finer grid, as shown in Figure 18, which further clarifies that the grid refinement has only a small impact on the simulation results. Additionally, the amount of the residually trapped gas was much smaller than the amount of the dissolved CO_2 shown in Table 4 and, moreover, this small amount of the trapped gas itself is subject to dissolution in medium over long time frames.

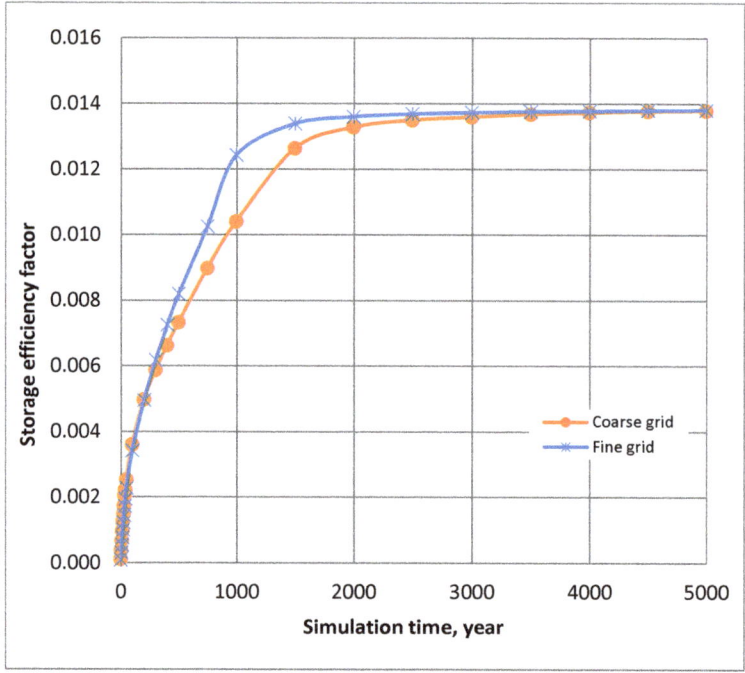

Figure 18. Grid refinement effect on CO_2 storage efficiency in a homogeneous domain. Cases 5 and 6 (Table 4).

Some preceding publications concluded only a slight or no impact of grid resolution on the simulation results. Conversely, ranking our simulation cases, according to the attribute of safer storage of the disposed gas in Table 4, the finer grid case was found to be on the top of the list. Therefore, and despite the excessive execution time required to conduct simulation runs with finer grids, it is important to magnify focus on the behaviour of the injected gas in different phases (i.e., dissolved, residually trapped and free-gas) within in situ pores network by using reasonably refined grids. The results from the finer mesh (see case 6 in Table 4) detected only about 0.497 MMT of free gas at the end of the simulation lifetime compared to about 0.514 MMT for the coarse grid in case 1, which reflects about 3.2% safer storage by the means of deploying the finer grid. This should be motivating for the researchers to refine their modelling grids for further focused and more credible accurate results. Nevertheless, it is recommended that a sensible balance between the grid refinement and the computational time required for calculations be embraced.

For the cylindrical domain modelled in this research work (3 km radius and 96 m thickness), it was found that discretizing it to 176 by 48 nodes in the lateral and vertical direction, respectively, was found to provide a reasonably effective level of refinement in terms of balancing between the accuracy of the achieved results and the computational time and requirements.

4. Conclusions

A set of numerical simulation cases was developed and conducted using the STOMP-CO_2 numerical simulation code to investigate the influence of various types of heterogeneity, injection schemes, grid resolution, anisotropy, and injection orientation on the CO_2–water flow system behaviour and storage efficiency in saline aquifers. It was found that heterogeneous formations amplify the residual trapping mechanism, while CO_2 dissolution (i.e., solubility trapping mechanism) shows higher trends in homogeneous formations. However, overall, the heterogeneous media were found to be more effective in storing CO_2 safely over long-time frames. Compared to the homogeneous media, cyclic injection methodology has shown more influence on the heterogeneous domains through which the injected gas spreads out further, leading to greater interfacial area with brine and, consequently, escalating CO_2 dissolution. However, further research work is required to investigate more details about optimizing the injection times and pausing intervals in long-term sequestration projects where mineralization trapping plays an important role.

We observed that, while a CO_2 gas plume extends further at higher k_v/k_h values, lower ratios enhance the solubility trapping of CO_2 at early stages of simulation. Additionally, stronger hysteresis at higher permeability ratios enhances the residual trapping mechanism. Overall, storage efficiency increases proportionally with the permeability ratio of geological formations because higher ratios facilitate the further extent of the gas plume and increase the solubility trapping of the integrated gas. Using the maximum length of the gas plume in Equation (21) to calculate the available porous volume in open-boundary domains requires more investigation because it produced some unrealistic results. Therefore, an optimization is suggested to set up the domain length according to the employed injection rate or pressure, so that the extended gas plume just reaches but does not pass it. This length can be called the effective length and can be used to calculate the volume available to host the potential injected gas. It is also concluded in this study that employing longer horizontal wells does not increase storage efficiency. However, more research work is recommended to optimize the length of the horizontal wells and the injection techniques, including injecting chase brine along with scCO_2.

Despite the excessive execution time required to run simulations on finer-grid domains, it is important to magnify focus on the behaviour of the injected gas in order to increase the accuracy of the logged results. Finer-resolution grids can slightly increase the calculated values of the storage efficiency factor, specifically in the medium terms of sequestration. However, practical balance should be maintained between the refinement level and the computational requirements along with the execution time needed.

Author Contributions: Conceptualization, K.J.K. and D.B.D.; Methodology, K.J.K. and D.B.D.; Software, K.J.K.; Validation, K.J.K.; Formal Analysis, K.J.K. and D.B.D.; Investigation, K.J.K. and D.B.D.; Resources, K.J.K. and D.B.D.; Data Curation, K.J.K.; Writing-Original Draft Preparation, K.J.K.; Writing-Review & Editing, K.J.K. and D.B.D.; Visualization, K.J.K.; Supervision, D.B.D.; Project Administration, K.J.K. and D.B.D.; Funding Acquisition, K.J.K. and D.B.D. All authors have read and agreed to the published version of the manuscript.

Funding: This research received no external funding.

Acknowledgments: This work was part of a Doctoral Thesis Completed by the first author of this paper, supported by the Chemical Engineering Department, Loughborough University. Both authors would like to thank Mark D. White from Pacific Northwest National Laboratory (PNNL), USA, for his insight and helpful comments in employing the STOMP-CO_2 simulation code in this study.

Conflicts of Interest: The authors declare no conflict of interest.

References

1. IPCC. *Carbon Dioxide Capture and Storage*; IPCC Special Report; Cambridge University Press: Cambridge, UK, 2005; Volume 217.
2. Said, A.; Laukkanen, T.; Järvinen, M. Pilot-scale experimental work on carbon dioxide sequestration using steelmaking slag. *Appl. Energy* **2016**, *177*, 602–611. [CrossRef]
3. Ren, F.; Ma, G.; Wang, Y.; Fan, L.; Zhu, H. Two-phase flow pipe network method for simulation of CO_2 sequestration in fractured saline aquifers. *Int. J. Rock Mech. Min. Sci.* **2017**, *98*, 39–53. [CrossRef]
4. Cui, G.; Wang, Y.; Rui, Z.; Chen, B.; Ren, S.; Zhang, L. Assessing the combined influence of fluid-rock interactions on reservoir properties and injectivity during CO_2 storage in saline aquifers. *Energy* **2018**, *155*, 281–296. [CrossRef]
5. Jiang, L.; Xue, Z.; Park, H. Enhancement of CO_2 dissolution and sweep efficiency in saline aquifer by micro bubble CO_2 injection. *Int. J. Heat Mass Transf.* **2019**, *138*, 1211–1221. [CrossRef]
6. Singh, M.; Chaudhuri, A.; Chu, S.P.; Stauffer, P.H.; Pawar, R.J. Analysis of evolving capillary transition, gravitational fingering, and dissolution trapping of CO_2 in deep saline aquifers during continuous injection of supercritical CO_2. *Int. J. Greenh. Gas Control* **2019**, *82*, 281–297. [CrossRef]
7. Seo, S.; Mastiani, M.; Hafez, M.; Kunkel, G.; Asfour, C.G.; Garcia-Ocampo, K.I.; Kim, M. Injection of in-situ generated CO_2 microbubbles into deep saline aquifers for enhanced carbon sequestration. *Int. J. Greenh. Gas Control* **2019**, *83*, 256–264. [CrossRef]
8. Rasheed, Z.; Raza, A.; Gholami, R.; Rabiei, M.; Ismail, A.; Rasouli, V. A numerical study to assess the effect of heterogeneity on CO_2 storage potential of saline aquifers. *Energy Geosci.* **2020**, *1*, 1–2, 20–27. [CrossRef]
9. Khudaida, K.J.; Das, D.B. A numerical study of capillary pressure–saturation relationship for supercritical carbon dioxide (CO_2) injection in deep saline aquifer. *Chem. Eng. Res. Des.* **2014**, *92*, 3017–3030. [CrossRef]
10. Abidoye, L.K.; Khudaida, K.J.; Das, D.B. Geological carbon sequestration in the context of two-phase flow in porous media: A review. *Crit. Rev. Environ. Sci. Technol.* **2015**, *45*, 1105–1147. [CrossRef]
11. Kopp, A.; Class, H.; Helmig, R. Investigations on CO_2 storage capacity in saline aquifers—Part 2: Estimation of storage capacity coefficients. *Int. J. Greenh. Gas Control* **2009**, *3*, 277–287. [CrossRef]
12. Gough, C.; Shakley, S.; Holloway, S.; Bentham, M.; Bulatov, I.; McLachlan, C.; Klemes, J.; Purdy, R.; Cockerill, T. An integrated assessment of carbon dioxide capture and storage in the UK. In Proceedings of the 8th International Conference on Greenhouse Gas Control Technologies, Trondheim, Norway, 19–22 June 2006.
13. Bachu, S. Review of CO_2 storage efficiency in deep saline aquifers. *Int. J. Greenh. Gas Control* **2015**, *40*, 188–202. [CrossRef]
14. Jin, C.; Liu, L.; Li, Y.; Zeng, R. Capacity assessment of CO_2 storage in deep saline aquifers by mineral trapping and the implications for Songliao Basin, Northeast China. *Energy Sci. Eng.* **2017**, *5*, 81–89. [CrossRef]
15. Vilarrasa, V.; Bolster, D.; Olivella, S.; Carrera, J. Coupled hydromechanical modeling of CO_2 sequestration in deep saline aquifers. *Int. J. Greenh. Gas Control* **2010**, *4*, 910–919. [CrossRef]
16. Potdar, R.S.; Vishal, V. Trapping Mechanism of CO_2 Storage in Deep Saline Aquifers: Brief Review. In *Geologic Carbon Sequestration*; Springer: Cham, Switzerland, 2016; pp. 47–58.
17. Khan, S.; Khulief, Y.A.; Al-Shuhail, A. Mitigating climate change via CO_2 sequestration into Biyadh reservoir: Geomechanical modeling and caprock integrity. *Mitig. Adapt. Strateg. Glob. Chang.* **2019**, *24*, 23–52. [CrossRef]
18. Goodarzi, S.; Settari, A.; Ghaderi, S.M.; Hawkes, C.; Leonenko, Y. The effect of site characterization data on injection capacity and cap rock integrity modeling during carbon dioxide storage in the Nisku saline aquifer at the Wabamun Lake area, Canada. *Environ. Geosci.* **2020**, *27*, 49–65. [CrossRef]
19. Gorecki, C.D.; Sorensen, J.A.; Bremer, J.M.; Ayash, S.C.; Knudsen, D.J.; Holubnyak, Y.I.; Harju, J.A. Development of Storage Coefficients for determining Carbon Dioxide Storage in Deep Saline Formations. In Proceedings of the SPE International Conference on CO_2 Capture, Storage, and Utilization, San Diego, CA, USA, 2–4 November 2009.
20. Knopf, S.; May, F. Comparing methods for the estimation of CO_2 storage capacity in saline aquifers in Germany: Regional aquifer based vs. structural trap based assessments. *Energy Procedia* **2017**, *114*, 4710–4721. [CrossRef]
21. Pingping, S.; Xinwei, L.; Qiujie, L. Methodology for estimation of CO_2 storage capacity in reservoirs. *Pet. Explor. Dev.* **2009**, *36*, 216–220. [CrossRef]

22. Kim, Y.; Jang, H.; Kim, J.; Lee, J. Prediction of storage efficiency on CO_2 sequestration in deep saline aquifers using artificial neural network. *Appl. Energy* **2017**, *185*, 916–928. [CrossRef]
23. Zhou, Q.; Birkholzer, J.T.; Tsang, C.; Rutqvist, J. A method for quick assessment of CO_2 storage capacity in closed and semi-closed saline formations. *Int. J. Greenh. Gas Control* **2008**, *2*, 626–639. [CrossRef]
24. Frailey, S.M. Methods for estimating CO_2 storage in saline reservoirs. *Energy Procedia* **2009**, *1*, 2769–2776. [CrossRef]
25. Ehlig-Economides, C.; Economides, M.J. Sequestering carbon dioxide in a closed underground volume. *J. Pet. Sci. Eng.* **2010**, *70*, 123–130. [CrossRef]
26. Macminn, C.W.; Juanes, R. A mathematical model of the footprint of the CO_2 plume during and after injection in deep saline aquifer systems. *Energy Procedia* **2009**, *1*, 3429–3436. [CrossRef]
27. Kopp, A.; Class, H.; Helmig, R. Investigations on CO_2 storage capacity in saline aquifers: Part 1. Dimensional analysis of flow processes and reservoir characteristics. *Int. J. Greenh. Gas Control* **2009**, *3*, 263–276. [CrossRef]
28. Okwen, R.T.; Stewart, M.T.; Cunningham, J.A. Analytical solution for estimating storage efficiency of geologic sequestration of CO_2. *Int. J. Greenh. Gas Control* **2010**, *4*, 102–107. [CrossRef]
29. Kopp, A.; Probst, P.; Class, H.; Hurter, S.; HELMIG, R. Estimation of CO_2 storage capacity coefficients in geologic formations. *Energy Procedia* **2009**, *1*, 2863–2870. [CrossRef]
30. Yang, F.; Bai, B.; Tang, D.; Shari, D.; David, W. Characteristics of CO_2 sequestration in saline aquifers. *Pet. Sci.* **2010**, *7*, 83–92. [CrossRef]
31. Goodman, A.; Bromhal, G.; Strazisar, B.; Rodosta, T.; Guthrie, W.F.; Allen, D.; Guthrie, G. Comparison of methods for geologic storage of carbon dioxide in saline formations. *Int. J. Greenh. Gas Control* **2013**, *18*, 329–342. [CrossRef]
32. Anderson, S.T.; Jahediesfanjani, H. Estimating the pressure-limited dynamic capacity and costs of basin-scale CO_2 storage in a saline formation. *Int. J. Greenh. Gas Control* **2019**, *88*, 156–167. [CrossRef]
33. Goodman, A.; Hakala, A.; Bromhal, G.; Deel, D.; Rodosta, T.; Frailey, S.; Guthrie, G. US DOE methodology for the development of geologic storage potential for carbon dioxide at the national and regional scale. *Int. J. Greenh. Gas Control* **2011**, *5*, 952–965. [CrossRef]
34. Szulczewski, M.L. The Subsurface Fluid Mechanics of Geologic Carbon Dioxide Storage. Ph.D. Thesis, Massachusetts Institute of Technology, Cambridge, MA, USA, 2013; pp. 32–38.
35. Chasset, C.; Jarsjö, J.; Erlström, M.; Cvetkovic, V.; Destouni, G. Scenario simulations of CO_2 injection feasibility, plume migration and storage in a saline aquifer, Scania, Sweden. *Int. J. Greenh. Gas Control* **2011**, *5*, 1303–1318. [CrossRef]
36. Birkholzer, J.T.; Zhou, Q.; Tsang, C. Large-scale impact of CO_2 storage in deep saline aquifers: A sensitivity study on pressure response in stratified systems. *Int. J. Greenh. Gas Control* **2009**, *3*, 181–194. [CrossRef]
37. Jikich, S.A.; Sams, W.N.; Bromhal, G.; Pope, G.; Gupta, N.; Smith, D.H. Carbon dioxide injectivity in brine reservoirs using horizontal wells. In Proceedings of the 2nd Annual Conference on Carbon Sequestration, Pittsburgh, PA, USA, 5–8 May 2003.
38. Ghaderi, S.; Keith, D.; Leonenko, Y. Feasibility of injecting large volumes of CO_2 into aquifers. *Energy Procedia* **2009**, *1*, 3113–3120. [CrossRef]
39. De Silva, P.N.K.; Ranjith, P.G. A study of methodologies for CO_2 storage capacity estimation of saline aquifers. *Fuel* **2012**, *93*, 13–27. [CrossRef]
40. Vilarrasa, V. Impact of CO_2 injection through horizontal and vertical wells on the caprock mechanical stability. *Int. J. Rock Mech. Min. Sci.* **2014**, *66*, 151–159. [CrossRef]
41. Hassanzadeh, H.; Pooladi-Darvish, M.; Keith, D. Accelerating CO_2 Dissolution in Saline Aquifers for Geological Storage—Mechanistic and Sensitivity Studies. *Energy Fuels* **2009**, *23*, 3328–3336. [CrossRef]
42. White, M.D.; Watson, D.J.; Bacon, D.H.; White, S.K.; Mcgrail, B.P.; Zhang, Z.F. *Stomp- Subsurface Transport Over Multiple Phases- STOMP- CO_2 and—CO_2e Guide*; V 1.1: 1.1–5.66; Pacific Northwest National Laboratory: Richland, WA, USA, 2013.
43. Holtz, M.H. Residual gas saturation to aquifer influx: A calculation method for 3-D computer reservoir model construction. In Proceedings of the SPE Gas Technology Symposium, Calgary, AB, Canada, 30 April–2 May 2002.
44. Khudaida, K.J. Modelling CO_2 Sequestration in Deep Saline Aquifers. Ph.D. Thesis, Loughborough University, Loughborough, UK, 2016; pp. 46–47.

45. Span, R.; Wagner, W. A new equation of state for carbon dioxide covering the fluid region from the triple-point temperature to 1100 K at pressures up to 800 MPa. *J. Phys. Chem. Ref. Data* **1996**, *25*, 1509–1596. [CrossRef]
46. Giljarhus, K.E.T.; Munkejord, S.T.; Skaugen, G. Solution of the Span–Wagner equation of state using a density–energy state function for fluid-dynamic simulation of carbon dioxide. *Ind. Eng. Chem. Res.* **2011**, *51*, 1006–1014. [CrossRef]
47. Spycher, N.; Pruess, K.; Ennis-King, J. CO_2-H_2O mixtures in the geological sequestration of CO_2. I. Assessment and calculation of mutual solubilities from 12 to 100 °C and up to 600 bar. *Geochim. Cosmochim. Acta* **2003**, *67*, 3015–3031. [CrossRef]
48. Spycher, N.; Pruess, K. A phase-partitioning model for CO_2–brine mixtures at elevated temperatures and pressures: Application to CO_2-enhanced geothermal systems. *Transp. Porous Media* **2010**, *82*, 173–196. [CrossRef]
49. Van Genuchten, M.T. A closed-form equation for predicting the hydraulic conductivity of unsaturated soils. *Soil Sci. Soc. Am. J.* **1980**, *44*, 892–898. [CrossRef]
50. Kaluarachchi, J.J.; Parker, J.C. Multiphase flow with a simplified model for oil entrapment. *Transp. Porous Media* **1992**, *7*, 1–14. [CrossRef]
51. Land, C.S. Calculation of imbibition relative permeability for two- and three-phase flow from rock properties. *Trans. Am. Inst. Min. Metall. Pet. Eng* **1968**, *243*, 149–156. [CrossRef]
52. Mualem, Y. A new model for predicting the hydraulic conductivity of unsaturated porous media. *Water Resour. Res.* **1976**, *12*, 513–522. [CrossRef]
53. White, M.D.; Oostrom, M.; Lenhard, R.J. A Practical Model for Mobile, Residual, and Entrapped NAPL in Water-Wet Porous Media. *Groundwater* **2004**, *42*, 734–746. [CrossRef] [PubMed]
54. Birkholzer, J.T.; Zhou, Q.; Rutqvist, J.; Jordan, P.; Zhang, K.; Tsang, C. *Research Project on CO2 Geological Storage and Groundwater Resources: Large-Scale Hydrogeological Evaluation and Impact on Groundwater Systems*; Report LBNL-63544; Lawrence Berkeley National Laboratory: Berkeley, CA, USA, 2007; pp. 20–31.
55. Szulczewski, M.L.; Macminn, C.W.; Juanes, R. How pressure buildup and CO_2 migration can both constrain storage capacity in deep saline aquifers. *Energy Procedia* **2011**, *4*, 4889–4896. [CrossRef]
56. Szulczewski, M.L. Storage Capacity and Injection Rate Estimates for CO_2 Sequestration in Deep Saline Aquifers in the Conterminous United States. Ph.D. Thesis, Massachusetts Institute of Technology, Cambridge, MA, USA, 2009; pp. 38–51.
57. Alabi, O.O. Fluid flow in homogeneous and heterogeneous porous media. *Electron. J. Geotech. Eng.* **2011**, *16*, 61–70.
58. Zhao, X.; Liao, X.; Wang, W.; Chen, C.; Rui, Z.; Wang, H. The CO_2 storage capacity evaluation: Methodology and determination of key factors. *J. Energy Inst.* **2014**, *87*, 297–305. [CrossRef]
59. Juanes, R.; Spiteri, E.J.; Orr, F.M.; Blunt, M.J. Impact of relative permeability hysteresis on geological CO_2 storage. *Water Resour. Res.* **2006**, *42*. [CrossRef]
60. Widarsono, B.; Muladi, A.; Jaya, I. Vertical–Horizontal Permeability Ratio in Indonesian Sanstone and Carbonate Reservoirs. In Proceedings of the Simposium Nasional IATMI, Yogyakarta, Indonesia, 25–28 July 2007.
61. Okwen, R.; Stewart, M.; Cunningham, J. Effect of well orientation (vertical vs. horizontal) and well length on the injection of CO_2 in deep saline aquifers. *Transp. Porous Media* **2011**, *90*, 219–232. [CrossRef]
62. Gonzalez-Nicolas, A.; Cody, B.; Baù, D. Numerical simulation of CO_2 injection into deep saline aquifers. In Proceedings of the AGU Hydrology Days, Fort Collins, CO, USA, 21–23 March 2011; pp. 107–113.
63. Bielinski, A. Numerical Simulation of CO_2 Sequestration in Geological Formations. Ph.D. Thesis, University of Stuttgart, Stuttgart, Germany, 2007.

 © 2020 by the authors. Licensee MDPI, Basel, Switzerland. This article is an open access article distributed under the terms and conditions of the Creative Commons Attribution (CC BY) license (http://creativecommons.org/licenses/by/4.0/).

Article

Model Development for Carbon Capture Cost Estimation

Tryfonas Pieri * and Athanasios Angelis-Dimakis

Department of Chemical Sciences, School of Applied Sciences, University of Huddersfield, Queensgate, Huddersfield HD1 3DH, UK; a.angelisdimakis@hud.ac.uk
* Correspondence: tryfonas.pieri@hud.ac.uk

Abstract: Carbon capture is the most critical stage for the implementation of a technically viable and economically feasible carbon capture and storage or utilization scheme. For that reason, carbon capture has been widely studied, with many published results on the technical performance, modelling and, on a smaller scale, the costing of carbon capture technologies. Our objective is to review a large set of published studies, which quantified and reported the CO_2 capture costs. The findings are grouped, homogenised and standardised, and statistical models are developed for each one of the categories. These models allow the estimation of the capture costs, based on the amount of CO_2 captured and the type of source/separation principle of the capture technology used.

Keywords: carbon capture; CCU; CCS; cost estimation

Citation: Pieri, T.; Angelis-Dimakis, A. Model Development for Carbon Capture Cost Estimation. *Clean Technol.* 2021, 3, 787–803. https://doi.org/10.3390/cleantechnol3040046

Academic Editors: Diganta B. Das and Sesha S. Srinivasan

Received: 30 June 2021
Accepted: 12 October 2021
Published: 20 October 2021

Publisher's Note: MDPI stays neutral with regard to jurisdictional claims in published maps and institutional affiliations.

Copyright: © 2021 by the authors. Licensee MDPI, Basel, Switzerland. This article is an open access article distributed under the terms and conditions of the Creative Commons Attribution (CC BY) license (https://creativecommons.org/licenses/by/4.0/).

1. Introduction

Global warming, i.e., the rapid and unusual increase in the earth's average surface temperature, is considered one of the major current environmental issues. It is caused by the increased amount of anthropogenic greenhouse gas emissions (carbon dioxide, methane, nitrous oxide and water), which can trap solar radiation in the form of heat. To respond to these environmental pressures, the target set by the European Union in the Roadmap for 2050 is the reduction of greenhouse gas emissions by 40% below the 1990 values by 2030, by 60% by 2040 and by 80% by 2050. The European Commission has thus defined three alternative approaches that could contribute positively towards achieving these targets: (a) wider implementation of renewable energy sources, (b) low carbon energy supply options, supported by carbon capture, and (c) energy-saving measures.

Carbon capture has been thus recognised by many as a mitigation tool for global warming. In terms of carbon capture and storage (CCS), it can reduce carbon dioxide (CO_2) emissions by capturing and storing CO_2 underground. Carbon capture and utilization (CCU) is an alternative way of reducing CO_2 emissions via recycling, by capturing CO_2 and purifying it to the required standards of industries. The purified CO_2 is transported by the available means of transportation to an industrial process to be sold for profit and reuse.

CCU value chains have not been widely commercialised yet because they face multiple technical, legislative and social barriers (e.g., utilization options, source-sink matching, lack of relevant policy and regulations, market, public acceptance, construction rate), but the most critical parameter towards their commercialization is their economic viability [1–3]. The economic components of a CCU value chain include capture cost, transportation cost, utilisation cost (which expresses the modification required in the production line of the receiver) and the profits from the carbon trading market and selling of captured CO_2. A CCU value chain is considered viable when the profits from selling captured the CO_2 and carbon trading market are higher than capture, transportation and utilization costs [4]. Estimations show that capture costs comprise 70–90% of costs of the whole value chain, making capture costs the component with the greatest importance and the critical economic barrier (or driver) in the development and commercialization of CCU value chains [5,6].

For this reason, and since capture is an integral part of CCS value chains as well, carbon capture has been given a lot of attention and many studies focused on the quantification and reporting of CO_2 capture costs. This paper will therefore aim to review such studies, where capture costs have been quantified, homogenise the approaches used and explain how their quantification can be incorporated in the optimisation of CCS or CCU value chains to facilitate its commercialisation. The paper also presents statistical models, and the methodology of developing such models, for the quantification of CO_2 capture costs using chemical absorption, physical absorption and oxyfuel combustion capture technologies.

2. Materials and Methods

To build the models for the estimation of carbon capture costs, a thorough literature review has been performed, collecting all the relevant published literature, and presented in Section 2.1. The collected data have been grouped based on the carbon source and technology used, and several models have been developed (Section 2.2).

2.1. Carbon Capture Costs

The importance of the economic viability of CCS and CCU value chains has led many authors and organisations to quantify and report them and as a result, different nomenclature and costing/reporting methods emerged [7,8]. Although there are many approaches and methods to estimate economic data, which are carried without specified boundary conditions or consistency, certain similarities exist across studies, which show consistency in the cost elements and metrics of CCS and CCU. Various previously published reviews highlighted the inconsistencies in nomenclature, costing and reporting methods and proposed a framework for the reporting of CCS and CCU cost data [7–10].

2.1.1. Components of Carbon Capture Costs

Carbon capture costs are divided into two categories; capital costs and operating and maintenance costs. Capital costs can be expressed in a number of ways where each expression covers the required costs for building and completing a project in increasing depth of detail considering more costs.

The Bare Elected Cost (BEC) of a carbon capture project is a value estimated by the contractor to complete the project and includes the cost of all the required equipment, materials and labour. BEC can be rated according to the level of detail ranging from simplified, least detailed to finalised, most detailed. BEC serves as the core for costing CCS projects as other cost elements are estimated as a percentage of this value [7,11].

The Engineering, Procurement and Construction (EPC) cost is the BEC cost increased by the cost of fees for additional engineering services, estimated as a percentage of BEC. EPC costs include direct and indirect costs related to project management, engineering, facilities, equipment and labour [7,12].

The Total Plant Cost (TPC) is a term that includes BEC, additional engineering services and contingency costs. The contingency costs of a project are included to account for the risks associated with technological maturity, performance and regulatory difficulties. Contingency costs can be estimated as a percentage of BEC or EPC according to the level of detail [7,11,12]. TPC is rarely used for reporting capital costs.

The Total Overnight Cost (TOC) equals the total plant cost increased by the owner's cost, which covers components that have not been taken into account (e.g., feasibility studies, surveys, land, insurance, permitting, finance transaction costs, pre-paid royalties, initial catalyst and chemicals, inventory capital, pre-production (start-up), other site-specific items unique to the project). The owner's costs do not include interest during construction [7,11,12]. TOC is rarely used for reporting capital costs.

Finally, the Total Capital Requirement (TCR) sometimes referred to as Total As Spent Cost (TASC) or Total Capital Cost (TCC) is the sum of all the previously mentioned costs, before including interest during construction [7,11,12]. TCR is the most common method of reporting capital costs.

Operating and maintenance costs are expressed as fixed and variable depending on whether a component has a fixed or variable cost and involve all costs of running a project. They are usually reported as a single value, but they can be broken down into more components if needed, using estimations provided from different organisations. O&M costs are estimated as a percentage of capital cost (usually between 3–15%) [7,13,14].

Fixed operating and maintenance costs are independent of plant size and consist of operating, maintenance, administrative and support labour, maintenance materials, property taxes and insurance [7]. Variable operating and maintenance costs include cost components that are directly proportional to the production of the product (usually the amount of electricity produced). Those components include fuel, other consumables, waste disposal, by-product sales or emissions tax [7].

2.1.2. Carbon Capture Metrics

A series of metrics have been introduced to express the economic viability of a carbon capture plant investment and allow the comparison between different configurations.

The Levelized Cost of Electricity (LCOE) is a metric that is widely applied but can only be applied to power producing industries. It expresses the price that electricity should be sold, for the capture investment to be profitable, based on a specified return on investment (ROI) and project lifetime. It is estimated by incorporating all expenses related to producing a certain amount of electricity per year, for a specified project lifetime and ROI [7,9,15]. Therefore, it serves as an indicator for the potential profitability of a specific project and allows comparison between projects with different plant sizes and electricity generation technologies, assuming project lifetime and ROI are the same and TCR costs are estimated in similar ways. Another similar metric is the first-year cost of electricity, which is identical to LCOE with the only difference that inflation rates and cost escalation rates are assumed to be zero for the first year of operation [7,9].

The cost of CO_2 avoided quantifies the average cost of avoiding a unit of CO_2 per unit of useful product by comparing a plant with capture to a reference plant of similar type and size, without a capture unit. This metric is equal to the CO_2 emission tax for which the cost of producing a unit of product for a plant without capture is equal to the same cost of a plant with capture and includes costs of capture (including transportation and storage/utilization, otherwise CO_2 will not be avoided) [7,9,10]. The cost of CO_2 captured is a similar indicator with the only difference that it covers only the cost of capturing and producing CO_2 as a chemical product, and unlike the cost of CO_2 avoided metric, it excludes transportation and storage/utilization. Both indicators are expressed in monetary units per tCO_2 [1,2,10].

The cost of CO_2 abated quantifies the cost of minimising CO2 emissions by changing the process of producing electricity, i.e., by modifying the process in any way, changing generators, fuel, region, country, and utility system, anything that changes the current situation to one with lower CO_2 emissions including CCS [10]. The energy penalty metric expresses the power output difference between a power plant with carbon capture and a similar reference power plant without capture.

From all the above-mentioned metrics, the only metric that quantifies the cost of CO_2 capture and is appropriate for our study is the "cost of CO2 captured". It includes only the stage of capturing and excludes the transportation and any potential storage or utilization.

2.2. Estimation of Carbon Capture Costs

The objective of this manuscript is to develop models for the estimation of capture costs (TCR and O&M), by extracting the relevant data from the published studies. These models will have as parameters the amount of CO_2 captured, the capture technology used and the carbon source type.

2.2.1. Data Collection

For that purpose, a literature review was performed to collect the cost of CO_2 captured from published studies. Table 1 summarises the studies used in the analysis and the parameters retrieved from each one of them. The data required for the analysis are the capital cost of capture (TCR, TOC or TPC), annual O&M costs and the annual amount of CO_2 captured. The capture cost elements reported usually include the capital costs of both the base plant and the CO_2 capture plant and, if a reference case is provided, then the capital cost of capture can be estimated by subtracting the cost of the reference case from the cost with capture. Annual O&M costs, if not reported, can be estimated by assuming to be equal to a percentage of capital costs. The annual amount of CO_2 captured also needs to be reported, which can be used to adjust the cost of capture based on the required size of the capture plant.

Table 1. Summary of Literature Sources Characteristics.

Parameters	[10]	[16]	[2]	[17]	[11]	[18]	[19]	[1]	[12]	[20]	[21]
Source type	NPR	PR	PR	PR	Both	PR	PR	Both	PR	PR	NPR
Separation principle	Y	Y	Y	Y	Y	Y	Y	N	Y	Y	N
Compression	Y	Y	Y	Y	Y	Y	N	N	Y	Y	Y
Amount of CO_2 captured	E	Y	Y	N	Y	E	N	N	E	Y	Y
Currency	Y	Y	Y	Y	Y	Y	Y	Y	Y	Y	Y
Year	Y	Y	Y	Y	Y	Y	Y	Y	Y	Y	N
Constant/Current	Co	Co	Co	Co	Co	N	N	Co	N	Co	Cu
Project lifetime	Y	Y	N	Y	Y	N	N	N	Y	Y	Y
Annual Working Hours	Y	Y	Y	Y	Y	Y	N	N	A	Y	Y
Reference Plant Capacity	N	Y	Y	Y	N	N	N	N	Y	Y	N
Reference Capital cost	N	N	Y	Y	N	Y	N	N	Y	Y	Y
Reference O&M cost	N	N	N	Y	N	Y	N	N	Y	Y	Y
With Capture Plant Capacity	Y	Y	Y	Y	N	N	Y	N	Y	Y	N
With Capture Capital cost	Y	Y	Y	Y	Y	Y	N	Y	Y	Y	Y
With Capture O&M cost	Y	Y	N	Y	Y	Y	N	N	Y	Y	Y
Cost of CO_2 captured	N	N	Y	N	N	N	N	N	N	N	N
Cost of CO_2 avoided	N	Y	Y	N	N	N	N	N	Y	Y	N
LCOE	N	Y	Y	N	N	Y	Y	N	Y	Y	N
Parameters	[22]	[23]	[24]	[25]	[15]	[26]	[13]	[27]	[28]	[29]	[30]
Source type	PR	PR	PR	PR	PR	NPR	PR	PR	Both	PR	Both
Separation principle	Y	Y	N	N	Y	Y	Y	N	Y	N	Y
Compression	Y	N	N	Y	Y	Y	Y	Y	N		Y
Amount of CO_2 captured	E	Y	N	Y	Y	Y	Y	Y	N		Y
Currency	Y	Y	Y	Y	Y	Y	Y	Y	Y		Y
Year	Y	Y	N	Y	Y	N	Y	Y	Y		Y
Constant/Current	Cu	Y	N	Co	Co	Co	N	Cu	N		Co
Project lifetime	N	Y	Y	N	Y	Y	Y	Y	N		N
Annual Working Hours	A	N	Y	N	N	N	N	N	N		N
Reference Plant Capacity	Y	Y	N	N	Y	N	Y	N	N	Y	Y
Reference Capital cost	Y	Y	N	N	Y	Y	Y	Y	N	Y	Y
Reference O&M cost	Y	Y	N	N	A	Y	N	N	N	N	N
With Capture Plant Capacity	Y	Y	N	Y	Y	Y	Y	Y	Y	Y	Y
With Capture Capital cost	Y	Y	Y	Y	Y	Y	Y	Y	N	Y	Y
With Capture O&M cost	Y	Y	N	Y	A	Y	N	N	N	N	N
Cost of CO_2 captured	Y	N	N	N	N	Y	Y	N	Y	Y	Y
Cost of CO_2 avoided	Y	N	N	N	Y	Y	Y	N	Y	Y	Y
LCOE	Y	N	N	N	Y	N	N	N	N	Y	Y

From the parameters reported in Table 1, the "source type" indicates whether the published study included data for power related sources, non-power related sources or both. The separation principle determines if the type of capture technology used was specified. The amount of CO_2 capture indicates if the annual amount of CO_2 captured is directly reported or if it can be estimated. Currency, base year, constant/current, project lifetime and working hours per year are the data required to standardise cost. The base year specifies the year of the reported costs, whereas constant or current indicates if inflation is included. The cost of CO_2 captured, cost of CO_2 avoided and LCOE are not required for this study, but they were included as reference values.

2.2.2. Data Standardisation

The extracted cost data were standardised to constant USD_{2018} prices [31]. The method adjusts for inflation of the reported currencies to 2018 prices using local CPI values and then converting currencies to USD_{2018} using market exchange rate data from [32]. The selected studies usually report cost data in US dollars, British pounds or euro. A base year was always provided, and it was specified whether inflation was included. For US dollars and British pounds, it was easy to find CPI values which are based on location because it is a country-based index, but for costs reported in Euro, it was not possible to find CPI values because there was no indication of country. Instead, the costs were converted to USD of the base year and then adjusted for inflation to USD_{2018} using the USD CPI. Cost data in current values were standardised for inflation of currencies to 2018 prices using local CPI values by dividing the CPI_{2018} by 100 and multiplying by the current price and then converting currencies to USD_{2018} using market exchange rate data from [32].

2.2.3. Data Aggregation

The extracted data include information on the source type and capture technology and ideally a model can be developed for each type of source using all available capture technologies. Because of the lack of data, this was not possible for all of them. Instead of developing a model for each capture technology applied on every source the data were sorted per source type and capture technology, and a model was developed for each source and capture technology (where enough data existed).

The extracted data were sorted according to source type for non-power related sources (metal industry, fluid catalytic cracking (FCC), combined stack, cement industry, hydrogen, ammonia, ethylene oxide production and synthetic fuel) and power-related sources (pulverised coal (PC), integrated gasification combined cycle (IGCC), supercritical pulverised coal (SCPC), ultra-supercritical pulverised coal (USCPC), natural gas combined cycle (NGCC), gas-fired furnaces, combined heating and power station (CHP), fluidised bed combustion (CFB)). To develop a model for each case, it was required to have at least 10 data points, which at the same time cover a reasonable range of flowrates (at least 2 Mt_{CO_2}/yr). The cost data from each source were also sorted according to the classification of capture technologies per separation principle, chemical absorption, physical absorption, oxyfuel combustion, chemical adsorption, chemical looping, cryogenics, inorganic membranes and hydrate crystallization). The criteria used to be able to develop a model were the same with the sorting per source. Therefore, it was not possible to develop models for chemical adsorption, chemical looping, cryogenics, inorganic membranes and hydrate crystallization CO_2 capture technologies because there are less than 10 pieces of data for each of them that cover a very small range of flows.

3. Results

Numerous regression analyses were performed and assessed to develop a model that best describes the investment cost and O&M cost based on the amount of CO_2 captured. Power regression analysis was chosen to develop a model that predicts the total capital requirement (TCR) and annual operating and maintenance cost (O&MC) (dependent variables) based on the annual amount of CO_2 captured (independent variable). The proposed models are presented in the following sections.

3.1. Capture Costs Based on the Separation Principle

Data were split based on the separation principle used in each case into three categories; chemical absorption, physical absorption and oxyfuel combustion. Regarding chemical absorption, many data points covered a decent range of flowrates between 0–6.7 Mt_{CO_2}/yr. This data group includes data from various CO_2 sources. The analysis showed that the curve which best described the data has the form of $y = ax^b$ where a and b are constants calculated from regression (Figures 1 and 2).

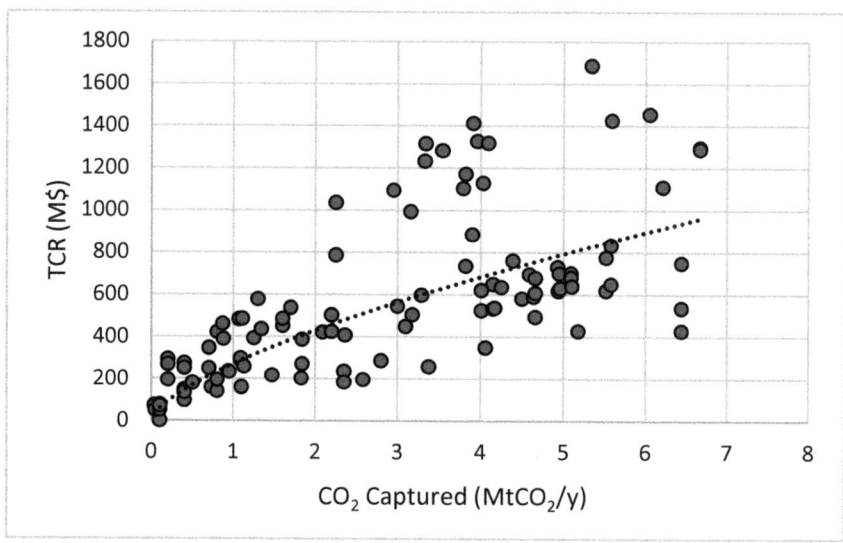

Figure 1. Estimation of TCR based on the amount of CO_2 captured for chemical absorption.

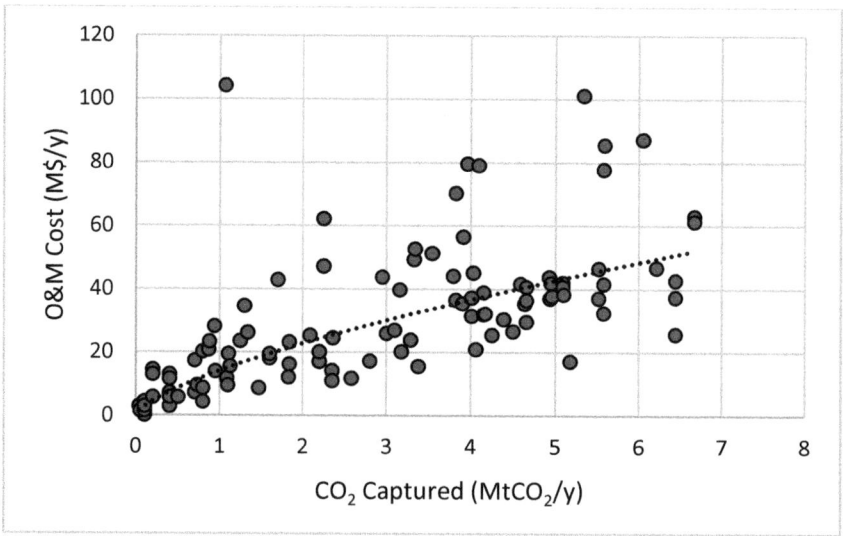

Figure 2. Estimation of O&M cost based on the amount of CO_2 captured for chemical absorption.

For physical absorption, the data points covered a decent range of flowrates between 0–6.4 Mt_{CO_2}/yr and included data from various non-power related CO_2 sources like the metal industry, cement industry, chemical and petrochemical industry and only IGCC from power-related sources. The analysis showed a curve of the shape of $y = ax^b$ where a and b are constants calculated from regression (Figures 3 and 4).

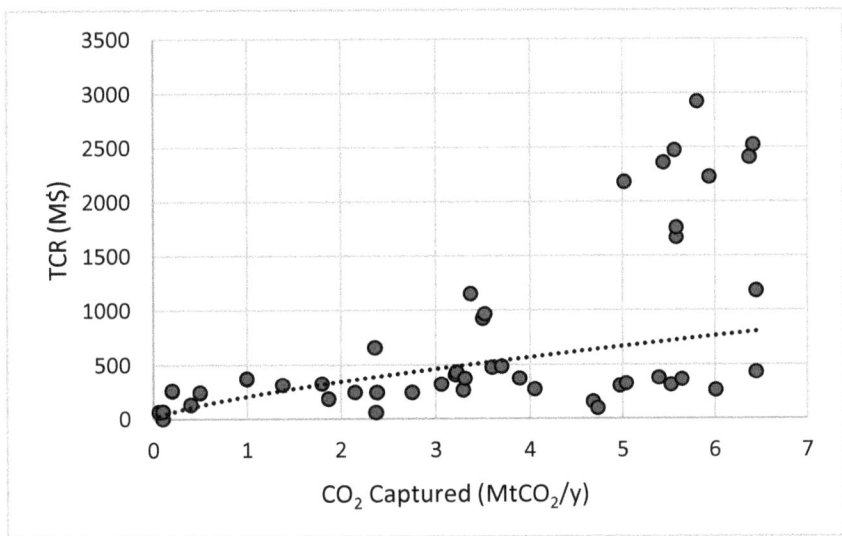

Figure 3. Estimation of TCR based on the amount of CO_2 captured for physical absorption.

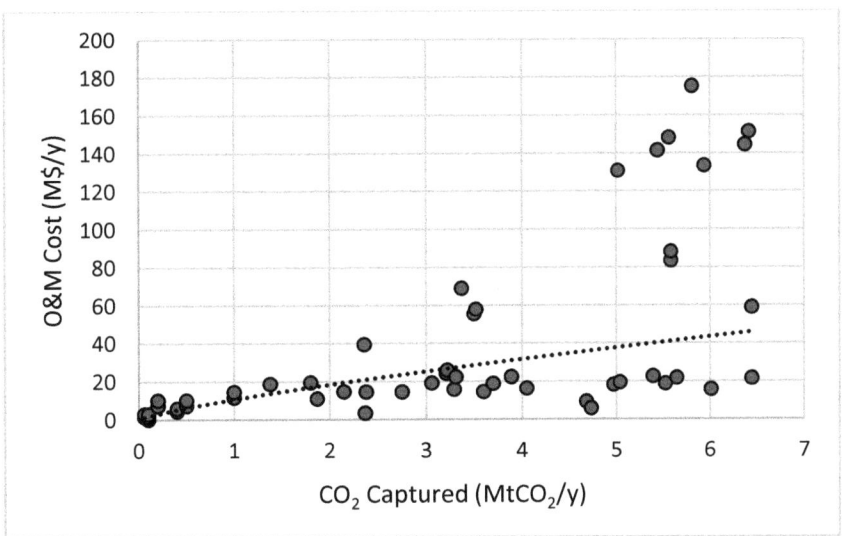

Figure 4. Estimation of O&M cost based on the amount of CO_2 captured for physical absorption.

For the oxy-fuel combustion capture technology, there were fewer data points, still covering a decent range of flowrates between 0–6 Mt_{CO_2}/yr. This data group includes data from sources that include combustion like SCPC, USCPC, CFB, PC&NGCC and chemical and petrochemical industry and cement industry. The power model that was proposed by the regression analysis is presented in Figures 5 and 6.

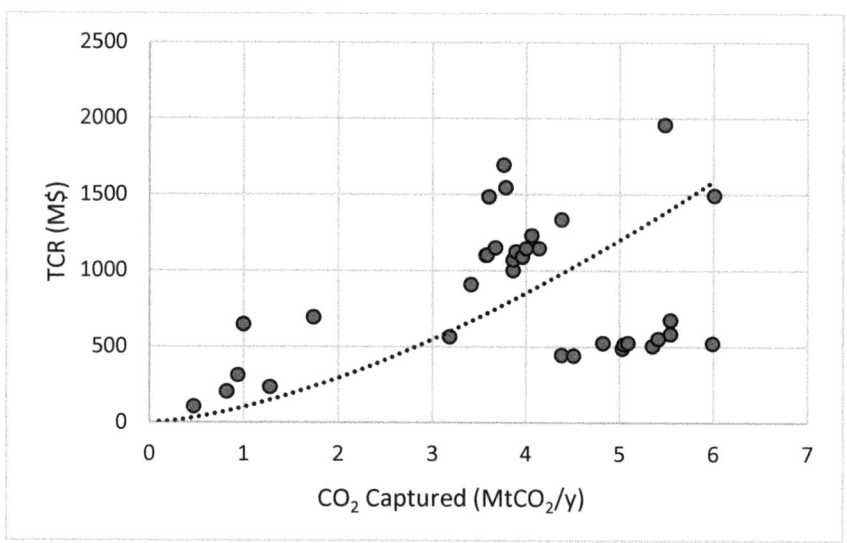

Figure 5. Estimation of TCR based on the amount of CO_2 captured for oxy-fuel combustion.

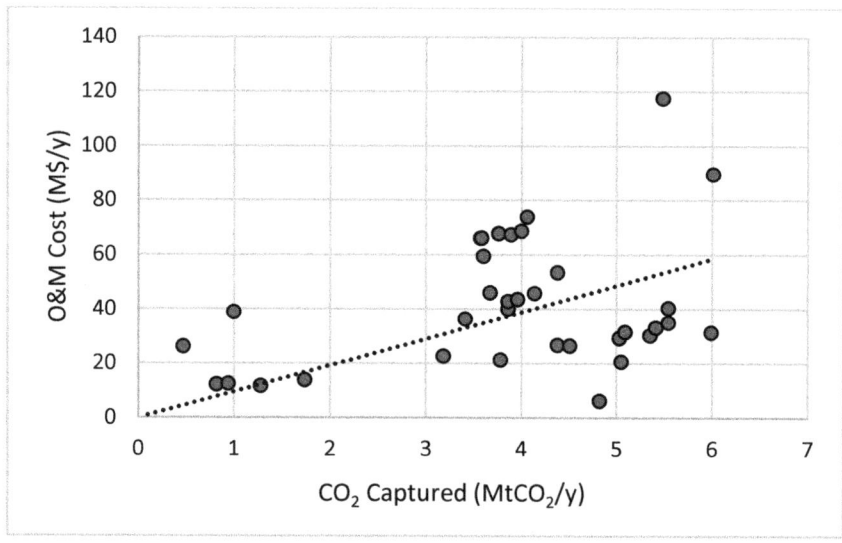

Figure 6. Estimation of O&M cost based on the amount of CO_2 captured for oxy-fuel combustion.

3.2. Capture Costs Based on the Source Type

It was also decided to categorise the data points based on the type of the source and, for those sources who had an appropriate number of data points, specify the model that best described their profile.

3.2.1. Metal Industry

The different CO_2 sources of the metal industry, including blast furnace, top gas recycling blast furnace (TGRBF), smelting reduction iron and raw material production, were all grouped under the metal industry because there were not enough data to develop a model for each one individually. There were 20 points of data that covered a range of flowrates between 0–6.4 Mt_{CO_2}/yr. Some data points are stacked, because the authors of the references cited, wanted to compare different capture technologies applied on the same CO_2 source. This data group includes various capture technologies like chemical absorption, physical absorption, inorganic porous membranes, physical adsorption, calcium looping and cryogenics. The power model that was proposed by the regression analysis is presented in Figures 7 and 8.

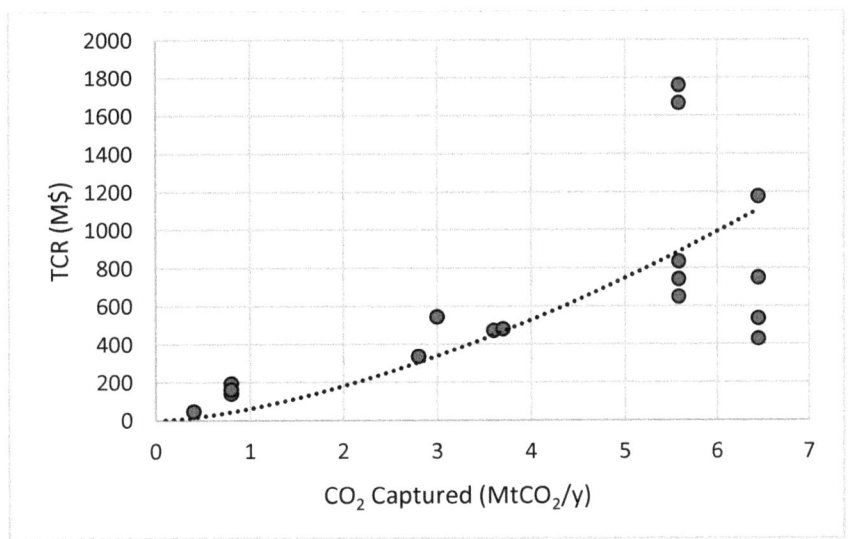

Figure 7. Estimation of TCR based on the amount of CO_2 captured for metal industry.

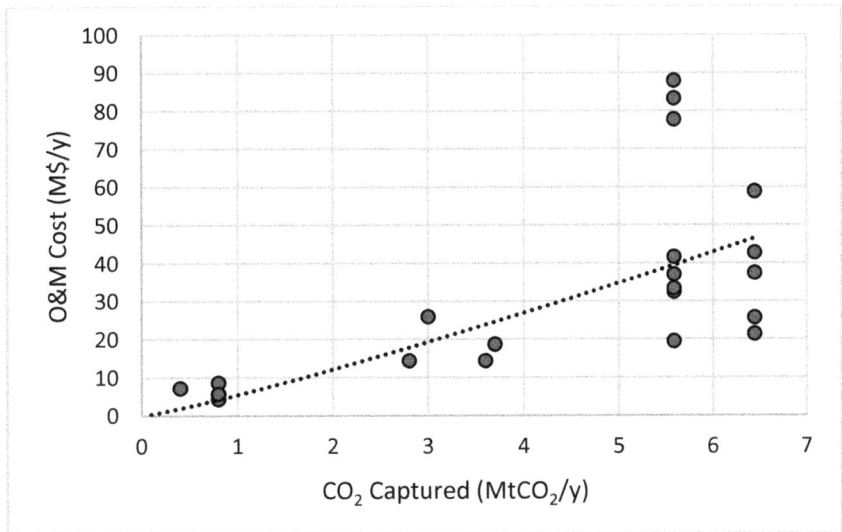

Figure 8. Estimation of O&M cost based on the amount of CO_2 captured for the metal industry.

3.2.2. Cement Industry

For the cement industry, data from the pre-calciner and the entire cement plant were grouped together, because there were not enough data to develop a model for each one individually. There were 13 points of data that covered flowrates between 0–1.4 Mt_{CO_2}/y although some data points are stacked. This data group includes various capture technologies like oxy-fuel combustion, chemical looping, chemical absorption, physical absorption and cryogenics. The power model that was proposed by the regression analysis is presented in Figures 9 and 10

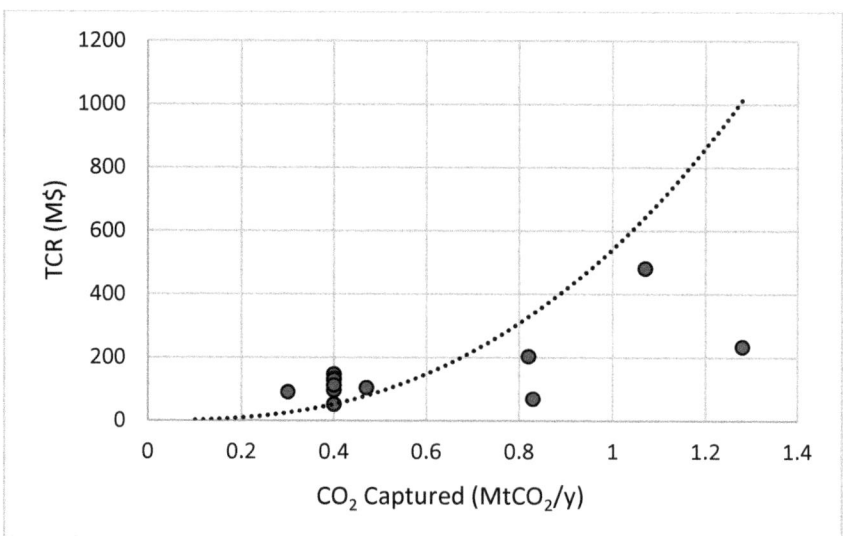

Figure 9. Estimation of TCR based on the amount of CO_2 captured for the cement industry.

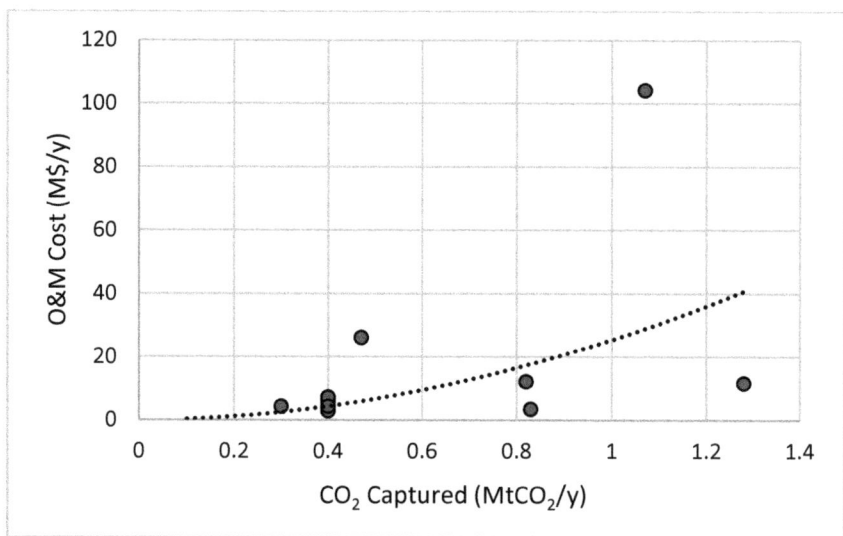

Figure 10. Estimation of O&M cost based on the amount of CO_2 captured for the cement industry.

3.2.3. Fluid Catalytic Cracking (FCC)

There were 18 points of data for FCC that covered flowrates between 0–1 Mt_{CO_2}/y. The range is relatively small, but representative of the size of the source, when compared to power related ones. The power model that was proposed by the regression analysis is presented in Figures 11 and 12.

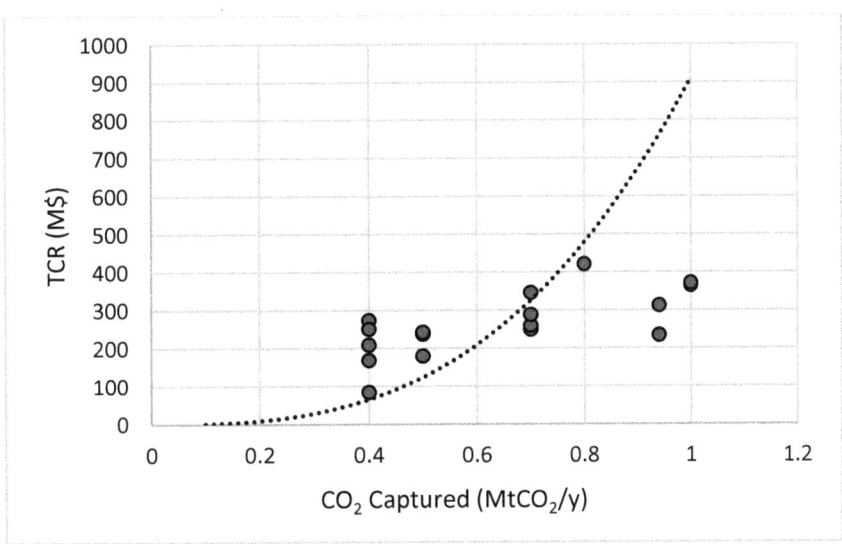

Figure 11. Estimation of TCR based on the amount of CO_2 captured for FCC.

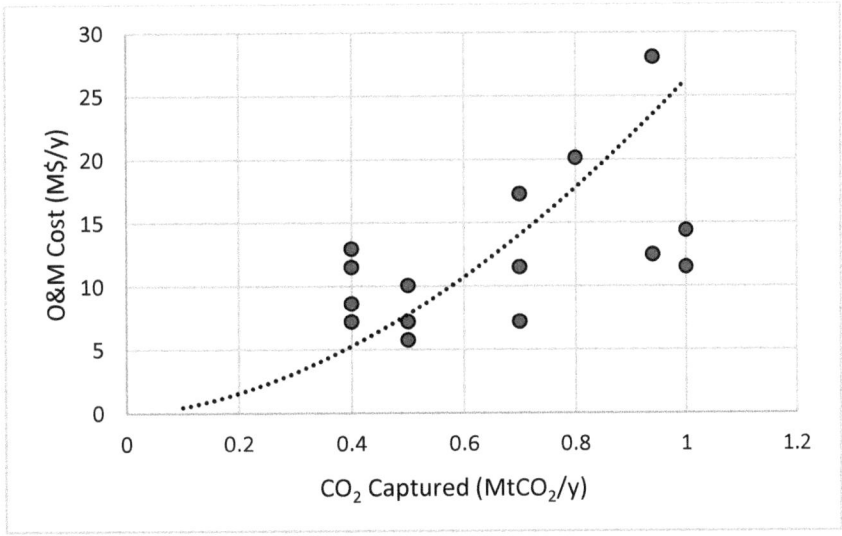

Figure 12. Estimation of O&M cost based on the amount of CO_2 captured for FCC.

3.2.4. Power Related Sources

The power-related sources were the category with the most available data. There were 57 data points for IGCC, 65 for SCPC, 23 points of data for NGCC and 16 for USCPC. All of them covered a range greater than 4.5 Mt_{CO_2} (from 0–4.5 Mt_{CO_2} to 0–6.7 Mt_{CO_2}). For IGCC, the data set includes only physical absorption capture technology, mainly (selexol), whereas for SCPC includes chemical absorption, oxy-fuel combustion and gas separation membranes capture technologies. For NGCC the data set include only chemical absorption (mostly MEA) and for USCPC it combines chemical absorption and oxy-fuel combustion capture technologies. The power model that was proposed by the regression analysis is presented in Figures 13–20.

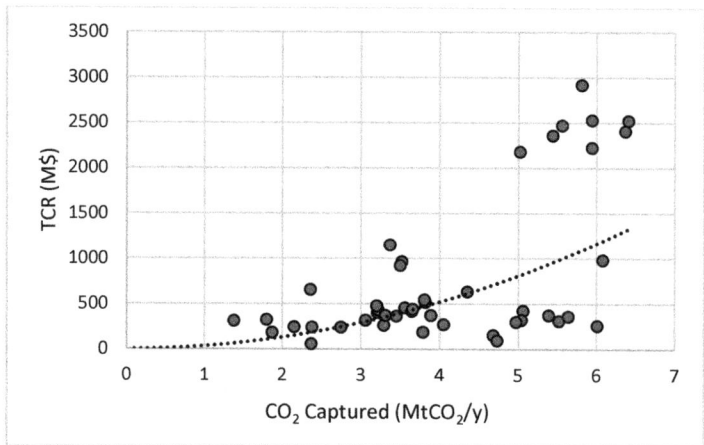

Figure 13. Estimation of TCR based on the amount of CO_2 captured for IGCC.

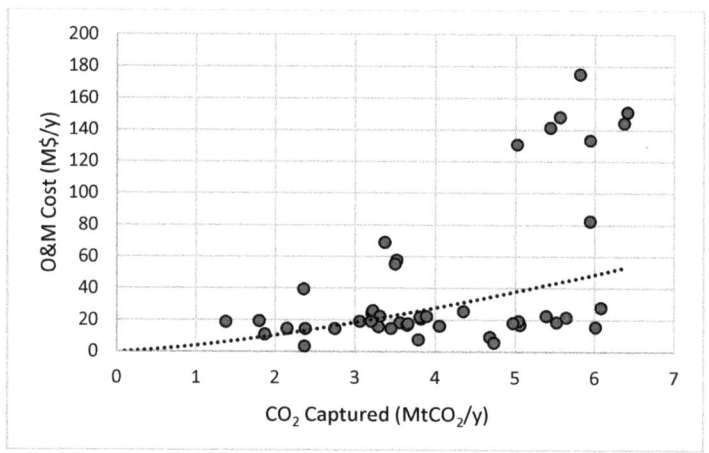

Figure 14. Estimation of O&M cost based on the amount of CO_2 captured for IGCC.

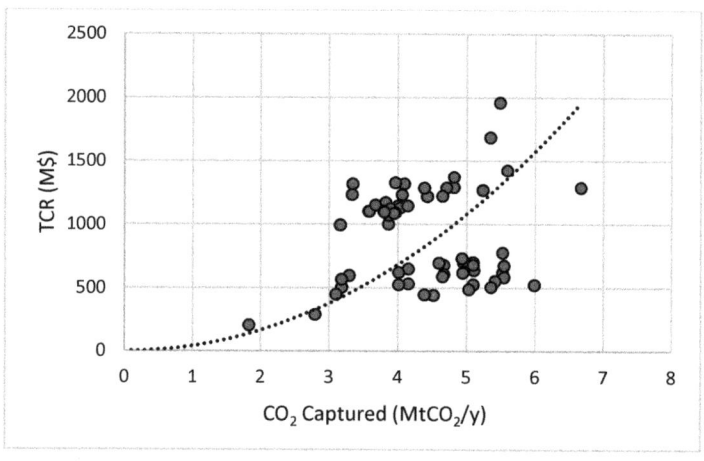

Figure 15. Estimation of TCR based on the amount of CO_2 captured for SCPC.

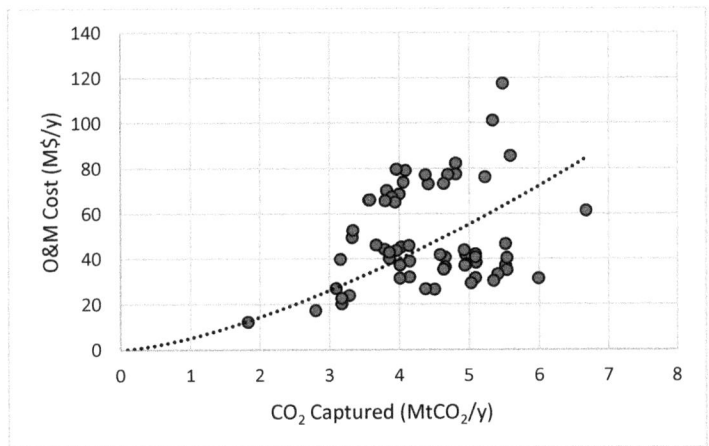

Figure 16. Estimation of O&M cost based on the amount of CO_2 captured for SCPC.

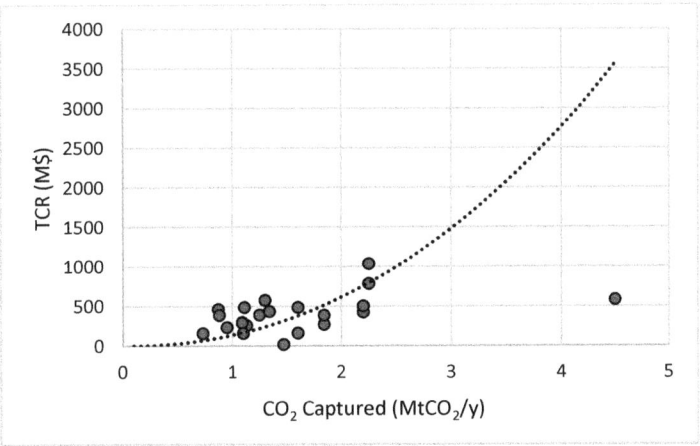

Figure 17. Estimation of TCR based on the amount of CO_2 captured for NGCC.

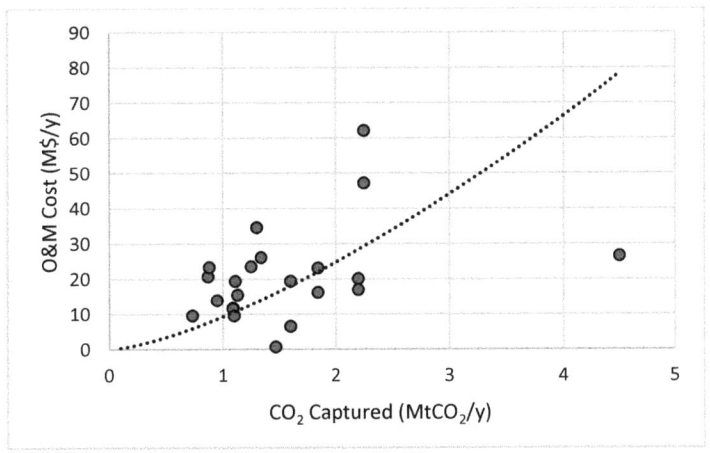

Figure 18. Estimation of O&M cost based on the amount of CO_2 captured for NGCC.

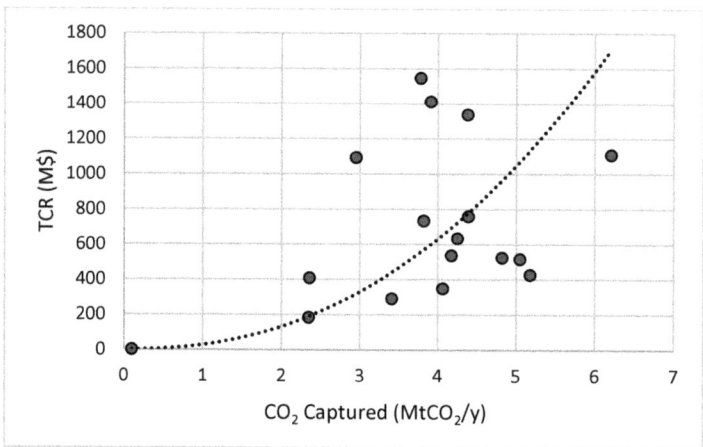

Figure 19. Estimation of TCR based on the amount of CO_2 captured for USCPC.

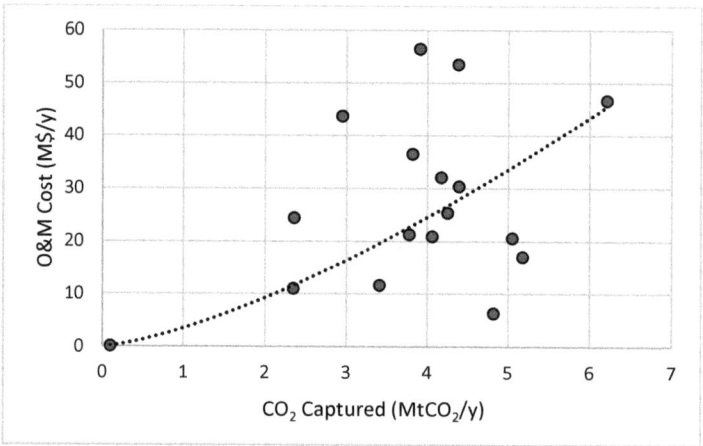

Figure 20. Estimation of O&M cost based on the amount of CO_2 captured for USCPC.

3.3. Model Validation

The regression analysis in all cases was forced to go through (0,0) because the cost to capture zero amount of CO_2 is zero. Since some of the models are not valid for (0,0), the data point (0.1, 0.1) was used instead. The model that best fitted the data in all cases was a power model, $y = ax^b$, where a and b are constants calculated from regression. Further analysis was carried out to the obtained model to determine its statistical characteristics like R^2 value and the p-value. The R^2 value which indicates the accuracy of the model is high in all cases with the lowest being 0.415 and the highest 0.908 and shows that most of the models would produce accurate predictions. The p-value, which signifies the statistical significance of the model, is significantly small in all cases and allows to demonstrate that the model is statistically significant by rejecting the null hypothesis.

The last characteristic "trend" indicated if the model follows economies of scale or reverse economies of scale. Economies of scale is a term that relates the cost of production to the amount produced. If economies of scale are followed, then a product would cost less if the production was increased. However, economies of scale are only observed up to a certain point. When that point is passed, reverse economies of scale describe the process. In the carbon capture case, if the amount of CO_2 captured is doubled and the cost less than doubles, then economies of scale are observed. In any other case, reverse economies of scale are followed. All data are summarised in Tables 2 and 3.

Table 2. Statistical analysis of TCR and O&M costs models per separation principle.

Type	Cost	Model	R^2	p-Value	Trend
Chemical Absorption	TCR	$y = 270.3x^{0.668}$	0.531	<0.001	Less than doubles
	O&M	$y = 14.089x^{0.690}$	0.669	<0.001	Less than doubles
Physical Absorption	TCR	$y = 206.2x^{0.731}$	0.415	<0.001	Less than doubles
	O&M	$y = 10.698x^{0.781}$	0.584	<0.001	Less than doubles
Oxyfuel Combustion	TCR	$y = 101.9x^{1.533}$	0.652	<0.001	More than doubles
	O&M	$y = 9.503x^{1.014}$	0.650	<0.001	More than doubles

Table 3. Statistical analysis of TCR and O&M costs models per source type.

Type	Cost	Model	R^2	p-Value	Trend
Metal Industry	TCR	$y = 61.629x^{1.550}$	0.777	<0.001	More than doubles
	O&M	$y = 5.464x^{1.151}$	0.817	<0.001	More than doubles
Cement Industry	TCR	$y = 543.5x^{2.538}$	0.637	<0.001	More than doubles
	O&M	$y = 25.382x^{1.913}$	0.637	<0.001	More than doubles
FCC	TCR	$y = 908.8x^{2.875}$	0.711	<0.001	More than doubles
	O&M	$y = 26.265x^{1.754}$	0.668	<0.001	More than doubles
IGCC	TCR	$y = 32.931x^{1.990}$	0.696	<0.001	More than doubles
	O&M	$y = 3.889x^{1.412}$	0.557	<0.001	More than doubles
SCPC	TCR	$y = 39.333x^{2.060}$	0.774	<0.001	More than doubles
	O&M	$y = 5.152x^{1.475}$	0.745	<0.001	More than doubles
NGCC	TCR	$y = 137.4x^{2.165}$	0.644	<0.001	More than doubles
	O&M	$y = 9.276x^{1.419}$	0.517	<0.001	More than doubles
USCPC	TCR	$y = 27.436x^{2.263}$	0.908	<0.001	More than doubles
	O&M	$y = 3.488x^{1.407}$	0.800	<0.001	More than doubles

4. Discussion

All models have been analysed for the type of economies they follow in the long term, by observing the effect on costs when the amount of CO_2 captured is doubled. If costs exactly double, then a linear relationship is observed, if the costs are more than double then reverse economies of scale are observed and if the costs are less than double then economies of scale are observed [33]. All models show to follow reverse economies of scale except for the models of chemical absorption and physical absorption, where economies of scale are observed. This is sensible because, when more CO_2 needs to be captured, then bigger capture equipment would be required to accommodate that flow. On first thought, this would justify only the reverse economies of scale for capital costs for the larger equipment and larger quantities of solvents, but it also justifies the O&M cost models following reverse economies of scale too because more energy would be required to regenerate the additional amount of solvent that is now required.

The required data used for the development of the models introduced some limitations to the models in the sense that the data used were gathered from various sources and standardised to process and use them. Capital costs were reported in different currencies and sometimes these were not expressed using the same metric. The reported capital costs did not always include the same cost components. Additionally, CO_2 compression is most of the time, but not always, included. Costs were reported in various currencies and base years. O&M costs were not always reported in the amount of money per year and had to be converted using assumptions. In other cases, they were not directly reported but instead reported as estimates using percentages of capital cost. The annual amount of CO_2 captured was not always reported and sometimes had to be estimated from plant capacity and annual working hours.

Nevertheless, the developed models are valid equations that only require one parameter (the annual amount of CO_2 to be captured) to estimate the capital cost of capture and the annual O&M cost of capture. In that sense, they can be used by industries, to estimate the total cost of carbon capture plants, having in mind the limitations. In terms of data validation, the models have been applied in real industrial cases as part of a pre-feasibility assessment of a potential carbon capture investment. The major observations from the application are that the models might not provide an accurate estimate in the boundary regions. Moreover, in some cases/categories the sources analysed have varying purities, thus requiring a different level of purification before being captured, which affects the overall capture costs, and thus the accuracy of the models.

Although the extracted data differed slightly and were standardised to allow for the development of the models, the models can provide robust, accurate estimations, with statistical significance. This allows for the cost estimation for any CO_2 source or any one of the most widely used capture technologies.

Author Contributions: Conceptualization, A.A.-D. and T.P.; methodology, T.P.; software, T.P.; validation, A.A.-D. and T.P.; formal analysis, T.P.; investigation, T.P.; resources, T.P.; data curation, T.P.; writing—original draft preparation, T.P.; writing—review and editing, A.A.-D.; visualization, A.A.-D. and T.P.; supervision, A.A.-D. All authors have read and agreed to the published version of the manuscript.

Funding: This research received no external funding.

Conflicts of Interest: The authors declare no conflict of interest.

References

1. Budinis, S.; Krevor, S.; Mac Dowell, N.; Brandon, N.; Hawkes, A. An assessment of CCS costs, barriers and potential. *Energy Strategy Rev.* **2018**, *22*, 61–81. [CrossRef]
2. Rubin, E.S.; Davison, J.E.; Herzog, J.H. The cost of CO_2 capture and storage. *Int. J. Greenh. Gas Control* **2015**, *40*, 378–400. [CrossRef]
3. Wu, N.; Parsons, J.E.; Polenske, K.R. The impact of future carbon prices on CCS investment for power generation in China. *Energy Policy* **2013**, *54*, 160–172. [CrossRef]
4. Xing, Y.; Ping, Z.; Xian, Z.; Lei, Z. Business model design for the carbon capture utilisation and storage (CCUS) project in China. *Energy Policy* **2018**, *121*, 519–533.
5. Spigarelli, B.P.; Kawatra, K.S. Opportunities and challenges in carbon dioxide capture. *J. CO_2 Util.* **2013**, *1*, 69–87. [CrossRef]
6. Wang, M.; Lawal, A.; Stephenson, P.; Sidders, J.; Ramshaw, C. Post-combustion CO_2 capture with chemical absorption: A state of the art review. *Chem. Eng. Res. Des.* **2011**, *89*, 1609–1624. [CrossRef]
7. Rubin, E.S.; Short, C.; Booras, G.; Davison, J.; Ekstrom, C.; Matuszewski, M.; McCoy, S. A proposed methodology for CO_2 capture and storage cost estimates. *Int. J. Greenh. Gas. Control* **2013**, *17*, 488–503. [CrossRef]
8. Hu, B.; Zhai, H. The cost of carbon capture and storage for coal-fired power plants in China. *Int. J. Greenh. Gas. Control* **2017**, *65*, 23–31. [CrossRef]
9. Rubin, E.S.; Booras, G.; Davison, J.; Ekstrom, C.; Matuszewski, M.; McCoy, S.; Short, C. *Towards a Common Method of Cost Estimation for CO_2 Capture and Storage at Fossil Fuel Power Plants*; Global CCS Institute: Canberra, Australia, 2013.
10. Rubin, E.S. Understanding the pitfalls of CCS cost estimates. *Int J. Greenh. Gas. Control* **2012**, *10*, 181–190. [CrossRef]
11. Element Energy. *Demonstrating CO_2 Capture in the UK Cement, Chemical, Iron and Steel and Oil Refining Sectors by 2012, A Techno-Economic Study*; Element Energy: Cambridge, UK, 2014.
12. Finkerath, M. *Cost and Performance of Carbon Dioxide Capture from Power Generation*; IEA: Paris, France, 2011.
13. SNC Lavalin. *Impact of Impurities on CO_2 Capture, Transport and Storage*; IEA: Paris, France, 2004.
14. Kuramochi, T.; Ramirez, A.; Turkenburg, W.; Faaij, A. A Comparative assessment of CO_2 capture technologies for carbon-intensive industrial processes. *Prog. Energy Combust.* **2012**, *38*, 87–112. [CrossRef]
15. Davison, J.; Mancuso, L.; Ferrari, N. Costs of CO_2 capture technologies in coal-fired power plants. *Energy Proced.* **2014**, *63*, 7598–7607. [CrossRef]
16. Porter, R.T.J.; Fairweather, M.; Kolster, C.; Mac Dowell, N.; Shah, N.; Woolley, R.M. Cost and performance of some carbon capture technology options for producing different quality CO_2 product streams. *Int. J. Greenh. Gas. Control* **2017**, *57*, 185–195. [CrossRef]
17. Kuramochi, T.; Ramirez, A.; Turkenburg, W.; Faaij, A. Techno-economic prospects for CO_2 capture from distributed energy systems. *Renew. Sus Energy Rev.* **2013**, *19*, 328–347. [CrossRef]
18. David, J.; Herzog, H. *The Cost of Carbon Capture*; MIT: Cambridge, MA, USA, 2011.
19. Adams, T.A.; Hoseinzade, L.; Bhaswanth Madabhushi, P.; Okeke, I.J. Comparison of CO_2 capture approaches for fossil-based power generation: Review and meta-study. *Processes* **2017**, *5*, 44. [CrossRef]
20. ZEP. *The Costs of CO2 Capture: Post-Demonstration CCS in the EU*; ZEP: Brussels, Belgium, 2011.
21. IEA GHG. *CO_2 Capture in Cement Industry*; IEA Greenhouse Gas R&D Program: Cheltenham, UK, 2008.
22. NETL. *Carbon Dioxide Capture Handbook*; National Energy Technology Laboratory: Albany, OR, USA, 2015.

23. Irlam, L. *Global Costs of Carbon Capture and Storage*; Global CCS Institute: Canberra, Australia, 2017.
24. Hendriks, G.; Crijns-Graus, W.H.J.; van Bergen, F. *Global Carbon Dioxide Storage Potential and Costs*; ECOFYS: Utrecht, The Netherlands, 2004.
25. Rotterdam Climate Initiative. *CO_2 Capture, Transport and Storage in Rotterdam*; DCMR: Schiedam, The Netherlands, 2009.
26. Barker, D.J.; Turner, S.A.; Napier-Moore, P.A.; Clark, M.; Davison, J.E. CO_2 Capture in the cement industry. *Energy Procedia* **2009**, *1*, 87–94. [CrossRef]
27. EPRI. *Program on Technology Innovation: Integrated Generation Technology Options*; EPRI: Washington, DC, USA, 2011.
28. IEA. *CO_2 Capture and Storage*; IEA: Paris, France, 2010.
29. NETL. *Current and Future Technologies for Natural Gas Combined Cycle (NGCC) Power Plants*; NETL: Albany, OR, USA, 2013.
30. IPCC. *Carbon Dioxide Capture and Storage: Capture of CO_2*; Cambridge University Press: Cambridge, UK; New York, NY, USA, 2005.
31. Turner, H.C.; Lauer, J.A.; Tran, B.X.; Teerawattananon, Y.; Jit, M. Adjusting for inflation and currency changes within health economic studies. *Value Health* **2019**, *22*, 1026–1032. [CrossRef] [PubMed]
32. Consumer Price Index (2010=100) United States. Available online: https://data.worldbank.org/indicator/FP.CPI.TOTL?end=2018&locations=US&start=1995&view=chart (accessed on 25 April 2020).
33. Knoope, M.M.J. Costs, Safety and Uncertainties of CO_2 Infrastructure Development. Ph.D. Thesis, University of Utrecht, Utrecht, The Netherlands, 4 September 2015.

Article

Techno-Economic Assessment of IGCC Power Plants Using Gas Switching Technology to Minimize the Energy Penalty of CO_2 Capture

Szabolcs Szima [1], Carlos Arnaiz del Pozo [2], Schalk Cloete [3,*], Szabolcs Fogarasi [1], Ángel Jiménez Álvaro [2], Ana-Maria Cormos [1], Calin-Cristian Cormos [1] and Shahriar Amini [3]

[1] Faculty of Chemistry and Chemical Engineering, Babes-Bolyai University, 400028 Cluj Napoca, Romania; szabolcs.szima@gmail.com (S.S.); fogarasi.szabolcs@yahoo.com (S.F.); ana.cormos@ubbcluj.ro (A.-M.C.); cormos@chem.ubbcluj.ro (C.-C.C.)

[2] Departamento de Ingeniería Energética, Escuela Técnica Superior de Ingenieros Industriales (ETSII), Universidad de Politécnica de Madrid, c/José Gutiérrez Abascal n°2, 28006 Madrid, Spain; cr.arnaiz@upm.es (C.A.d.P.); a.jimenez@upm.es (Á.J.Á.)

[3] Flow Technology Group, SINTEF Industry, 7031 Trondheim, Norway; shahriar.amini@sintef.no

* Correspondence: schalk.cloete@sintef.no

Citation: Szima, S.; Arnaiz del Pozo, C.; Cloete, S.; Fogarasi, S.; Jiménez Álvaro, Á.; Cormos, A.-M.; Cormos, C.-C.; Amini, S. Techno-Economic Assessment of IGCC Power Plants Using Gas Switching Technology to Minimize the Energy Penalty of CO_2 Capture. *Clean Technol.* **2021**, *3*, 594–617. https://doi.org/10.3390/cleantechnol3030036

Academic Editor: Diganta B. Das

Received: 27 June 2021
Accepted: 3 August 2021
Published: 10 August 2021

Publisher's Note: MDPI stays neutral with regard to jurisdictional claims in published maps and institutional affiliations.

Copyright: © 2021 by the authors. Licensee MDPI, Basel, Switzerland. This article is an open access article distributed under the terms and conditions of the Creative Commons Attribution (CC BY) license (https://creativecommons.org/licenses/by/4.0/).

Abstract: Cost-effective CO_2 capture and storage (CCS) is critical for the rapid global decarbonization effort recommended by climate science. The increase in levelized cost of electricity (LCOE) of plants with CCS is primarily associated to the large energy penalty involved in CO_2 capture. This study therefore evaluates three high-efficiency CCS concepts based on integrated gasification combined cycles (IGCC): (1) gas switching combustion (GSC), (2) GSC with added natural gas firing (GSC-AF) to increase the turbine inlet temperature, and (3) oxygen production pre-combustion (OPPC) that replaces the air separation unit (ASU) with more efficient gas switching oxygen production (GSOP) reactors. Relative to a supercritical pulverized coal benchmark, these options returned CO_2 avoidance costs of 37.8, 22.4 and 37.5 €/ton (including CO_2 transport and storage), respectively. Thus, despite the higher fuel cost and emissions associated with added natural gas firing, the GSC-AF configuration emerged as the most promising solution. This advantage is maintained even at CO_2 prices of 100 €/ton, after which hydrogen firing can be used to avoid further CO_2 cost escalations. The GSC-AF case also shows lower sensitivity to uncertain economic parameters such as discount rate and capacity factor, outperforms other clean energy benchmarks, offers flexibility benefits for balancing wind and solar power, and can achieve significant further performance gains from the use of more advanced gas turbine technology. Based on all these insights, the GSC-AF configuration is identified as a promising solution for further development.

Keywords: gas switching combustion; gas switching oxygen production; integrated gasification combined cycle; chemical looping combustion; CCS

1. Introduction

The global power sector faces a key challenge in the 21st century: achieving rapid emissions reductions despite strong demand growth [1]. The target set at the Paris Climate Agreement [2] is to limit the global average temperature increase to "well below 2 °C" by the end of the century. The models presented by the Intergovernmental Panel on Climate Change (IPCC) requires zero or even negative emissions from the power sector to comply with the 2 °C target [3].

Several options are available to reduce CO_2 emissions depending on the source of origin, including energy efficiency, renewable energy, nuclear energy, fuel switching, and CO_2 capture and storage (CCS). Among these pathways, CCS is arguably the most promising for drastic emissions reduction for three main reasons: (1) CCS retrofits can achieve emissions reductions from plants that have already been built, (2) CCS can be applied to sectors other than electricity such as direct industrial emissions or clean fuels, and (3) CCS can achieve negative emissions through BECCS or direct air capture. Unfortunately, the deployment of CCS is lagging far behind the trajectory required by the Paris Climate Accord [4], mostly because of economic and political challenges. Capturing and storing CO_2 will always be more expensive than simply emitting it to the atmosphere, and, to date, there have been limited policy incentives for covering these added costs.

However, the added cost of CCS can be minimized through more advanced CO_2 capture processes. Lowering the energy demand for the CO_2 separation process presents one promising pathway towards lower operating and capital costs of CCS plants. In the case of power production from solid fuels, the integration of a chemical looping combustion (CLC) [5,6] unit in an integrated gasification combined cycle (IGCC) system offers a promising pathway to a lower energy penalty [7]. In the IGCC system, the fuel is gasified and burned in a combined cycle gas turbine for power production. In general, the net electric efficiency of an IGCC power plant is around 47% without capture, whereas, if the conventional pre-combustion CO_2 capture is added, the efficiency drops as low as 36% [8]. This substantial energy penalty presents the major obstacle to CCS deployment.

CLC offers a way to substantially reduce this energy penalty, leading to considerable reductions in the CO_2 avoidance cost [9]. The CLC process relies on the basic idea of supplying oxygen in the combustion media via a solid oxygen carrier, as presented in Figure 1 (left). Oxygen is separated from air in the air reactor and transported to the fuel reactor via the OC, where it reacts with the fuel. Combustion occurs in a nitrogen-free zone, thus requiring only water condensation for delivering a high-purity CO_2 stream. Spallina et al. [10] compared the performance of several packed bed CLC-IGCC power plant strategies, obtaining an electrical efficiency of 41%, lowering the energy penalty relative to a pre-combustion capture benchmark with 5.7%-points. Hamers et al. [11] compared the performance of packed and fluidized bed reactors in CLC-IGCC systems but found no significant effect on the efficiency of the plant. In this case, net efficiency as high as 42% was obtained, further reducing the energy penalty with 6.92% points relative to pre-combustion. Cloete et al. [12] replaced the air separation unit within the IGCC plant with a chemical looping oxygen production unit reducing the energy penalty by 8.1%-points for an efficiency of 43.4% and reaching 45.4% if hot gas clean-up technology is employed.

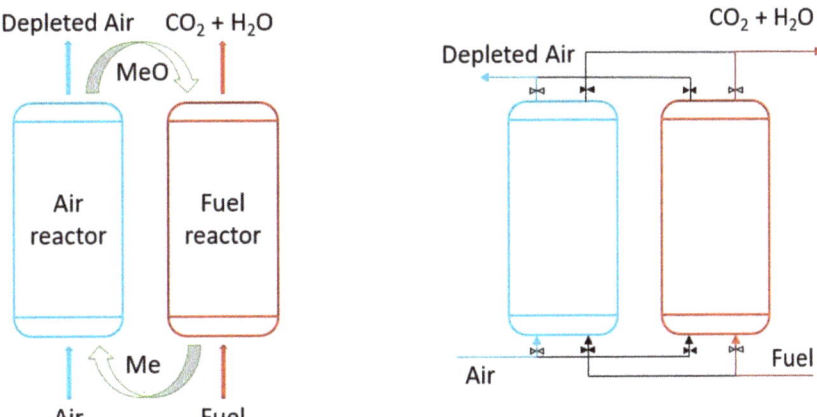

Figure 1. The chemical looping combustion process (**left**) and the gas switching combustion variant (**right**) that was investigated in this study.

Alternatively, a 3-step chemical looping combustion configuration can be employed where an extra reactor is used to partially oxidize an iron-based oxygen carrier using steam to produce hydrogen for driving a combined cycle. Sorgenfrei et al. [13] present the design and evaluation of a CLC IGCC system based on this configuration, achieving a net electric efficiency of 44.8% using a British Gas/Lurgi gasifier. Wu et al. [14] evaluated a similar configuration to obtain net efficiencies as high as 45.6%. This three-step configuration achieves efficiency benefits because the hydrogen firing can achieve considerably higher turbine inlet temperatures (TIT) than the hot depleted air stream from the conventional CLC configuration. However, the extra reactor makes the configuration considerably more complex than conventional two-reactor CLC, and equilibrium limitations enforce the use of moving beds, adding further complexity and increasing reactor size.

One important challenge with CLC is scale-up under pressurized conditions. To overcome this challenge, gas switching combustion (GSC) [15] was proposed. As shown in Figure 1 (right), the GSC concept keeps the solid OC in a single reactor where it is alternately oxidized with air and reduced by the fuel. The alternating feed gas streams are fed to the reactor using inlet switching valves. Similar switching valves are needed at the reactor outlet to separate the alternating depleted air and CO_2 streams emerging from each gas switching reactor. Such a simple standalone bubbling fluidized bed reactor promises to be substantially easier to scale up and pressurize than the interconnected dual circulating fluidized bed CLC configuration. To maintain continuous operation, a coordinated cluster of several dynamically operated GSC reactors can be used. Figure 1 (right) illustrates a simple cluster of two reactors where the reactor on the left is being oxidized and the one on the right is being reduced. When the desired degree of oxygen carrier conversion is achieved, the feed valves will switch to start reducing the reactor on the left and oxidizing the one on the right. A cluster of only two reactors is shown here for simplicity, but, since the air flowrate is much larger than the fuel flowrate, it is necessary to split the air feed between a larger number of reactors to maintain a similar fluidization velocity in all reactors [16].

The GSC-IGCC configuration was recently investigated with the aim of maximizing the process efficiency by circumventing two main efficiency challenges [17]. First, an additional combustor fired by natural gas was added after the GSC reactors to increase the TIT, thereby increasing the power cycle efficiency. Second, a recuperator was implemented to recover heat from the reduction outlet gases and transfer this thermal energy through the topping power cycle for more efficient electricity production. In addition, the condensation enthalpy in the steam originating from fuel combustion could be partially recovered at suitable temperatures in the steam cycle due to the high pressure of the GSC reduction outlet gases. Combined, these features succeeded in eliminating the energy penalty of CO_2 capture from an IGCC power plant, reaching efficiencies as high as 50%.

A major contributor to auxiliary consumption in an IGCC power plant is the air separation unit (ASU), the unit providing the necessary oxygen for the gasification of the fuel. The chemical looping process can be successfully applied for the separation of oxygen from the other constituents of air [18] using several metal oxides. Shi et al. [19] investigated several chemical looping air separation layouts, both continuous and batch types, and concluded that batch operation is more cost-effective for oxygen production. Deng et al. [20] modelled a chemical looping air separation unit using a fluidized bed reactor and optimized the process. A gas switching variant of this principle, called gas switching oxygen production (GSOP), was recently proposed to displace the ASU in a pre-combustion CO_2 capture IGCC configuration [21]. This oxygen production pre-combustion (OPPC) plant could achieve a net efficiency of more than 45%, albeit with a somewhat lower CO_2 avoidance of around 80%. Another benefit is that the relatively low operating temperature of the GSOP reactors will circumvent possible technical challenges with downstream valves and filters after GSC reactors.

The present study will investigate the effects of these large efficiency gains from an economic point of view. For the GSC configuration with added natural gas firing, greater efficiency will decrease levelized costs related to coal fuel and CO_2 transport and storage. Extracting more power from the syngas by means of a higher TIT will also substantially reduce the levelized costs of the expensive gasification train (coal and ash handling, gasifier, air separation unit, and gas clean-up). On the other hand, the use of natural gas for added firing will increase fuel costs because natural gas is more expensive than coal and reduce CO_2 avoidance because the CO_2 from natural gas combustion is not captured. For the OPPC configuration, levelized cost reductions can also be expected due to the high efficiency, but the relatively diluted syngas produced by this configuration will substantially increase the capital cost of the gasifier and gas clean-up units.

To quantify these trade-offs, this study presents a bottom-up economic assessment of GSC-IGCC plants with and without added natural gas firing and the OPPC plant. These results are compared to several benchmarks, including IGCC plants with and without conventional pre-combustion CO_2 capture. The plant performance will be quantified in terms of the levelized cost of electricity and CO_2 avoidance cost, relative to the IGCC and supercritical pulverized coal plant without CO_2 capture. In addition, the sensitivity of these performance measures to key economic assumptions such as fuel costs and discount rate will be identified. Finally, the economic performance of these advanced IGCC plants will be benchmarked against other clean energy technologies, including nuclear, wind, and solar PV, in a future energy system with high CO_2 prices.

2. Methodology

In this paper, five coal-fired IGCC power plant layouts are compared from a techno-economic point of view:

- Case 1: IGCC power plant without CO_2 capture (IGCC);
- Case 2: IGCC power plant with pre-combustion CO_2 capture using Selexol^{TM} liquid-gas absorption (IGCC-PCC);
- Case 3: GSC-IGCC power plant with inherent CO_2 capture (GSC);
- Case 4: GSC-IGCC power plant with added natural gas firing (GSC-AF);
- Case 5: GSOP-IGCC power plant with pre-combustion CO_2 capture using Selexol^{TM} liquid-gas absorption (OPPC).

The results are also compared to a supercritical pulverized coal power plant [8] as this technology is widely deployed in the power sector today. Simplified schematics of the power plants for Cases 3, 4, and 5 are shown in Figures 2–4, respectively. More detailed schematics can be found in previously published technical assessments [17,21]. The most important differences between the GSC (Figure 2) and GSC-AF (Figure 3) plants are (1) the GSC-AF plant fires natural gas after the GSC oxidation step to increase the TIT, and (2) the GSC-AF plant transfers heat from the CO_2 rich reduction step outlet gases to the compressed air stream using a recuperator. In contrast, the GSC plant must use the relatively high-grade heat in the GSC reduction step outlet gases to superheat steam for the bottoming cycle because insufficient high-grade heat is available from the gas turbine outlet gases, resulting from the lower GT firing temperature.

The OPPC plant (Figure 4) does not use GSC reactors, relying on a conventional pre-combustion CO_2 capture train to separate CO_2. However, large efficiency gains are achieved by using GSOP reactors to produce an N_2-free oxidant stream (17%mol of O_2) for the gasifier and pre-heating the air to 900 °C. In this way, the OPPC plant avoids the energy penalty of an ASU and greatly reduces the amount of H_2 required (and the associated steam consumption) to reach the desired TIT relative to a conventional pre-combustion plant. The process flowsheets of the reference plants (Cases 1 & 2) are similar to the layouts shown in Gazzani et al. [22].

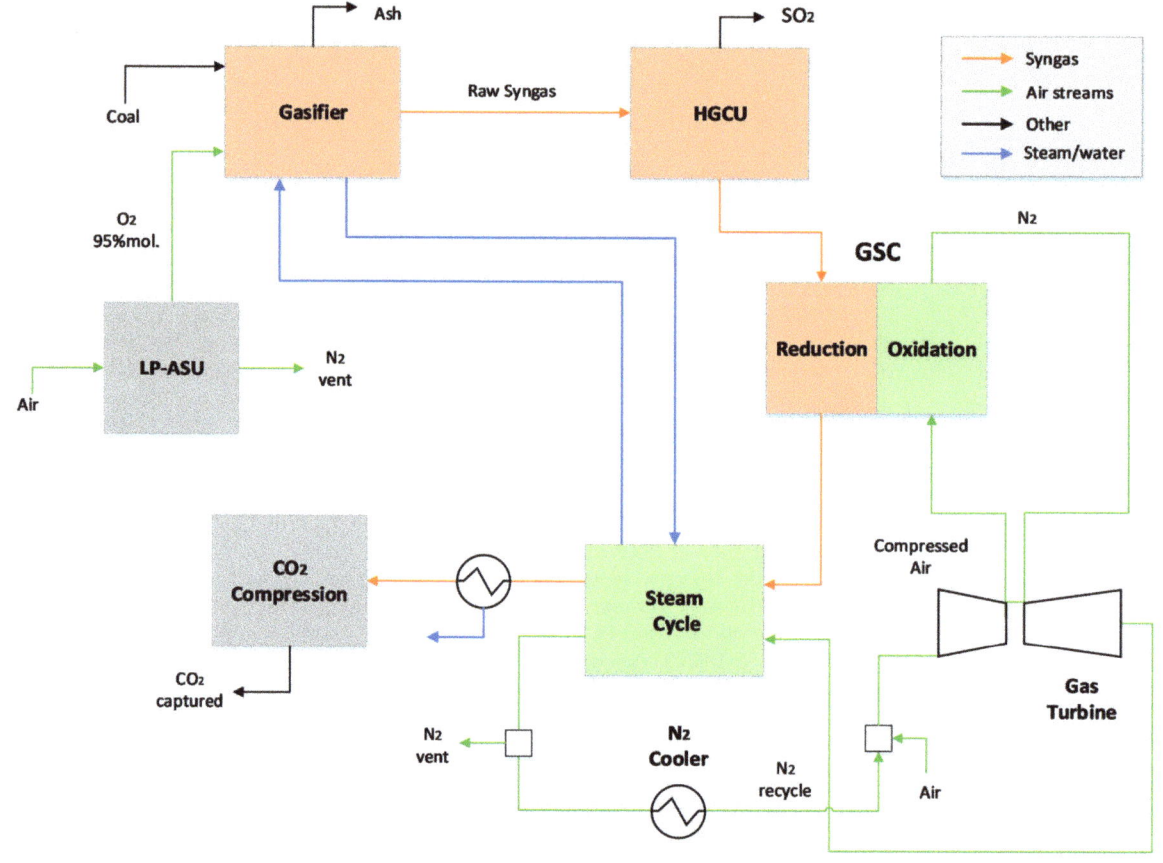

Figure 2. Schematic of the GSC power plant (Case 3) [17].

2.1. Process Simulation

Two benchmark IGCC plants are considered in this work: the unabated IGCC model (Case 1) consists of a dry fed entrained flow gasifier (Shell Type), syngas scrubbing, and heat recovery with cold gas desulphurization unit. O_2 is delivered by a high-pressure air separation unit, and coal is loaded with high purity N_2. The ASU is 50% integrated with the gas turbine compressor, while all available N_2 is mixed with the syngas fuel to minimize NOx emissions for complying with regulations. The power island assumptions considered in this work are similar to the ones in Spallina et al. [10], assuming an F-class turbine adapted to operate with syngas instead of natural gas.

The pre-combustion CO_2 capture model (Case 2) has a similar setup to the unabated IGCC plant, but a low-pressure ASU is used instead (no integration is advised for H_2 co-production and reliability), while coal is loaded with CO_2, resulting in slightly higher cold gas efficiency. After syngas scrubbing and steam addition from the HP stage steam turbine outlet (reaching a steam to CO ratio of 1.9 to avoid catalyst deterioration), the water-gas shift (WGS) reaction is carried out in two intercooled adiabatic reactors. CO_2 is removed with Selexol^{TM} absorption, modelled based on the work of Kapetaki [23] for component solubility, and compressed in a five-stage intercooled compressor. H_2-rich fuel is saturated and mixed with N_2 from the ASU for NOx abatement and fired in the gas turbine (GT).

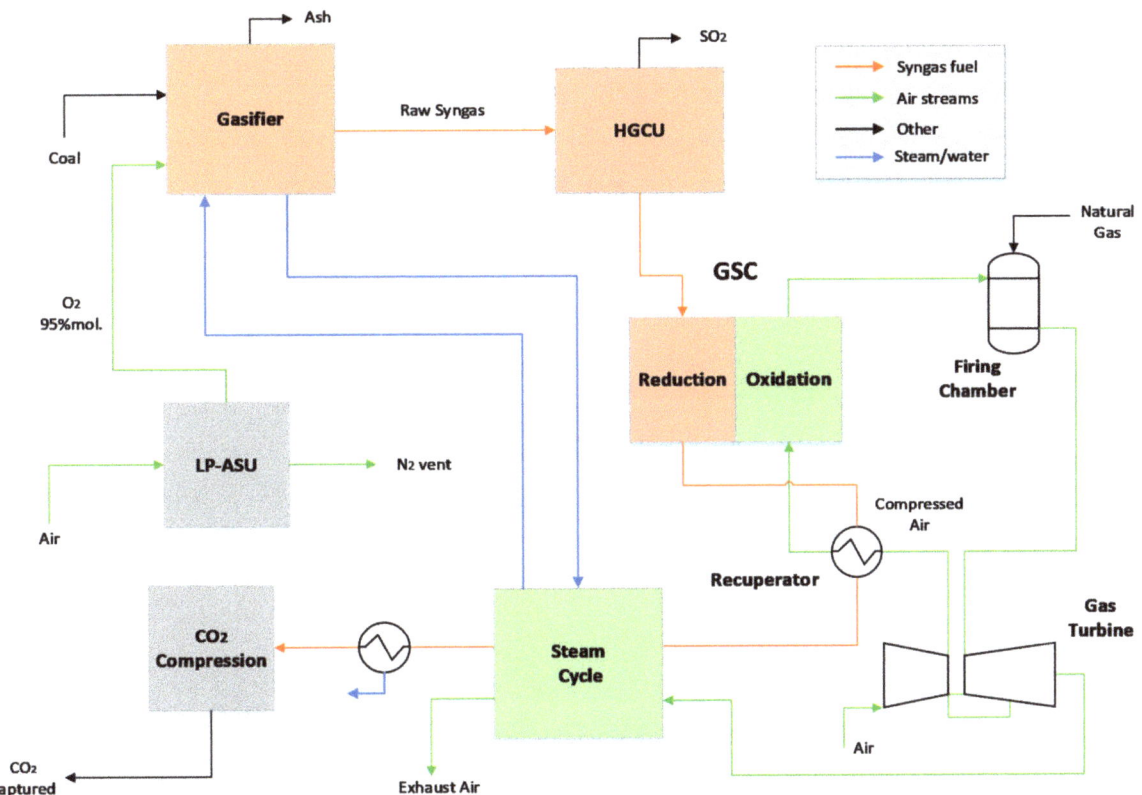

Figure 3. Schematic of GSC+AF power plant (Case 4) [17].

The GSC plants (Cases 3 and 4) are modelled with a similar approach as in Arnaiz del Pozo et al. [17], with the notable difference of employing an NiO oxygen carrier instead of Ilmenite, which shows higher feasibility to operate under the assumed maximum temperatures (1200 °C) [24] and has a better performance in terms of undesired mixing, achieving higher capture ratios due to the higher oxygen carrying capacity that facilitates longer reactor cycles. The same component efficiencies for the power island are taken and, considering the reduced turbine inlet temperature resulting from the mechanical limits of the oxygen carrier, a simple correlation by Horlock [25] is taken to determine stator cooling, neglecting cooling of the rotor (Case 3). For the GSC plant with natural gas extra firing, the plant simulations performed in the present study consider a GT cooling flow model resulting a small decrease in efficiency, a lower capture rate, and a higher heat input provided by the extra natural gas relative to the results reported in Arnaiz del Pozo et al. [17] (Case 4). The latter study reveals that carrying out extra firing with a portion of syngas results in significantly lower (more than 15%-points below) carbon capture relative to natural gas, because of its larger carbon intensity, while lower electrical efficiency is attained, due to thermal losses of syngas production and treating, which curtails the attractiveness of this option. Similar to the benchmark IGCC plants, the configurations integrating GSC technology produce syngas with a Shell gasifier, but also include hot has desulphurization as an additional efficiency enhancement.

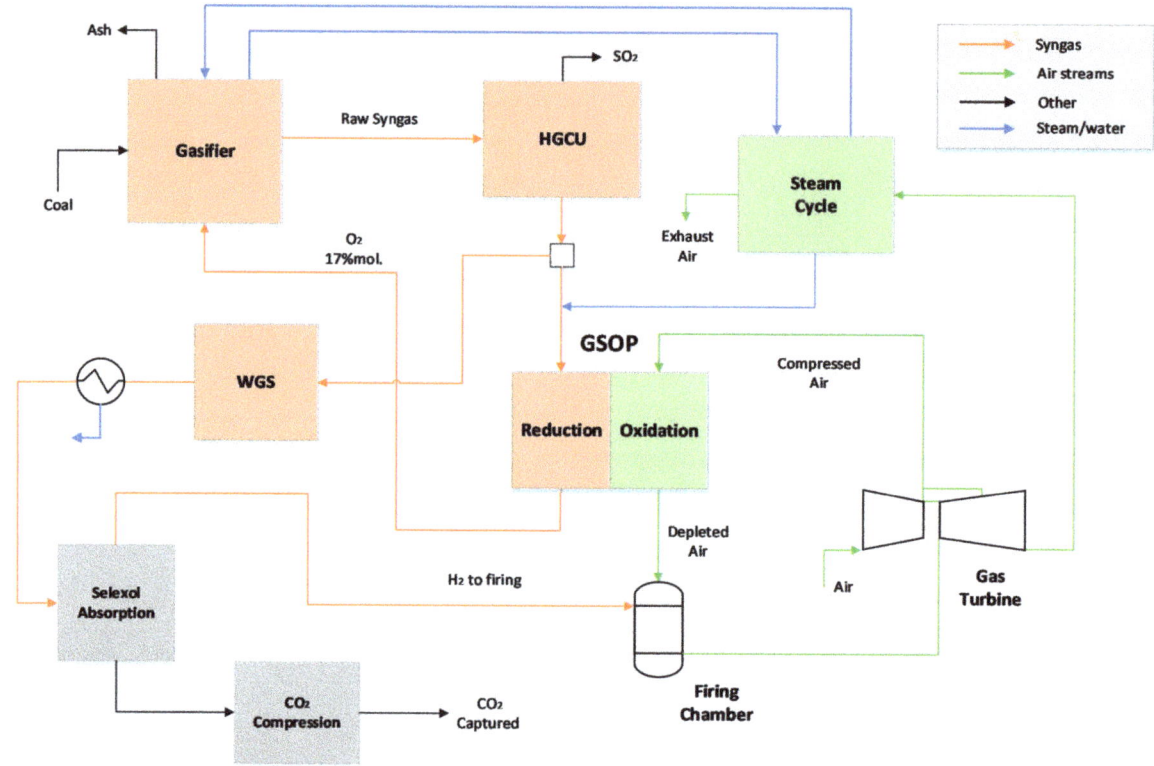

Figure 4. Schematic of the OPPC power plant (Case 5) [21].

The OPPC plant presented here (Case 5) has a similar configuration to the one shown in Arnaiz del Pozo et al. [21], where a GSOP cluster delivers an oxidant stream to a Winkler gasifier to produce syngas. After hot gas desulphurization and contaminant removal, a portion of the syngas, together with some intermediate pressure steam from the bottoming cycle, is routed to the GSOP cluster. The remaining syngas is sent to a WGS unit similarly to the pre-combustion capture model. Subsequent CO_2 sequestration is performed with a simplified Selexol^{TM} unit (as H_2S has already been removed). The compressed air is firstly heated in the GSOP oxidation stage to 900 °C, and then it enters an extra firing chamber to reach higher temperatures by combustion of the H_2-rich fuel produced in the WGS unit. When incorporating the coolant flows in the GT model, a smaller portion of air passes through the GSOP cluster relative to Arnaiz del Pozo et al. [21] and, in parallel, a larger fraction of syngas must be sent to the WGS unit to generate sufficient H_2 to reach the required combustor outlet temperature (COT).

The calibrated natural gas-fired turbine has a COT of 1440 °C and a TIT of 1360 °C with a turbine outlet temperature of 603 °C, operating with a pressure ratio of 18.1 and a simple cycle efficiency of 39%. When applied to the syngas fired models, it is assumed that the turbine operates at its nominal design point (equal pressure ratio and polytropic efficiencies of compressor and expansion stages) and that the cooling flows are adjusted to operate at the same TIT with the same cooling fraction to the rotor. This assumes a higher level of blade cooling technology and an appropriate compressor design to account for the higher flow rate of lower energy density fuel relative to the natural gas case. Furthermore, the coal flow rate to the plant is fixed, resulting in a different size of the GT for each case. This is consistent with the fact that the gasification island is the major cost component of the plant, with a constant heat input for all cases, and since GSC technology has a long deployment horizon, it is safe to assume some flexibility in GT design. The steam cycle consists of a three-pressure level with a reheat heat recovery steam generator (HRSG).

For NOx control, a large amount of N_2 from the ASU is mixed with the fuel in the IGCC-PCC case, while the GSC case requires no special measures due to the flameless combustion in the GSC reactors. For the GSC-AF case, it is assumed that the spontaneous combustion of natural gas in the hot depleted air stream from the GSC reactors can be carried out in a manner approaching the behavior of a premixed combustor by employing many fuel injectors and high turbulence [26]. A similar approach is followed in the OPPC case.

The power plant models were built with the process simulator Unisim Design R451 using the Peng Robinson equation of state and the ASME steam tables for thermodynamic property calculations. Detailed modelling assumptions of the plant units are provided in the Appendix A. The time-averaged operating points of the gas switching reactors as input for the power plant were determined with a transient 0-D model in Matlab, described in more detail in the technical assessments of the GSC and GSC-AF plants [17] and the OPPC plant [21]. This model assumes ideal gas behavior of the gaseous species, which is acceptable due to the high temperature and relatively low-pressure values encountered in the reactors. The reactions included in the models for the GSC (Equations (1)–(4)) and GSOP (Equations (5)–(8)) processes are summarized below. Equations (1)–(7) are assumed to proceed to completion, whereas Equation (8) is assumed to reach equilibrium as defined in Arnaiz del Pozo et al. [21]:

$$CH_4 + 4NiO \rightarrow 4Ni + CO_2 + 2H_2O \quad (1)$$

$$H_2 + NiO \rightarrow Ni + H_2O \quad (2)$$

$$CO + NiO \rightarrow Ni + CO_2 \quad (3)$$

$$O_2 + 2Ni \rightarrow 2NiO \quad (4)$$

$$CH_4 + 8Ca_2AlMnO_{5.5} \rightarrow 8Ca_2AlMnO_5 + CO_2 + 2H_2O \quad (5)$$

$$H_2 + 2Ca_2AlMnO_{5.5} \rightarrow 2Ca_2AlMnO_5 + H_2O \quad (6)$$

$$CO + 2Ca_2AlMnO_{5.5} \rightarrow 2Ca_2AlMnO_5 + CO_2 \quad (7)$$

$$O_2 + 4Ca_2AlMnO_5 \leftrightarrow 4Ca_2AlMnO_{5.5} \quad (8)$$

2.2. Economic Assessment

The economic assessment methodology is presented in four parts: (1) the design and cost assessment of gas switching reactors and heat exchangers, (2) other capital cost assumptions, (3) operating and maintenance cost assumptions, and (4) the methodology for calculating the levelized cost of electricity and the cost of CO_2 avoidance.

2.2.1. Reactor and Heat Exchanger Design

The reactor cost was estimated by assuming the wall structure presented in Figure 5 where, from left to right, the layers represent the inner Ni-alloy to withstand the temperature, corrosion and abrasion loads, the middle 0.54 m thick layer of thermal insulation for an outer wall temperature of 80 °C, and the outer carbon steel shell to carry the pressure load. The cost of the reactor strongly depends on the cost of the shell, which depends on the insulation thickness employed. This is investigated in a sensitivity analysis in the results section. Each reactor was assumed to consist of two process vessels: an inner Ni-alloy vessel and an outer carbon steel vessel. The fully installed cost of these vessels is estimated using the correlations given by Turton [27], with the cost of the inner vessel being doubled to account for elements such as the gas distributor and downstream particle filters. However, the cost of high-temperature outlet valves is included following Hamers et al. [11]. The cost of the initial load of OC is added to the capital cost of the reactor.

The fluidization velocity in the reactor is assumed to be 1 m/s, which will be on the upper edge of the bubbling fluidization regime (shortly before the transition to turbulent fluidization) when 150 μm particles are used according to the correlations of Bi and Grace [28]. This assumed fluidization velocity requires the total cross-sectional area of all the reactors to be 191.5 m^2. The reactors in the cluster are 1.84 m in diameter, 3.68 m in height, and a total number of 72. Costs are updated for the year 2018 using the Chemical Engineering Plant Cost Index [29].

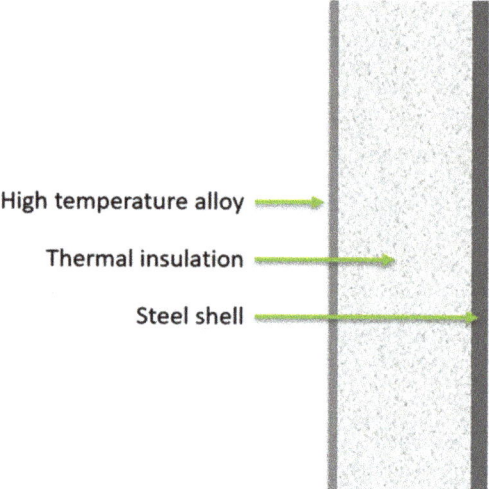

Figure 5. Assumed reactor wall structure.

For the GSOP reactors, a similar methodology is applied as in the case of the GSC reactors with the following differences: the required cross-section of the reactors is 105.4 m^2, the height is 3.66 m, the diameter is 1.83 m, and the number of units is 40 to maintain the desired fluidization velocity of 1 m/s. Since the GSOP reactors operate at a considerably lower temperature than the GSC reactors, a thinner insulation layer of 0.31 m could be used to maintain the outer wall temperature of 80 °C.

The cost estimation methodology for the heat exchangers involved a similar methodology. Shell-and-tube heat exchangers are selected with stainless steel used for both the shell and the tubes. Information about the heat transfer duty and log mean temperature difference from the process simulation is combined with calculated film and overall heat transfer coefficients necessary to determine the heat transfer area required in each heat exchanger. This heat transfer area is then used in the cost functions presented in Turton [27]. The overall heat transfer coefficient is calculated as a function of film coefficients of the cold and hot streams using Nusselt number correlations from the literature [30].

2.2.2. Capital Cost Estimation

Capital costs are estimated using the costs from Franco et al. [8] and scaled to a chosen modeling parameter as presented in the general form of the cost (Equation (9)). C_0 and Q_0 are the reference cost and capacity of the unit, and M is an exponent that depends on the equipment type. The parameters for the cost calculation are presented in Tables 1 and 2 for the cases without CO_2 capture and with CO_2 capture, respectively. The obtained capital cost is updated using the Chemical Engineering Plant Cost Index [29] for the year 2018:

$$C = C_0 * \left(\frac{Q}{Q_0}\right)^M \tag{9}$$

Table 1. Reference costs, capacities and scaling exponents for the case without CO_2 capture used in Equation (9).

Equipment	Scaling Parameter	Reference Cost (M€)	Reference Capacity	Scaling Exponent	Year	Ref.
ASU	Oxygen produced [kg/s]	64.48	26.54	0.67	2011	[8]
Coal handling	Coal input [kg/s]	49.50	32.90	0.67	2011	[8]
Ash handling	Ash flowrate [kg/s]	16.00	4.65	0.60	2011	[8]
HRSG	ST gross power [MW]	35.46	182.36	0.67	2011	[8]
Gas turbine	Net power output [MW]	88.60	254.42	1	2011	[8]
Steam turbine	ST gross power [MW]	55.00	182.36	0.67	2011	[8]
Condenser	ST gross power [MW]	40.56	182.36	0.67	2011	[8]
Gasifier	Coal thermal input [MW]	162.00	828.02	0.67	2011	[8]
Gas clean-up	Syngas flowrate [kg/s]	58.03	75.26	0.67	2011	[8]

The capital cost estimation for the base case IGCC power plant without CO_2 capture is performed using the reference data presented in Table 1 and applied in Equation (9).

The capital cost estimations of Cases 2–5 are performed using the parameters presented in Table 2. Case 2 involves standard technologies for gas clean-up, whereas the other three cases use hot gas clean-up as this offers significant efficiency improvements for IGCC systems [31]. The standard gas clean-up is assumed to consist of the following units: acid-gas removal, gas cleaning, water treatment, and the Claus burner. The cost correlation parameters for the hot gas clean-up are obtained as 75% of the standard gas clean-up unit presented by Franco et al. [8] as estimated from an RTI report [32]. The cost of the WGS unit used in Cases 2 and 5 is obtained from the work of Spallina et al. [33]. All other costs are taken from Franco et al. [8]. A scaling exponent of 1 was employed for the CO_2 compression because the lower costs for the cases with GSC that generate already pressurized CO_2 streams stem from fewer compression stages and not from smaller compressors.

Table 2. Reference costs, capacities and scaling exponents for the cases with CO_2 capture used in Equation (9).

Equipment	Scaling Parameter	Reference Cost (M€)	Reference Capacity	Scaling Exponent	Year	Ref.
ASU	Oxygen produced [kg/s]	72.80	31.45	0.67	2011	[8]
Coal handling	Coal input [kg/s]	53.89	38.72	0.67	2011	[8]
Ash handling	Ash flowrate [kg/s]	17.42	5.48	0.67	2011	[8]
HRSG	ST gross power [MW]	34.10	168.46	0.67	2011	[8]
Gas turbine	Net power output [MW]	93.32	282.87	1	2011	[8]
Steam turbine	ST gross power [MW]	52.00	168.46	0.67	2011	[8]
Condenser	ST gross power [MW]	39.00	168.46	0.67	2011	[8]
Gasifier	Thermal input [MW]	180.00	954.08	0.67	2011	[8]
Gasifier for Case 5	Raw syngas flowrate [kg/s]	167.1	65.60	0.67	2011	[8]
Gas clean-up	Syngas flowrate [kg/s]	61.49	89.21	0.67	2011	[8]
Hot gas clean-up	Syngas flowrate [kg/s]	46.12	89.21	0.67	2011	[8]
Selexol™ CO_2 capture unit	Shifted syngas flowrate [kg/s]	45.00	111.04	0.67	2011	[8]
WGS unit	Syngas flowrate [kg/s]	21.12	89.21	0.67	2011	[33]
CO_2 compression	Compressor power [MW]	30.00	20.69	1	2011	[8]

One important uncertainty is the gasifier cost assessment for Case 5. First, a different gasification technology is used (fluidized bed in Case 5 vs. entrained flow in the other cases). Second, the produced syngas has a much lower heating value, because the O_2 diluted oxidant stream (17%mol) from the GSOP employed for gasification, resulting in more than double the raw syngas flowrate relative to the other cases. This higher syngas flowrate can be expected to increase the required gasifier cross-sectional area, but it is also reasonable to expect that the gasification reactions will proceed faster due to the high concentration of CO_2 and H_2O in the oxidant stream and the high temperature at which this stream enters the gasifier, thus mitigating the required gasifier volume increase. In addition, elements like lock hoppers will be cheaper because the coal feed rate is the same, but the gasifier operating pressure is lower. To account for these conflicting effects, two scaling parameters were used for the gasifier cost: (1) the thermal input like the other cases and (2) the raw syngas flowrate that resulted in a much higher cost. In Table 2, scaling with the raw syngas flowrate (before water addition in the scrubber) is done from the reference cost of the gasifier in the pre-combustion plant in this study, which is slightly smaller than the one in Franco et al. [8]. The gasifier cost was then taken as the average of these two cost estimations. The effect of this uncertainty on the LCOE will be quantified in the results section.

The total investment cost was calculated as outlined in Table 3. A process contingency of 30% was added to the GSC reactor cluster due to its low level of technological maturity, while a 10% contingency was added to the hot gas clean-up unit which is near commercial readiness [34]. A project contingency of 18% and owner's cost of 12% are applied in line with our previous work [9]. These relatively high values are assumed to account for the technological uncertainty involved in IGCC technology.

Table 3. Estimation methodology for the total overnight cost of the plant.

Component	Definition
Total install cost (TIC)	Installed cost of each unit
Process contingency (PS)	30% of install cost for GSC reactors
	10% of install cost for the hot gas clean-up
Engineering procurement and construction costs (EPCC)	14% of (TIC + PS)
Project contingency (PT)	18% of (TIC + PS + EPCC)
Total plant costs (TPC)	TIC + PS + EPCC + PT
Owners cost	12% of TPC
Total overnight costs	TPC + Owners costs

2.2.3. Operating and Maintenance Costs

Table 4 presents the assumptions for the fixed and variable operating and maintenance (O&M) costs used in every case. The operating labour cost is included in the maintenance cost, according to Franco et al. [8], in both without and with carbon capture cases. The maintenance cost is estimated based on the gross power output of the plant. References are provided in the table for the estimations, and the fuel costs are varied in a sensitivity assessment in the results section.

Table 4. Fixed and variable operating & maintenance cost assumptions for the GSC plant.

Fixed O&M Costs		
Operating labour	Included in maintenance	
Maintenance and administrative costs	56 [8]	€/kW/year
Cost of coal	2.5 [32]	€/GJ LHV
Cost of ash disposal	9.73 [32]	€/t
Cost of NG	6.5 [8]	€/GJ LHV
Variable O&M Costs		
Process water costs	6 [8]	€/t
Cooling water make up costs	0.325 [8]	€/t
Catalyst Replacement		
Oxygen carrier	12,500 [35]	€/t
Selexol^{TM} replacement	5000 [8]	€/t
CO_2 Costs		
Transport and storage	10 [35]	€/t
Chemicals		
Cooling water chemical treatment	0.0025 [35]	€/m^3
Process water chemical treatment	45,000 [35]	€/mo.

The oxygen carrier replacement period is selected as two years (also varied in a sensitivity analysis later), and the Selexol^{TM} absorbent loss in the system is assumed to be 7 g lost/MWh gross power generated [8]. The economic parameters used for the OC in the GSOP reactors are the same as in the case of the GSC option.

2.2.4. Cash Flow Analysis

The levelized cost of electricity (LCOE) is calculated as the electricity price that would yield a net present value (NPV) of zero at the end of the plant's economic lifetime, according to Equation (10). Here, i is the discount rate, and ACF is the annual cash flow in every year over the construction and operating periods specified in Table 5. The annual cash flow combines revenues from electricity sales and expenditures from capital, fuel, and O&M costs. The construction period for the reference case without CO_2 capture is assumed to be lower, 3 years. A sensitivity analysis to the discount rate and capacity factor is presented in the results section.

Table 5. Cash flow analysis assumptions.

Economic lifetime	25 years
Discount rate	8%
Construction period	4 years
Capacity factor	85%
First year capacity factor	65%

The cost of CO_2 avoidance (COCA) is calculated using Equation (11), where LCOE represents the levelized cost of electricity and E the specific CO_2 emissions of the plant, respectively. Subscript CC denotes the plant with CO_2 capture and ref the reference plant without CO_2 capture, respectively:

$$NPV = \sum_{t=0}^{n} \frac{ACF_t}{(1+i)^t} \tag{10}$$

$$COCA\left(\frac{\text{€}}{tCO_2}\right) = \frac{LCOE_{cc} - LCOE_{ref}}{E_{ref} - E_{cc}} \tag{11}$$

COCA is calculated based on two references: the IGCC plant evaluated in this study ($COCA_{IGCC}$) and the supercritical pulverized coal plant from previous work [9] ($COCA_{SCPC}$). The supercritical pulverized coal plant has an LCOE of €55.7/MWh and an emission intensity (E) of 763 kg/MWh.

3. Results

The results will be presented in four parts. First, a brief outline of the revised plant performance will be given. Second, the economic performance of the different plants under base-case assumptions will be presented. Third, a sensitivity analysis to the most uncertain assumptions will be presented. And finally, the economic performance of these plants will be compared to other clean energy supply technologies.

3.1. Power Plant Performance Summary

The model results shown in Table 6 reveal similar values to those presented in previous work [17,21] for the plants using gas switching technology, while the reference IGCC plants with and without CCS show a comparable performance to Franco et al. [8]. A few small differences from these previous works can be highlighted:

- The lower heating value of coal was adjusted with an increase of 181 kJ/kg (0.72%) to match Franco et al. [8], relative to the property estimation value from Unisim Design R451 used in our previous studies [17,21].

- The GSC-AF case shows around 0.5%-points lower efficiency relative to our previous study [17] because the GT model was improved to consider cooling flows, with a pressure ratio of 18.1 compared to 20 in the previous assessment. A larger natural gas heat input is required due to the increased air flow rate across the expander, leading to a small decrease in the CO_2 capture rate of 2%-points and a net power output increase by 8%.
- The CO_2 compression for the plants integrating GSC technology consists of two intercooled stages and a supercritical CO_2 pump instead of the CO_2 purification unit used previously [17]. This simplification results from the improvement in CO_2 purity enabled by the larger oxygen carrying capacity of NiO (8.6 wt% [36]) relative to the previously simulated ilmenite (3.3 wt% [37]), which allows for an almost $3\times$ longer time between valve switches. Such a reduced switching frequency reduces the amount of undesired N_2/CO_2 mixing taking place after the feed streams are switched [16], improving CO_2 purity. Lower N_2/CO_2 mixing also facilitates a 1%-point increase in capture rate. However, the longer cycles cause a slightly lower reactor temperature, reducing the efficiency by 0.1%-points.
- The OPPC results given in Table 6 represent the case from our previous work [21] with the GSOP cluster operating at 900 °C, employing Selexol^{TM} for CO_2 capture and no H_2 fuel dilution (only saturation with water with low temperature residual heat). The CO_2 capture ratio is 1%-point lower than the value reported in [21], as a result of the lower partial pressure of CO_2 in the syngas, which reduces the capture performance of the Selexol unit. The syngas is produced at lower pressure as the gasification pressure is fixed by the GT pressure ratio, which in this study is fixed to 18.1, relative to the value of 20 assumed in the earlier work.

Table 6. Power plant performance summary.

Item/Plant	IGCC	IGCC-PCC	GSC	GSC-AF	OPPC
Gas Turbine Net (MW)	283.3 *	268.5	209.7	369.5	244.0
Steam Turbine Net (MW)	189.3	157.9	220.4	277.1	193.8
Heat Input (MW)	854.0	854.0	854.0	1176.8	854.0
Total Auxiliaries (MW)	66.0	104.1	62.1	63.8	42.4
Gross Plant (MW)	472.7	426.3	430.1	646.6	437.9
Net Plant (MW)	406.7	322.2	368.0	582.8	395.5
Gross Efficiency (LHV %)	55.4	49.9	50.4	55.0	51.3
Net Efficiency (LHV %)	47.6	37.7	43.1	49.5	46.3
Specific Emissions (kgCO_2/MWh)	727.3	86.4	46.6	135.1	123.3
Capture Rate (%)	0.0	90.6	94.2	78.1	83.2

* includes air expander.

Table 6 shows that the IGCC-PCC benchmark plant suffers a large 9.9%-point energy penalty relative to the unabated IGCC case, while the advanced process configurations greatly reduce this penalty. In the case of GSC-AF, the efficiency is even higher than the unabated IGCC plant due to the added firing with natural gas. However, this added natural gas firing causes 56% higher specific emissions than the IGCC-PCC benchmark, although emissions remain $5.4\times$ lower than the IGCC plant. The OPPC model achieves only slightly lower specific emissions relative to the GSC-AF plant despite a 5%-point higher capture rate, given that its thermal efficiency is 3%-points lower and all combusted fuel is derived from carbon-intensive coal syngas. The GSC case is the only advanced plant that achieves lower specific emissions than the IGCC-PCC benchmark, although there is a tradeoff in terms of lower efficiency relative to the GSC-AF and OPPC configurations.

3.2. Base Case Economic Assessment

Capital costs generally represent the largest component of the LCOE of coal-fired plants with CCS. Figure 6 presents the capital cost breakdown in the GSC case. The gasifier and gas switching reactor island are the most expensive components of the plant. The rest of the units' share is at 10% or lower. It is also noteworthy that the power cycle represents only a third of the capital costs of the plant. The units involved in the chemical transformation of coal and CO_2 compression represent the other two thirds of the plant cost. This implies that any measures to get more useful electricity from the hot depleted air stream in the power cycle (such as the added firing with natural gas) can offer substantial reductions in the levelized capital cost of the plant.

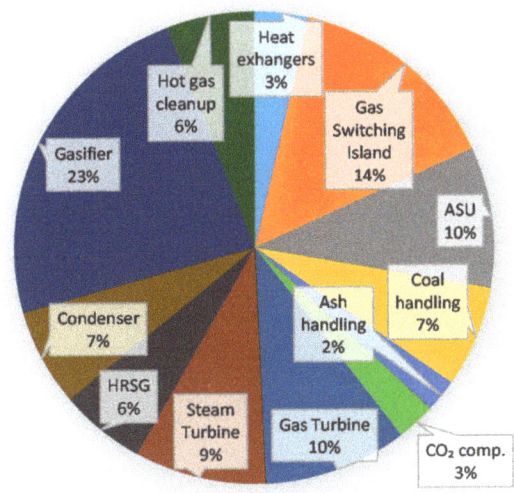

Figure 6. Total installed cost breakdown for the GSC case.

As presented in Table 7, the gasifier has the highest cost in all cases, with the gas turbine, gas switching island, and ASU also representing major shares of the plant capital cost. It is also interesting to note that the GSC plant relies more on the expensive steam cycle components (steam turbine, HRSG, and condenser) rather than the cheaper gas turbine relative to the other plants because of the relatively low TIT of this case. More power production from the gas turbine facilitated by the extra firing results in a more cost-effective power cycle. The OPPC plant suffers from a high gasifier cost due to the syngas flowrate that is more than double the size of the other plants. This high syngas flowrate also increases the gas clean-up cost.

The maintenance cost for the plant includes the labor cost, and it is calculated as a function of the gross power output of the plant, this explains the substantial difference between the two GSC models, the GSC-AF plant having a significantly higher output, as presented in Table 7. Variable O&M costs depend on the capacity factor, and this could change from year to year and can be expected to drop by the end of the economic lifetime. Table 8 presents O&M costs for the evaluated cases assuming a capacity factor of 85%, as used in the economic model. For the GSC-AF case, the high cost of natural gas is clearly shown, given that it represents only about a quarter of the LHV fuel input to the plant. Besides fuel costs, the costs associated with CO_2 storage have the highest impact on the economics of the plant. In the GSC plants, oxygen carrier replacement costs are also significant. These plants achieve a small process water revenue because of the water recovered from the high-pressure CO_2-rich stream from the GSC reactors.

Table 7. Installed costs for the main process components in each case.

Unit	IGCC	IGCC-PCC	GSC	GSC-AF	OPPC
Heat exchangers			26.63	13.14	33.37
Gas Switching Island			106.52	106.52	52.11
ASU	70.07	70.60	70.60	70.60	
Coal handling	52.03	50.79	50.79	50.79	50.79
Ash handling	16.78	16.42	16.42	16.42	17.71
CO_2 compression		34.86	19.32	19.56	31.56
Gas Turbine	99.62	91.21	71.24	125.53	82.90
Steam Turbine	58.73	51.86	64.84	75.61	59.58
HRSG	37.87	34.01	42.52	49.58	39.07
Condenser	43.31	38.89	48.63	56.71	44.68
Gasifier	170.30	172.08	172.08	172.08	232.28
Hot gas clean-up			43.26	43.28	72.15
WGS		19.47			20.18
Gas clean-up	57.21	56.70			
Selexol^{TM} plant		42.71			39.29
Total Install cost (M€)	605.93	679.60	732.86	799.83	776.70
Total overnight cost (M€)	912.91	1023.91	1104.15	1205.04	1170.20
Net power output (MW)	406.69	322.19	367.95	582.80	395.48
Specific investment cost (€/kWe)	2244.71	3178.00	3000.79	2067.68	2958.89

Table 8. O&M costs for the different cases.

Fixed O&M Costs (M€/Year)	IGCC	IGCC-PCC	GSC	GSC-AF	OPPC
Maintenance incl. labour	23.64	23.87	24.09	32.33	24.52
Variable O&M costs at 85% capacity factor (M€/Year)					
Cost of coal	57.27	57.27	57.27	57.27	57.27
Cost of NG				56.28	
Cost of ash disposal	1.25	1.25	1.25	1.25	1.40
Process water	3.20	6.67	−1.86	−1.87	2.24
Cooling water consumption	1.22	1.30	1.35	1.68	1.33
Oxygen carrier replacement			3.82	3.82	2.09
WGS catalyst replacement		0.44			0.44
Selexol^{TM} make up	0.12	1.03			0.94
CO_2 transport and storage		19.99	21.61	21.65	18.01
Total cost (M€/Year)	86.71	111.82	107.53	172.41	108.24

The main economic performance indicators are presented in Table 9 for all cases. The conventional pre-combustion capture plant has the highest LCOE, followed by the GSC and OPPC plants that reduce LCOE by 10 and 13 €/MWh, respectively. Added natural gas firing reduces the LCOE by an additional 13 €/MWh relative to the standard GSC plant. As discussed earlier, the gasifier cost is an important uncertainty in the estimation of the OPPC cost. For perspective, the LCOE of this case reduces to 76.37 €/MWh if the gasifier costs are scaled only by the thermal input and increases to 83.00 €/MWh if scaled only by the raw syngas flowrate. Trends in the COCA indicators are similar to those in the LCOE, although the COCA of the GSC-AF and OPPC plants are increased by their higher CO_2 emissions intensities (Table 6).

Table 9. LCOE and COCA indicators for each case.

	IGCC	IGCC-PCC	GSC	GSC-AF	OPPC
LCOE [€/MWh]	61.23	92.74	82.79	69.75	79.68
COCA$_{IGCC}$ [€/ton]	-	49.16	31.67	14.39	30.55
COCA$_{SCPC}$ [€/ton]	-	54.74	37.82	22.38	37.50

Figure 7 shows the breakdown of the LCOE for all cases considered in this paper. Fuel cost and O&M costs have similar ratios in the cost breakdown of the LCOE for the four carbon capture cases, capital cost being the one that varies from technology to technology. In the IGCC-AF case, the capital cost reduction obtained is counteracted to some extent by the higher cost of the NG. Even so, the overall cost is substantially reduced relative to the base GSC case and the OPPC case.

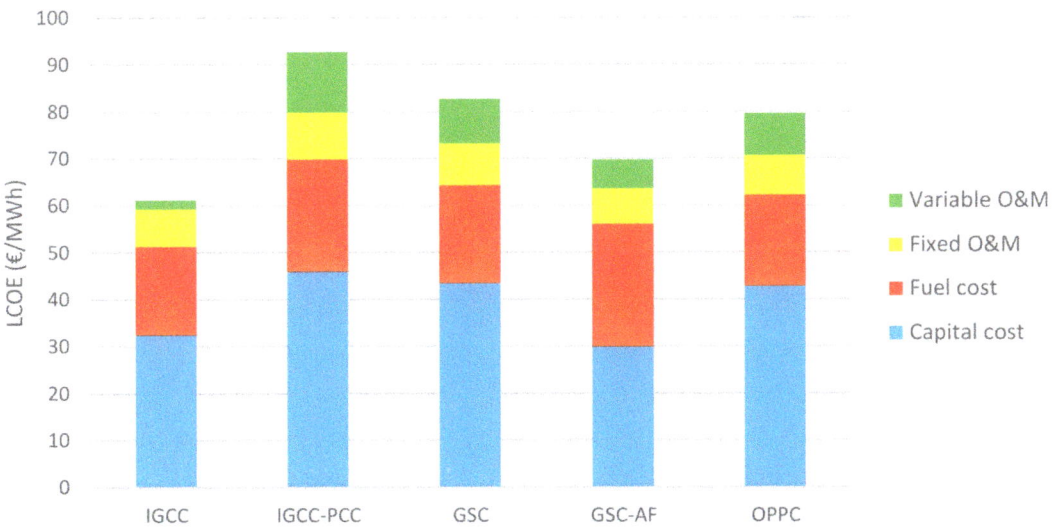

Figure 7. LCOE breakdown for the five IGCC configurations.

3.3. Sensitivity Analysis

The LCOE is sensitive to the cost of the fuel, as presented in Figure 8a,b, respectively. In all cases, aside from the GSC-AF case, the slopes of the lines in Figure 8a are inversely proportional to the plant efficiency. The GSC-AF plant has the lowest degree of dependency on the cost of coal because, in addition to having the highest efficiency, about a quarter of its fuel input is NG.

When the natural gas price is varied, a high degree of dependency is observed in the GSC-AF case, because of the high cost of natural gas when compared to coal. It is noteworthy that the LCOE of the GSC-AF plant remains lower than the GSC plant even at a natural gas cost of €10/GJ (4× higher than the coal cost). This illustrates the large benefit of using natural gas to raise the TIT so that the syngas produced by the costly gasification train can be converted to electricity more efficiently.

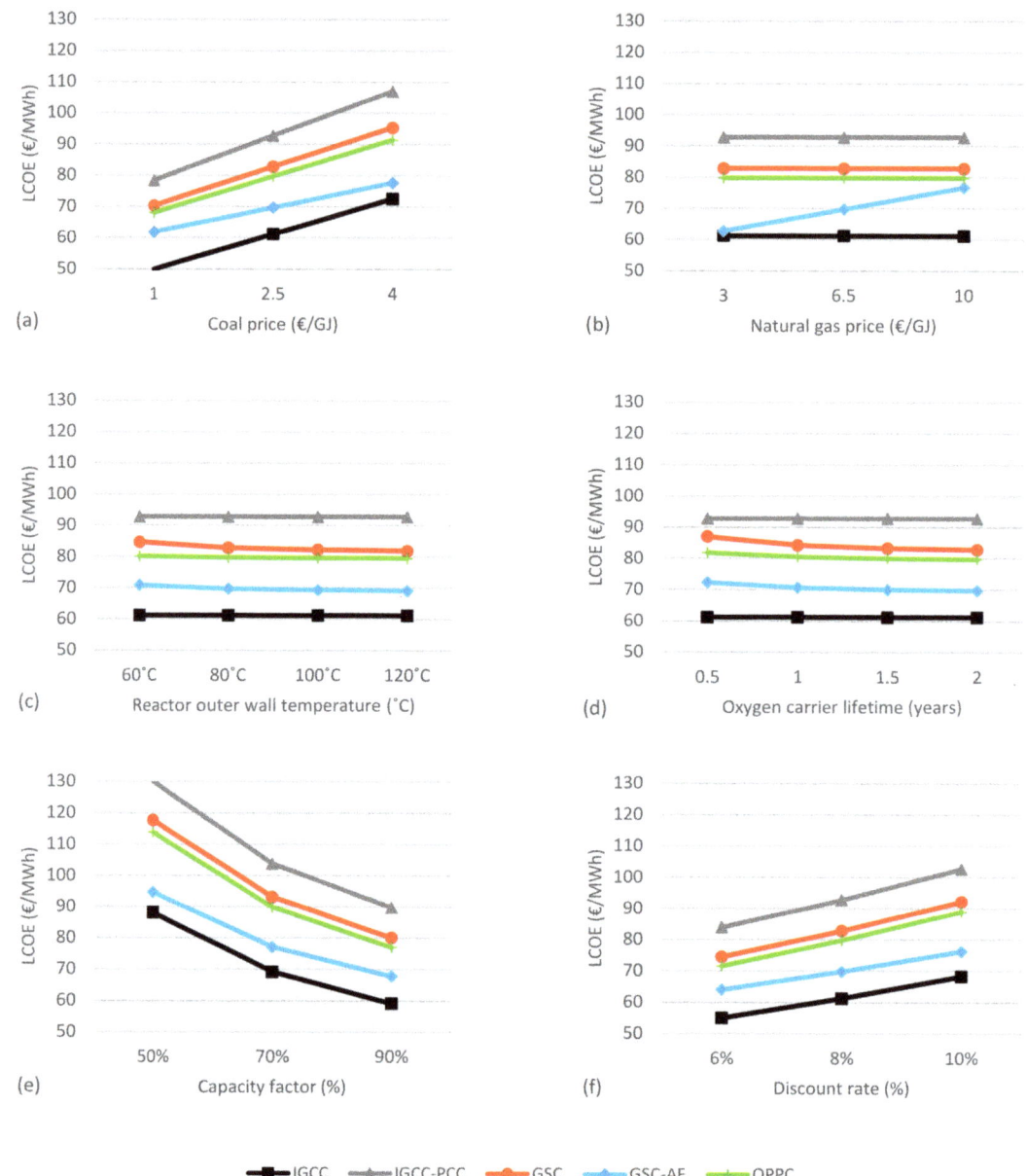

Figure 8. Sensitivity analysis to coal (**a**) and natural gas (**b**) prices, reactor outer wall temperature (**c**), oxygen carrier lifetime (**d**), capacity factor (**e**), and discount rate (**f**).

The outer carbon steel shell, carrying the pressure load in the GSC reactors, is the component showing the highest sensitivity because an increase in the insulation layer thickness increases both the shell volume and its required thickness, thus strongly increasing its cost. Increasing the insulation thickness from 0.54 m to 0.88 m in the GSC case lowers the shell temperature by 20 °C, but increases the LCOE by 1.86 €/MWh (Figure 8c). Allowing the shell temperature to reach 100 °C reduces the insulation thickness to 0.38 m while the LCOE drops with 0.6 €/MWh. Thus, even though the total reactor cost increased by 43%

from the 100 °C wall temperature to the 60 °C wall temperature, the effect on the LCOE is relatively small. The effect is even smaller in the GSC-AF and OPPC cases where the gas switching reactors represent a smaller fraction of total plant costs. The calculated heat losses for the three temperatures on the total surface of the reactors in the case of the GSC plant are 893.8 kW, 1113.5 kW, and 1365.9 kW, representing a bit more than 0.1% of the heat input.

Oxygen carrier lifetime is another important uncertainty for all concepts based on chemical looping technology. For the base case, a two-year replacement period is assumed for both GSC and GSOP reactors. As presented in Figure 8d, the LCOE would increase in all cases if the OC lifetime reduces. The GSC case is the most sensitive to the OC lifetime, showing a 4.2 €/MWh increase in LCOE if the OC lifetime reduces from 2 years to 0.5 years.

Given the capital-intensive nature of these plants, capacity factor and discount rate have the highest effect on the LCOE. A reduced capacity factor strongly increases the LCOE, as presented in Figure 8e. With the rapid growth of wind and solar power, thermal power plants are increasingly expected to act as balancing capacity, operating at lower capacity factors. In this respect, the GSC-AF plant offers some additional benefits because it is the least capital intensive, and, under part-load operation, it will reduce the fraction of fuel input required from more expensive natural gas. For example, when the F-class gas turbine output reduces by a little more than 50%, the TIT falls to the GSC outlet temperature [38], thus requiring no more natural gas firing. Under these conditions, the plant can operate with only a mild turndown of the relatively inflexible gasification train, but a substantial turndown in overall plant output, saving the high natural gas fuel costs and associated CO_2 emissions. The variation of the discount rate also has a large effect on the LCOE for all cases (Figure 8f), with the GSC-AF case being the least sensitive due to its relatively low specific capital cost.

3.4. Benchmarking against Other Clean Energy Technologies

In today's energy market, the COCA relative to unabated fossil fuel plants is not the most important indicator of the competitiveness of CCS technologies. Alternative clean energy technologies represent a more relevant benchmark. For this reason, the power plants assessed in this paper will be benchmarked against nuclear, wind, and solar technologies with cost data outlined in Table 10. Technology costs are taken from the IEA World Energy Outlook [39] for the year 2040 in the European Union. Wind and solar power integration costs, resulting from their large temporal and spatial variability, are taken from Hirth et al. [40] and are appropriate to the European Union for a wind and solar market share of 30–40%, increasing further for higher shares. Although nuclear and CCS plants would generally have longer operating lifetimes, all plants are assumed to have a 25-year economic lifetime. This assumption will give a conservative estimate of the competitiveness of the CCS plants evaluated in this study.

Table 10. Cost assumptions for nuclear, wind and solar benchmarks.

	Nuclear	Onshore Wind	Solar PV
Capital cost (€/kW)	3750	1417	508
Construction period (years)	6	1	1
Capacity factor	85%	30%	14%
O&M costs (€/MWh)	20	15	10
Fuel costs (€/MWh)	15		
Integration costs (€/MWh)		25–35	25–35

Figure 9 shows the results of this benchmarking exercise. Clearly, the conventional CO_2 capture plant (IGCC-PCC) is not well positioned in the competitive clean energy landscape. It is significantly more expensive than nuclear, only on par with wind and considerably more expensive than solar. Given the negligible air pollution and general green appeal of wind and solar energy, these clean technologies will be preferred over CCS if costs are similar. The GSC and OPPC plants achieve a better competitive position, being significantly cheaper than wind and nuclear and on par with solar with the higher integration costs bound. Only the GSC-AF plant outperforms other clean energy benchmarks, although only slightly in the case of solar. However, solar in Europe is subject to substantial seasonal variations that are misaligned with the seasonal electricity demand profile. Thus, Europe will continue relying strongly on wind despite the lower future LCOE projected for solar. It is noteworthy that the capital cost portion of the LCOE of the GSC-AF plant is lower even than that of solar PV, which has a 4× lower investment cost. This results from the 6x lower capacity factor of solar PV.

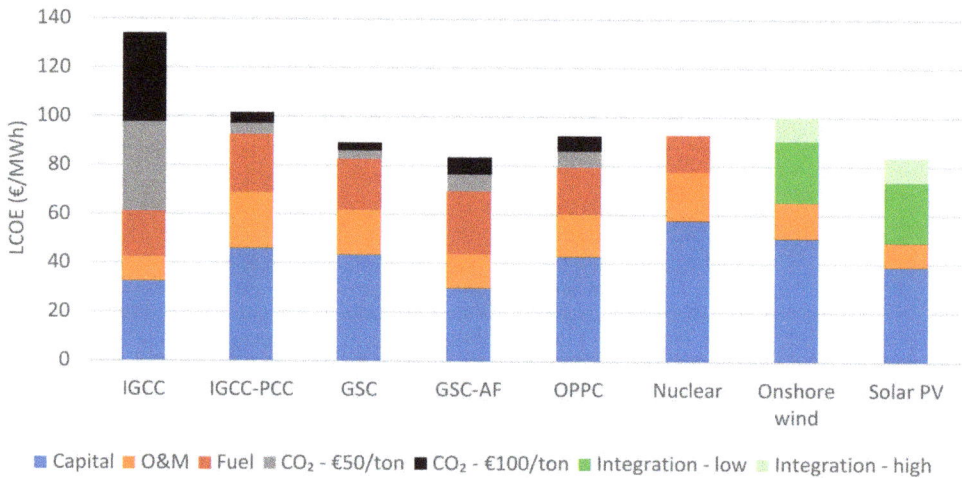

Figure 9. Benchmarking of the five IGCC-based power plants evaluated in this study against nuclear, wind, and solar power using costs relevant to the year 2040 when the CO_2 price is set to 50–100 €/ton.

Figure 9 also shows that the significant CO_2 emissions resulting from the combustion of natural gas after the GSC reactors in the GSC-AF plant reduces its competitiveness if CO_2 prices become very high. When the CO_2 price approaches €100/ton, it will become economical to do the added firing with clean hydrogen instead as recently calculated for a CLC-NGCC plant [41]. This possibility means CO_2 prices higher than €100/ton will not further increase the LCOE of the GSC-AF plant. Furthermore, a moderate fraction of biomass co-firing has the potential to bring CO_2 emissions below zero to achieve ultra-low emission targets while avoiding most of the technical challenges associated with biomass gasification and combustion.

This result suggests that highly efficient plants like the GSC-AF configuration will be required for CCS to be competitive in the clean energy landscape of the future. It should be noted, however, that the GSC-AF and OPPC configurations can benefit from using more advanced gas turbines with higher TITs to further increase efficiency and reduce costs. Flexibility is also an important criterion for the attractiveness of new CCS plants as the expansion of variable renewables continues [42]. The higher degree of flexibility offered by the GSC-AF case further increases its competitive position relative to the other CCS plants evaluated in this study.

4. Summary and Conclusions

This study compared the economic performance of five different IGCC power plant configurations: a benchmark IGCC plant without CCS, conventional pre-combustion CCS, gas switching combustion (GSC), GSC with added firing with natural gas (GSC-AF) to increase the TIT, and the oxygen production pre-combustion (OPPC) configuration that replaces the air separation unit (ASU) with more efficient gas switching oxygen production (GSOP) reactors.

The GSC plant returned a 10.7% lower LCOE than the conventional pre-combustion benchmark (82.7 €/MWh vs. 92.7 €/MWh) while maintaining a CO_2 capture rate of over 94%. Despite the higher cost of natural gas relative to coal, the high efficiency of the GSC-AF plant reduced the LCOE by another 15.7% to 69.8 €/MWh, reducing the cost of CO_2 avoidance as low as 22.4 €/ton when compared to a supercritical pulverized coal power plant. The large efficiency benefit of replacing the ASU with GSOP reactors in the OPPC configuration was partially counteracted by an increase in the gasifier cost and a lower CO_2 capture rate, resulting in a similar CO_2 avoidance cost to the GSC plant, despite achieving a 3.8% lower LCOE.

These results reveal that the GSC-AF configuration holds the most promise. In the sensitivity analysis, this case also showed reduced risk from several sources of uncertainty. Fuel costs are split evenly between coal and natural gas, limiting the sensitivity to price variations in either fuel. Uncertainties related to the GSC reactor cost and oxygen carrier lifetime are also limited since the added firing makes these components a smaller fraction of the LCOE. Added natural gas firing also makes the GSC-AF case less capital intensive (30% lower specific capital cost than GSC), limiting the cost increase related to lower capacity factors and higher discount rates. This plant could also hold benefits related to flexible operation for balancing wind and solar power since the expensive natural gas consumption can be ramped down first during part-load operation, requiring only a modest turndown of the relatively inflexible gasification train. The GSC-AF plant faces some risk from very high CO_2 prices due to the emissions from added natural gas firing, but this risk is mitigated by the possibility to do the added firing with clean hydrogen instead.

The good performance of the GSC-AF case was confirmed in comparisons to nuclear, wind, and solar power, where it emerged as the only CCS technology consistently less expensive than other clean energy benchmarks. Among the advanced IGCC power plant configurations investigated in this study, the GSC-AF configuration therefore emerges as the preferred option for further development. Future work will investigate the possibility of further performance gains using more advanced gas turbine technology and the potential to do the added firing with hydrogen extracted from the syngas stream.

Author Contributions: Conceptualization S.C.; methodology S.S., C.A.d.P. and S.C.; formal analysis, S.S., C.A.d.P., S.C. and S.F.; investigation, S.S. and C.A.d.P.; writing—original draft preparation, S.S., C.A.d.P. and S.C.; writing—review and editing, Á.J.Á., A.-M.C., C.-C.C., and S.A.; supervision, S.C., Á.J.Á. A.-M.C., and C.-C.C.; project administration, S.A.; funding acquisition, S.C, Á.J.Á., A.-M.C. and S.A. All authors have read and agreed to the published version of the manuscript.

Funding: This research was funded by the European Commission Horizon 2020 program, the Romanian National Authority for Scientific Research and Innovation, the Ministerio de Economía y Competitividad, and the Research Council of Norway under the ACT Grant Agreement No 691712.

Institutional Review Board Statement: Not applicable.

Informed Consent Statement: Not applicable.

Data Availability Statement: Data can be provided by the first author (szabolcs.szima@gmail.com) upon reasonable request.

Acknowledgments: The authors would also like to acknowledge Honeywell for the free academic license of Unisim Design R451 for the modelling of the power plants.

Conflicts of Interest: The author Szabolcs Szima is an employee of MDPI, however he does not work for the journal Clean Technologies at the time of submission and publication.

List of Abbreviations

ASU	Air separation unit
CCS	CO_2 capture and storage
CLC	Chemical looping combustion
COCA	Cost of CO_2 avoidance
COT	Combustor inlet temperature
EPCC	Engineering, procurement and construction cost
GS	Gas Switching
GSC	Gas switching combustion
GSC-AF	GSC power plant with added natural gas firing
GSOP	Gas switching oxygen production
GT	Gas Turbine
HGCU	Hot gas clean up
HRSG	Heat recovery steam generator
IGCC	Integrated gasification combined cycle
LCOE	Levelized cost of electricity
NPV	Net present value
O&M	Operating and maintenance
OPPC	Oxygen production pre-combustion power plant
PS	Process contingency
PT	Project contingency
TIC	Total install cost
TIT	Turbine inlet temperature
TPC	Total plant cost
WGS	Water-gas shift

Appendix A

Table A1. Gasification island assumptions.

Winkler Gasifier		
Item	Value	Units
Freeboard Temperature	900	°C
%w. CO_2 for coal loading	15	%
% LHV CH_4 in syngas	11.3	%
Oxidizer Overpressure	50	kPa
HP steam superheat	450	°C
Fixed carbon conversion	97	%
%w. Vented CO_2 in lock hoppers	10	%
Coal milling & handling	40	MJ/kg coal
Ash handling	200	MJ/kg ash
Shell Gasifier		
Item	Value	Units
Moderator (steam) to dry coal ratio	0.09	kg/kg
Oxygen to dry coal ratio	0.873	kg/kg
Moisture in Coal after drying	2	%
Syngas for coal drying %LHV	0.9	%
Fixed carbon conversion	99.3	%
Gasifier operating pressure	44	bar
Steam moderator pressure	54	bar
Heat loss as %LHV	0.7	%
Heat to membrane wall as %LHV	2	%
CO_2 HP/HHP Pressure	56/88	bar
CO_2 Temperature	80	°C
CO_2 to dry coal ratio	0.83	kg/kg

Table A1. *Cont.*

Winkler Gasifier		
Air Separation Unit		
Item	Value	Units
Main air compressor polytropic efficiency	89	%
Booster air compressor polytropic efficiency	87	%
Reboiler-condenser pinch	1.5	°C
Heat exchanger minimum approach temperature	2	°C
Process stream temperature after heat rejection	25	°C
Oxygen purity	95	%
Oxygen delivery pressure	48	bar
Oxygen pump efficiency	80	%
Exchanger pressure losses/side	10	kPa
Intercooler pressure loss	10	kPa

Table A2. Syngas treating unit assumptions.

HGCU		
Item	Value	Units
Adsorption temperature	400	°C
Regeneration temperature	750	°C
Filter pressure drop	5	%
Auxiliary consumption	5.34	MJe/kgH_2S
Compander polytropic efficiency	90	%
Syngas blower polytropic efficiency	80	%
O_2 mol fraction in regeneration stream	2	%
CGCU		
Item	Value	Units
Absorption temperature	30	°C
Auxiliary consumption	3	MJe/kg H_2S
LP steam requirement	50	MJth/kg H_2S
Syngas blower polytropic efficiency	80	%
Selexol pump efficiency	80	%
% H_2S to Claus unit	>25	%

Table A3. Power cycle assumptions.

Gas Turbine		
Item	Value	Units
GT compressor polytropic efficiency	91.5	%
GT turbine polytropic efficiency	87	%
GT pressure ratio	18.1	-
GT Electromechanical efficiency	98.6	%

Table A3. Power cycle assumptions.

Item	Value	Units
Gas Turbine Steam Cycle		
Steam turbine low pressure stage isentropic efficiency	88	%
Steam turbine intermediate pressure stage isentropic efficiency	94	%
Steam turbine high pressure stage isentropic efficiency	92	%
Steam turbine electromechanical efficiency	98.1	%
Pressure levels HP/IP/LP	144/36/4	bar
Auxiliaries for heat rejection	0.008	MJe/MJth
Pump isentropic efficiency	80	%
Live steam temperature	565	°C
CO_2 Compression		
CO_2 Compressor stage isentropic efficiency	80	%
Process stream temperature after cooler	25	°C

References

1. IEA. World Energy Outlook. International Energy Agency, 2018. Available online: https://www.iea.org/reports/world-energy-outlook-2018 (accessed on 7 October 2019).
2. The Paris Agreement, n.d. Available online: https://unfccc.int/process-and-meetings/the-paris-agreement/the-paris-agreement (accessed on 7 October 2019).
3. AR5 Synthesis Report: Climate Change. 2014. Available online: https://www.ipcc.ch/report/ar5/syr/ (accessed on 7 October 2019).
4. Tracking Clean Energy Progress. 2020. Available online: https://www.iea.org/topics/tracking-clean-energy-progress (accessed on 1 October 2020).
5. Ishida, M.; Zheng, D.; Akehata, T. Evaluation of a chemical-looping-combustion power-generation system by graphic exergy analysis. *Energy* **1987**, *12*, 147–154. [CrossRef]
6. Lyngfelt, A.; Leckner, B.; Mattisson, T. A fluidized-bed combustion process with inherent CO_2 separation; Application of chemical-looping combustion. *Chem. Eng. Sci.* **2001**, *56*, 3101–3113. [CrossRef]
7. Adánez, J.; Abad, A.; Mendiara, T.; Gayán, P.; de Diego, L.F.; García-Labiano, F. Chemical looping combustion of solid fuels. *Prog. Energy Combust. Sci.* **2018**, *65*, 6–66. [CrossRef]
8. Anantharaman, R.; Bolland, O.; Booth, N.; van Dorst, E.; Sanchez Fernandez, E.; Franco, F.; Macchi, E.; Manzolini, G.; Nikolic, D.; Pfeffer, A.; et al. *Cesar Deliverable D2.4.3. European Best Practice Guidelines for Assessment of CO_2 Capture Technologies*; Technical Report Number: CESAR-D2.4.3; Zenodo Array: Geneve, Switzerland, 2011. [CrossRef]
9. Cloete, S.; Tobiesen, A.; Morud, J.; Romano, M.; Chiesa, P.; Giuffrida, A.; Larring, Y. Economic assessment of chemical looping oxygen production and chemical looping combustion in integrated gasification combined cycles. *Int. J. Greenh. Gas Control* **2018**, *78*, 354–363. [CrossRef]
10. Spallina, V.; Romano, M.C.; Chiesa, P.; Gallucci, F.; van Sint Annaland, M.; Lozza, G. Integration of coal gasification and packed bed CLC for high efficiency and near-zero emission power generation. *Int. J. Greenh. Gas Control* **2014**, *27*, 28–41. [CrossRef]
11. Hamers, H.P.; Romano, M.C.; Spallina, V.; Chiesa, P.; Gallucci, F.; van Annaland, M.S. Comparison on process efficiency for CLC of syngas operated in packed bed and fluidized bed reactors. *Int. J. Greenh. Gas Control* **2014**, *28*, 65–78. [CrossRef]
12. Cloete, S.; Giuffrida, A.; Romano, M.; Chiesa, P.; Pishahang, M.; Larring, Y. Integration of chemical looping oxygen production and chemical looping combustion in integrated gasification combined cycles. *Fuel* **2018**, *220*, 725–743. [CrossRef]
13. Sorgenfrei, M.; Tsatsaronis, G. Design and evaluation of an IGCC power plant using iron-based syngas chemical-looping (SCL) combustion. *Appl. Energy* **2014**, *113*, 1958–1964. [CrossRef]
14. Wu, W.; Wen, F.; Chen, J.R.; Kuo, P.C.; Shi, B. Comparisons of a class of IGCC polygeneration/power plants using calcium/chemical looping combinations. *J. Taiwan Inst. Chem. Eng.* **2019**, *96*, 193–204. [CrossRef]

15. Zaabout, A.; Cloete, S.; Johansen, S.T.; van Sint Annaland, M.; Gallucci, F.; Amini, S. Experimental Demonstration of a Novel Gas Switching Combustion Reactor for Power Production with Integrated CO_2 Capture. *Ind. Eng. Chem. Res.* **2013**, *52*, 14241–14250. [CrossRef]
16. Cloete, S.; Romano, M.C.; Chiesa, P.; Lozza, G.; Amini, S. Integration of a Gas Switching Combustion (GSC) system in integrated gasification combined cycles. *Int. J. Greenh. Gas Control* **2015**, *42*, 340–356. [CrossRef]
17. Arnaiz del Pozo, C.; Cloete, S.; Cloete, J.H.; Jiménez Álvaro, Á.; Amini, S. The potential of chemical looping combustion using the gas switching concept to eliminate the energy penalty of CO_2 capture. *Int. J. Greenh. Gas Control* **2019**, *83*, 265–281. [CrossRef]
18. Moghtaderi, B. Application of chemical looping concept for air separation at high temperatures. *Energy Fuels* **2010**, *24*, 190–198. [CrossRef]
19. Shi, B.; Wu, E.; Wu, W. Novel design of chemical looping air separation process for generating electricity and oxygen. *Energy* **2017**, *134*, 449–457. [CrossRef]
20. Deng, Z.; Jin, B.; Zhao, Y.; Gao, H.; Huang, Y.; Luo, X.; Liang, Z. Process simulation and thermodynamic evaluation for chemical looping air separation using fluidized bed reactors. *Energy Convers. Manag.* **2018**, *160*, 289–301. [CrossRef]
21. Arnaiz del Pozo, C.; Cloete, S.; Hendrik Cloete, J.; Jiménez Álvaro, Á.; Amini, S. The oxygen production pre-combustion (OPPC) IGCC plant for efficient power production with CO_2 capture. *Energy Convers. Manag.* **2019**, *201*. [CrossRef]
22. Gazzani, M.; MacChi, E.; Manzolini, G. CO_2 capture in integrated gasification combined cycle with SEWGS-Part A: Thermodynamic performances. *Fuel* **2013**, *105*, 206–219. [CrossRef]
23. Kapetaki, Z.; Brandani, P.; Brandani, S.; Ahn, H. Process simulation of a dual-stage Selexol process for 95% carbon capture efficiency at an integrated gasification combined cycle power plant. *Int. J. Greenh. Gas Control* **2015**, *39*, 17–26. [CrossRef]
24. Kuusik, R.; Trikkel, A.; Lyngfelt, A.; Mattisson, T. High temperature behavior of NiO-based oxygen carriers for Chemical Looping Combustion. *Energy Procedia* **2009**, *1*, 3885–3892. [CrossRef]
25. Horlock, J.H. Cycle Efficiency with Turbine Cooling (Cooling Flow Rates Specified). *Adv. Gas. Turbine Cycles Pergamon* **2003**, 47–69. [CrossRef]
26. Khan, M.N.; Cloete, S.; Amini, S. Efficiency Improvement of Chemical Looping Combustion Combined Cycle Power Plants. *Energy Technol.* **2019**, *7*, 1900567. [CrossRef]
27. Turton, R.; Bailie, R.C.; Whiting, W.B.; Shaeiwitz, J.A. *Analysis, Synthesis, and Design of Chemical Processes*, 3rd ed.; Pearson Education: Boston, MA, USA, 2008.
28. Bi, H.T.; Grace, J.R. Flow regime diagrams for gas-solid fluidization and upward transport. *Int. J. Multiph. Flow* **1995**, *21*, 1229–1236. [CrossRef]
29. Plant Cost Index Archives-Chemical Engineering. Available online: http://www.chemengonline.com/ (accessed on 2 November 2017).
30. Bergman, T.L.; Lavine, A.S.; Incropera, F.S.; DeWitt, D.P. *Fundamentals of Heat and Mass Transfer*, 8th ed.; Wiley: Danvers, MA, USA, 2017.
31. Giuffrida, A.; Romano, M.C.; Lozza, G. Efficiency enhancement in IGCC power plants with air-blown gasification and hot gas clean-up. *Energy* **2013**, *53*, 221–229. [CrossRef]
32. Nexant. Preliminary Feasibility Analysis of RTI Warm Gas Clean Up (WGCU) Technology. 2007. Available online: https://fdocuments.in/document/preliminary-feasibility-analysis-of-rti-warm-gas-cleanup-wgcu-.html (accessed on 8 August 2021).
33. Spallina, V.; Pandolfo, D.; Battistella, A.; Romano, M.C.; Van Sint Annaland, M.; Gallucci, F. Techno-economic assessment of membrane assisted fluidized bed reactors for pure H_2 production with CO_2 capture. *Energy Convers. Manag.* **2016**, *120*, 257–273. [CrossRef]
34. Rubin, E.; Booras, G.; Davison, J.; Ekstrom, C.; Matuszewski, M.; Mccoy, S.; Short, C. Toward a Common Method of Cost Estimation for CO_2 Capture and Storage at Fossil Fuel Power Plants A White Paper Prepared by the Task Force on CCS Costing Methods. 2013. Available online: https://www.globalccsinstitute.com/resources/publications-reports-research/toward-a-common-method-of-cost-estimation-for-co2-capture-and-storage-at-fossil-fuel-power-plants/ (accessed on 15 April 2020).
35. Szima, S.; Nazir, S.M.; Cloete, S.; Amini, S.; Fogarasi, S.; Cormos, A.-M.; Cormos, C.C. Gas switching reforming for flexible power and hydrogen production to balance variable renewables. *Renew. Sustain. Energy Rev.* **2019**, *110*, 207–219. [CrossRef]
36. Abad, A.; Adánez, J.; García-Labiano, F.; de Diego, L.F.; Gayán, P.; Celaya, J. Mapping of the range of operational conditions for Cu-, Fe-, and Ni-based oxygen carriers in chemical-looping combustion. *Chem. Eng. Sci.* **2007**, *62*, 533–549. [CrossRef]
37. Abad, A.; Adánez, J.; Cuadrat, A.; García-Labiano, F.; Gayán, P.; de Diego, L.F. Kinetics of redox reactions of ilmenite for chemical-looping combustion. *Chem. Eng. Sci.* **2011**, *66*, 689–702. [CrossRef]
38. Gülen, S.C. *Gas Turbines for Electric Power Generation*; Cambridge University Press: Cambridge, UK, 2019. [CrossRef]
39. IEA. World Energy Outlook. International Energy Agency, 2019. Available online: https://www.iea.org/reports/world-energy-outlook-2019 (accessed on 1 October 2020).
40. Hirth, L.; Ueckerdt, F.; Edenhofer, O. Integration costs revisited-An economic framework for wind and solar variability. *Renew. Energy* **2015**, *74*, 925–939. [CrossRef]
41. Khan, M.N.; Chiesa, P.; Cloete, S.; Amini, S. Integration of chemical looping combustion for cost-effective CO_2 capture from state-of-the-art natural gas combined cycles. *Energy Convers. Manag. X* **2020**, *7*, 100044. [CrossRef]
42. Cloete, S.; Hirth, L. Flexible power and hydrogen production: Finding synergy between CCS and variable renewables. *Energy* **2020**, *192*, 116671. [CrossRef]

Article

Advanced Steam Reforming of Bio-Oil with Carbon Capture: A Techno-Economic and CO_2 Emissions Analysis

Jennifer Reeve, Oliver Grasham *, Tariq Mahmud and Valerie Dupont *

School of Chemical and Process Engineering, University of Leeds, Leeds LS2 9JT, UK; jspragg90@gmail.com (J.R.); t.mahmud@leeds.ac.uk (T.M.)
* Correspondence: o.r.grasham@leeds.ac.uk (O.G.); v.dupont@leeds.ac.uk (V.D.)

Abstract: A techno-economic analysis has been used to evaluate three processes for hydrogen production from advanced steam reforming (SR) of bio-oil, as an alternative route to hydrogen with BECCS: conventional steam reforming (C-SR), C-SR with CO_2 capture (C-SR-CCS), and sorption-enhanced chemical looping (SE-CLSR). The impacts of feed molar steam to carbon ratio (S/C), temperature, pressure, the use of hydrodesulphurisation pretreatment, and plant production capacity were examined in an economic evaluation and direct CO_2 emissions analysis. Bio-oil C-SR-CC or SE-CLSR may be feasible routes to hydrogen production, with potential to provide negative emissions. SE-CLSR can improve process thermal efficiency compared to C-SR-CCS. At the feed molar steam to carbon ratio (S/C) of 2, the levelised cost of hydrogen (USD 3.8 to 4.6 per kg) and cost of carbon avoided are less than those of a C-SR process with amine-based CCS. However, at higher S/C ratios, SE-CLSR does not have a strong economic advantage, and there is a need to better understand the viability of operating SE-CLSR of bio-oil at high temperatures (>850 °C) with a low S/C ratio (e.g., 2), and whether the SE-CLSR cycle can sustain low carbon deposition levels over a long operating period.

Keywords: sorption enhancement; chemical looping; hydrogen; bio-oil; carbon capture; techno-economics

Citation: Reeve, J.; Grasham, O.; Mahmud, T.; Dupont, V. Advanced Steam Reforming of Bio-Oil with Carbon Capture: A Techno-Economic and CO_2 Emissions Analysis. *Clean Technol.* **2022**, *4*, 309–328. https://doi.org/10.3390/cleantechnol4020018

Academic Editor: Diganta B. Das

Received: 11 March 2022
Accepted: 18 April 2022
Published: 26 April 2022

Publisher's Note: MDPI stays neutral with regard to jurisdictional claims in published maps and institutional affiliations.

Copyright: © 2022 by the authors. Licensee MDPI, Basel, Switzerland. This article is an open access article distributed under the terms and conditions of the Creative Commons Attribution (CC BY) license (https://creativecommons.org/licenses/by/4.0/).

1. Introduction

With ever-increasing global energy demand and calls for all-sector decarbonisation, interest in green and blue hydrogen is swelling. Hydrogen is and will continue to be a vital component for chemical and fertiliser manufacturing [1]. Whilst hydrogen is flexible and able to provide for a range of energy applications such as transport and energy storage, it is also highly attractive for heat in future energy landscapes [2]. Hydrogen production is currently dominated by steam methane reforming (SMR), which uses fossil-based natural gas as its feedstock [1]. The streamlined SMR process, which has benefitted from decades of optimisation, produces a cost-effective product, and the tailored global infrastructure has led multiple SMR plant operators to be open to operation with carbon capture utilisation and storage (CCUS) [3]. Combining fossil-derived hydrogen with CCUS has been termed "blue hydrogen".

Green hydrogen includes that derived from renewably fueled electrolysis of water or from biogenic feedstock [3]. Biogenic hydrogen is of particular interest from an environmental perspective due to the potential of introducing CCUS and therefore providing negative CO_2 emissions. Bioenergy with carbon capture and storage (BECCS) is one of the most promising options in not just limiting but reducing emissions according to the IPCC [4]. One encouraging method of generating H_2 from biomass is via the steam reforming of bio-oil. Bio-oil is the energy-dense liquid formed from pyrolysis of biomass-derived feedstocks, and its steam reforming has shown advantages in yield for hydrogen production compared to alternatives such as biomass gasification with shift conversion [5]. H_2 production may be an effective method to upgrade bio-oil, which suffers from medium–low heating value, high acidity, and chemical instability [6]. However, bio-oil is far easier to transport than H_2 and provides the potential for centralised plants that can benefit from economies of scale.

Recent advancements in reforming techniques such as sorption enhancement and chemical looping may provide the spark to bring hydrogen from biomass and BECCS into future energy markets [7]. Sorption-enhanced steam reforming (SE-SR) performs in situ CO_2 removal with a high-temperature sorbent in the reformer, providing a product stream of high H_2 purity. Moreover, the in situ CO_2 removal provides a favourable chemical equilibrium shift, aiding yields, meeting high temperature requirements, and forming an ideal foundation for CCUS. CaO is the most popular sorbent choice due to its low cost and availability, whilst demonstrating strong affinity for CO_2 sorption and capture [7].

Chemical looping steam reforming (CLSR) uses oxygen transfer material (OTM) for partial oxidation of the feedstock, which provides heat for autothermal conditions. The partial oxidation produces CO_2 as a by-product which also lends itself to CCU opportunities. The OTM is normally formed of a metal oxide such as Cu, Fe_2O_3, NiO, or Mn_3O_4 supported on an inert material such as Al_2O_3, $MgAl_2O_4$, SiO_2, TiO_2, or ZrO [8]. The OTM not only provides the oxygen for partial oxidation, but also often acts as a catalyst for steam reforming or water gas shift. As such, OTM analysis and selection make up the bulk of literature on CLSR, with nickel-based options being the most extensively researched [9]. Ni-based OTMs not only show high reactivity, high temperature stability and high selectivity to syngas production [8,10–12], but also are relatively low-cost and commercially widespread.

Sorption-enhanced chemical looping steam reforming (SE-CLSR) integrates the technical aspects behind both CLSR and SESR to provide autothermal operation and a high-purity product with in situ carbon capture [7,13–15]. SE-CLSR is characterised by at least two-stage cycling, where saturated sorbent is regenerated at higher temperatures by heat generated by OTM re-oxidation. The thermodynamic study presented by Spragg et al. [15] showcased the benefits of bio-oil SE-CLSR in purity, yield, carbon deposition, and process efficiency.

Because of the predominance of experimental and thermodynamic studies on bio-oil reforming, there is a need for techno-economic investigation to assess the potential for commercialisation and widespread implementation. Previous studies on steam reforming of bio-oil have revealed it can produce cost-competitive H_2. In 2010, Sarkar and Kumar [6] showed H_2 from autothermal bio-oil steam reforming from whole-tree biomass, forest residue, and agricultural biomass could be costed at USD 2.40, USD 3.00, and USD 4.55 per kg H_2, respectively. In 2014, Brown et al. [16] calculated conventional steam reforming (CSR) of bio-oil to produce H_2 at USD 3.25 to USD 5 per kg.

There is also scope to produce a techno-economic evaluation of the CO_2 capture potential in line with bio-oil reforming to H_2. Numerous studies have investigated CO_2 capture with steam methane reforming [17–19]. In 2021, a review by Yang et al. [20] detailed the avoidance costs for SMR plants ranging from EUR 40 to EUR 130 per t CO_2. In the same review, CLSR avoidance costs of EUR 86 per t CO_2 and advanced autothermal reforming systems as low as EUR 18 per t CO_2 were showcased. To the authors' knowledge, this paper will be the first of its kind to perform techno-economic studies on the CO_2 capture from bio-oil steam reforming, which can be compared to alternatives. This is of particular interest due to the negative emission potential of using a biogenic feedstock such as bio-oil.

2. Materials and Methods

The methods used for the techno-economic analysis operate on the basis that the plant is located in an industrial area such as Teesside (United Kingdom), where H_2 pipeline infrastructure can be taken advantage of. It is proposed that the H_2 is prepared under the same conditions as the H21 Leeds City Gate project [21] which also sets its plant location at Teesside. A H_2 export pressure of 40 bar was therefore assumed, at 25 °C and hydrogen purity greater than 99.98%.

Teesside was also chosen as the location for the case study due to the region's inevitable participation in future CO_2 capture and storage (CCS), where CO_2 will be piped to empty North Sea oil fields [22]. Where CO_2 capture was considered, a set of purity conditions were applied to the separated CO_2 to maintain transportation and storage infrastructure integrity. Given a lack of standardised CO_2 purity specifications, those used in this study and presented in Table 1 were based on those generated by CCS stakeholders in the CO_2 Europipe project [23]. For supercritical phase transportation, 110 bar was assumed as the specified CO_2 pressure.

Table 1. CO_2 specifications.

Component	Limit in CO_2
CO_2	>95 vol%
Ar CH_4 H_2 N_2 O_2	Total noncondensables <5 vol%
H_2O	No free water (<500 ppm_v)

The discussed facility is assumed as a centralised reforming plant that receives feedstock from multiple pyrolysis sites. This combines the benefits of economies of scale for the reforming stage with providing realistic capacities for pyrolysis from bio-compounds and associated feedstock limitations. A range of 5000 to 100,000 $Nm^3\ h^{-1}$ of bio-oil from 1 to 20 pyrolysis plants, to feed a central reforming facility, was used to analyse the impact of scale on the techno-economics.

2.1. Bio-Oil Feedstock

Bio-oil was modelled using a surrogate mixture, as in the work of Spragg et al. [15], closely resembling the elemental composition and differential thermogravimetric (DTG) curve of a real palm empty fruit bunch (PEFB) bio-oil [24]. Sensitivity analysis on PEFB bio-oil model mixtures shows equilibrium results are not sensitive to the exact mixture composition, provided a known elemental composition [25]. The bio-oil surrogate mixture was based on the work of Dupont et al. [26], and the bio-oil has been represented with a mixture of 6 macro-families following the methodology of García-Pérez et al. [27]. The mass fraction of each compound is described by Spragg et al. [15] and in the Supplementary Materials (S1). In this study, it is assumed that the bio-oil is mixed with 10 wt% methanol to reduce its viscosity and density [6]. Stainless steel tanks are used to store the bio-oil due to its corrosive nature [28].

2.2. Desulphurisation

Many existing techno-economic studies on bio-oil reforming, have assumed sufficiently low sulphur content in bio-oils to avoid the requirement for desulphurisation [6,29,30]. However, as this is a potentially important sensitivity for reforming catalysts, the impact of desulphurisation is considered and compared to a base case without. Assumptions for desulphurisation are based upon data available for naphtha hydrodesulphurisation (HDS), a common approach in refining [31]. Transition metal catalysts, such as sulphided $CoMo/Al_2O_3$ and $NiMo/Al_2O_3$, convert sulphur compounds in the liquid feedstock into H_2S, via reaction with hydrogen [32]. As well as consuming hydrogen, the process is a net consumer of power and steam, as well as fuel gas for a fired heater.

Sulphur levels are assumed equivalent to those used for the inlet to naphtha reforming, around 0.5 to 1 ppmwt [31,33]. Detailed process design was not performed for desulphurisation, rather order of magnitude estimates were used for techno-economic considerations based on data from Maples [31], such as the utilities presented in Table 2 and single point cost data in Table 3. Hydrogen consumption for a given wt% sulphur in the feed was derived from a correlation within the same work. The analysis performed details only the costs associated with desulphurisation and does not illustrate the potential benefits of improving catalyst lifetime and performance.

Table 2. Utilities consumption for desulphurisation.

Utility	Requirement per m^3 Bio-Oil/Methanol Feed
Power	12.58 kWh
Steam	42.79 kg
Fuel gas	55.30 kWh

2.3. Economic Costing

Levelised cost of hydrogen (LCOH) was used for a consistent comparison between the processes and the comparative systems in the literature. LCOH estimates the H_2 product value required to recover lifetime project costs, as calculated in Equation (1):

$$\text{LCOH} = \frac{\sum_{t=1}^{n} \frac{TCI_t + COM_{d,t}}{(1+r)^t}}{\sum_{t=1}^{n} \frac{H_t}{(1+r)^t}} \quad (1)$$

where n is the lifetime of the project, TCI_t is the capital investment, and $COM_{d,t}$ is the cost of manufacture in year t. H_t is the hydrogen generated in year t. The time value of money is accounted for by the discount rate (r), which discounts costs to the present value over the plant's lifetime.

For economic quantification of CO_2 capture, the cost of CO_2 avoided (CCA) was calculated; the CCA can be defined as the required carbon tax value for competitive CO_2 capture against a benchmark plant [18], as calculated in Equation (2):

$$CCA = \frac{\text{LCOH} - \text{LCOH}_{ref}}{E_{H_2,ref} - E_{H_2}} \quad (2)$$

where LCOH and LCOH$_{ref}$ are the levelised cost of hydrogen (USD kg$_{H_2}^{-1}$) in the plant with and without CO_2 capture, respectively. E_{H_2} and $E_{H_2,ref}$ are the specific emissions per unit production of H_2 (kg$_{CO_2}$ kg$_{H_2}^{-1}$) of the plant with and without CO_2 capture, respectively.

The total emissions include not only CO_2 emissions in the flue gases, but also those associated with imports and exports of electricity and steam. The specific emissions (kg$_{CO_2}$ kg$_{H_2}^{-1}$) were calculated as in Equation (3):

$$E_{H_2} = \frac{\dot{m}_{CO_2} + \left(\dot{Q}_{th}^+ - \dot{Q}_{th}^-\right)E_{th} + \left(\dot{P}_{el}^+ - \dot{P}_{el}^-\right)E_{el}}{\dot{m}_{H_2}} \quad (3)$$

where $\dot{m}_{CO_2}$ is flue gas CO_2 mass flow rate and $\dot{m}_{H_2}$ is the H_2 mass flow rate. E_{th} and E_{el} are the thermal and electrical emissions factors, respectively. $\dot{Q}_{th}$ and $\dot{P}_{el}$ are the thermal energy and electrical power, with + and − subscripts to signify imports and exports, respectively. Emission factors are taken from European Union data [34], where E_{el} is 0.391 kg kWh^{-1} and E_{th} is 0.224 kg kWh^{-1}, assuming 90% natural gas boiler efficiency. Any emissions of biogenic origin have been accounted as carbon-neutral.

Bare module costs were taken from Turton et al. [35] as much as possible, and the size factor was accounted for using Equation (4):

$$log_{10}C_P^o = K_1 + K_2 log_{10}(A) + K_3[log_{10}(A)]^2 \quad (4)$$

where C_P^o is the purchased cost of equipment at base case conditions (ambient operating pressure and carbon steel construction) and A is the size parameter. Aspen Plus–derived size parameters were used to calculate equipment cost under base conditions.

The purchased cost (C_p^o) was then multiplied by a series of factors that account for deviations from the base conditions, including specific equipment type, system pressure, and materials of construction, as outlined by Turton et al. [35,36] and described in the Supplementary Materials (S2). Due to the corrosiveness of bio-oil, exposed process parts were assumed as stainless steel. Remaining parts were assumed as carbon steel.

For systems and processing units where data was unavailable from Turton et al. [35], bare module costs (C_{mod}) were acquired from the literature and scale-adjusted using Equation (5):

$$C_{mod} = C_{mod,0} \left(\frac{S}{S_0}\right)^f \times I \qquad (5)$$

where f is the scaling exponent, S_{mod}, and $C_{mod,0}$ and $S_{mod,0}$ are the cost and size of the reference case, respectively. The value I is the installation factor (where given). Table 3 details the process units costed using this method with associated data for Equation (5).

Table 3. Single point cost data for bare module cost.

Unit	Base Size	Base Cost (mUSD)	f	Installation Factor	Year	Ref.
WGS	15.6 Mmol h^{-1} CO + H$_2$	36.9	0.85	1	2001	[37]
PSA	9600 kmol h^{-1} throughput	28	0.7	1.69	2001	[37]
CO$_2$ capture (MDEA)	62.59 kg s^{-1} CO$_2$ captured	104.2	0.8	-	2017	[37]
CO$_2$ compression and drying	13 MW compressor power	17.9	0.67	-	2017	[18]
High temperature three-way valve	2 m^3 s^{-1}	0.1695	0.6	-	2014	[18]
HDS plant	30,000 BPD	16	0.65	-	1991	[31]

The refractory-lined reactor vessels in the SE-CLSR study were designed in more detail, based on the methods of Peters et al. [38] and Hamers et al. [39]. This was performed due to the identification of their influence on overall plant cost. Reactor volumes were estimated via catalyst weight hourly space velocity (WHSV) and sorbent quantity. On the basis of this reactor volume, the masses of steel and refractory material were calculated, providing cost. Full details of calculations are available in the Supplementary Materials (S3).

To account for inflation, all costs were aligned to the year 2018 using the Chemical Engineering Plant Cost Index (CEPCI):

$$C_{BM,2018} = C_{BM,base} \left(\frac{CEPCI_{2018}}{CEPCI_{base}}\right) \qquad (6)$$

where $CEPCI_{2018}$ is the index value from 2018 (603.1) and $CEPCI_{base}$ is the index value from the year of the source cost. $C_{BM,base}$ is the cost in the base year and $C_{BM,2018}$ is the 2018 adjusted cost.

Total capital investment (TCI) was calculated taking into account a number of additional factors. Firstly, fees were assumed at 3% of bare module costs C_{BM} [35], a contingency of 30% of C_{BM}, as recommended by NETL for a concept with bench-scale data [40]. Consideration for auxiliary facilities, such as site development and buildings, was accounted as 50% of C_{BM}. These generated a figure for total fixed capital investment (FCI). Working capital was assumed at 15% of FCI, which when applied, created the TCI.

Operating costs were determined using the method from Turton et al. [35]. When each operating factor is accounted for, the total cost of manufacture without depreciation (COM_d) is as shown in Equation (7):

$$COM_d = 0.18FCI + 2.73C_{OL} + 1.23(C_{UT} + C_{WT} + C_{RM}) \qquad (7)$$

where C_{OL} is operating labour costs, C_{UT} is utility costs, C_{WT} is waste treatment costs, and C_{RM} is costs of raw materials. Table 4 details further assumptions made for the calculation of COM_d.

Table 4. Operating cost calculation assumptions.

	Materials		
Bio-oil price		0.2 USD kg^{-1}	[41–43]
Methanol price		0.37 USD kg^{-1}	[44]
Reforming catalyst/oxygen carrier price		20 USD kg^{-1}	[45,46]
WGS catalyst price		60 USD kg^{-1}	[47]
CaO sorbent		1.1 USD kg^{-1}	[48]
WHSV for steam reforming		1 h^{-1}	[49]
GHSV for WGS		3000 h^{-1}	[45,50]
WHSV for reforming stage of SE-CLSR		0.8 h^{-1}	[51]
Reforming catalyst lifetime (C-SR)		1 year	Assumed
Oxygen carrier lifetime (SE-CLSR)		2 years	Assumed
WGS catalyst lifetime		5 years	[52]
CaO sorbent lifetime		2 years	Assumed
MDEA solvent [a]		0.04 mUSD/year per kg$_{CO_2}$/s	[18]
	Waste treatment		
Waste water disposal		0.538 USD t^{-1}	[53]
Catalyst recovery		−0.11 USD/kg	[54]
	Utilities		
Process water		2 USD m^{-3}	[55]
Electricity (purchase)		100 USD MWh^{-1}	[14]
Electricity (export)		50 USD MWh^{-1}	[14]
Steam (purchase/export)		20.9 USD MWh^{-1}	Calculated [b]
Natural gas		25 USD MWh^{-1}	[14]
Cooling water		0.4 USD m^{-3}	[55]
	Other assumptions		
Plant availability		360 days per year	-
Conversion GBP to USD		1.29	[56]
Conversion EUR to USD		1.13	[56]
Labour cost for workers in UK industry		GBP 40,000 per year	[57]
Shifts worked per worker per week		5	-
Shifts per day		3	-
Weeks worked per year		47	-

[a] MDEA solvent cost estimated from [18] prorated to process size. [b] Based on natural gas boiler with 90% efficiency [18].

2.4. Process Modelling Methodology

As previous studies on bio-oil steam reforming have achieved bio-oil conversion and hydrogen yields close to 100% [49], an equilibrium-based approach has been used throughout the Aspen Plus model. Peng–Robinson property method was selected as suggested for hydrogen-rich applications [58] and used for similar applications in the literature [51,59,60].

For all models, RGibbs reactors were used to simulate each reforming reactor. For C-SR an isothermal reforming reactor is connected to an isothermal burner by an energy stream, representing external firing on reformer tubes. For SE-CLSR, adiabatic RGibbs blocks were utilised, where the outlet temperature is determined by Aspen Plus on a chemical equilibrium basis. The WGS reactor was an adiabatic REquil reactor in which the WGS reaction is specified, instead of an RGibbs block, to represent a reaction supported by a catalyst selective to CO_2 rather than CH_4 production.

SE-CLSR reactors are difficult to simulate due to their packed bed design, in which different stages (time intervals with different feed streams) are initiated by switching gas inputs. Figure 1a illustrates how this is implemented in a packed bed, showing each SE-CLSR stage with the catalyst redox changes. Figure 1b demonstrates the autothermal cycle formed by each stage with a temperature–pressure diagram.

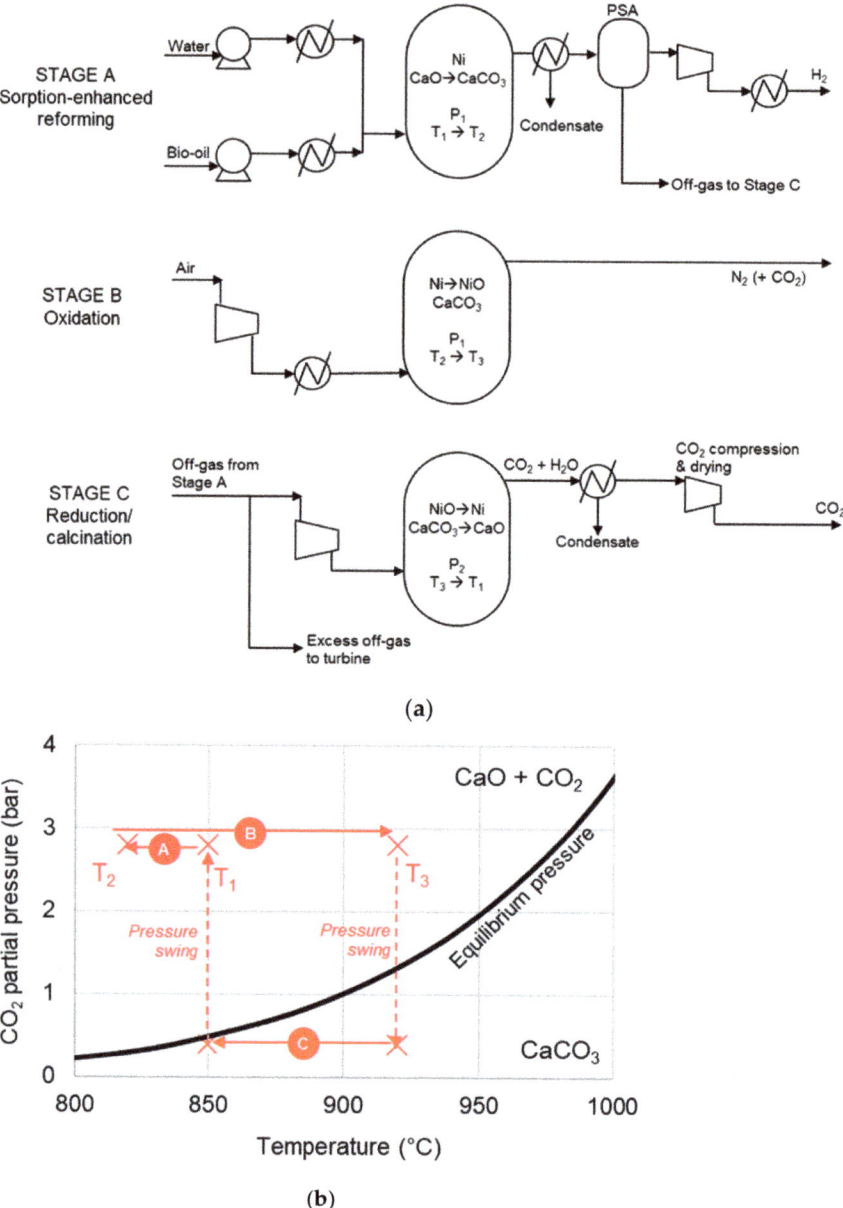

Figure 1. (a) Simplified process flow diagram of bio-oil SE-CLSR. (b) Example SE-CLSR operating conditions on CO_2 equilibrium partial-pressure diagram. CaO/CO_2 equilibrium properties from [61].

To model this gas switching process in Aspen Plus, each stage is represented by a different reactor block in the process flowsheet, despite them being of a singular vessel design in reality. Conceptual separator blocks isolate solids from the outlet of the reactors and are copied by transfer blocks as inputs to the next stage, rather than representing the physical movement of material between reactors. This approach instead simulates the retention of solids in the same reactor like a type of semi-batch process. Meanwhile, the C-SR is modelled as a continuous, steady-state process.

Separator blocks were used to simulate the PSA with a 90% H_2 recovery and the absorption-based capture process with 95% CO_2 recovery. Energy demand for capture and compression was taken from the work of Meerman et al. [62], who modelled an activated MDEA process in syngas at similar conditions. Pressure drops in heat exchangers were used as in the work of Seider et al. [63], and efficiencies of turbomachines were used as in the work of Spallina et al. [47]. Other assumptions were as follows:

- Where gas volumes are given in Nm^3, normal conditions are 20 °C and 1.01325 bar;
- Air is composed of 79% N_2 and 21% O_2;
- To ensure storage in liquid form, the bio-oil/methanol mixture is stored above its vapour pressure (around 3 bar);
- All other fluid inputs enter the system at 25 °C and 1.01325 bar;
- Reactor pressure drop is 5% of inlet pressure;
- Heat exchanger minimum approach of 10 °C.

Flue gases from the furnace and gas turbine are cooled to 180 °C before emission to the atmosphere [64]. Low-pressure (LP) steam at 6 bar and 160 °C is produced using process excess heat and sold as a by-product [47]. The only system heat imports are fuel gas for the net demands of C-SR and C-SR-CCS. The process flow diagrams for the Aspen Plus models for C-SR without CO_2 capture, C-SR with CO_2 capture, and the SE-CLSR process can be found in Figures 2–4, respectively. Reactions involved in each process are as in the work of Spragg et al. [15].

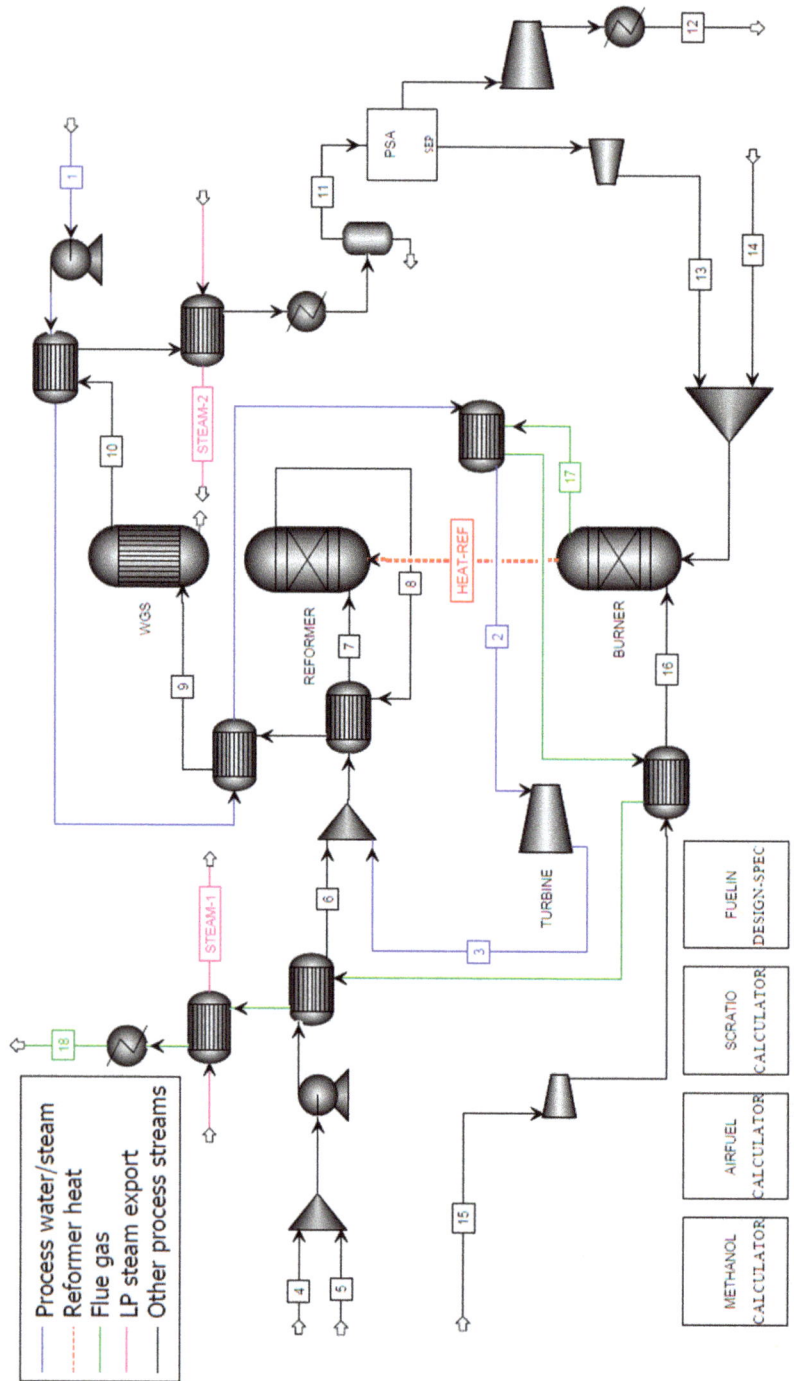

Figure 2. Aspen Plus flowsheet for C-SR of bio-oil without CO_2 capture.

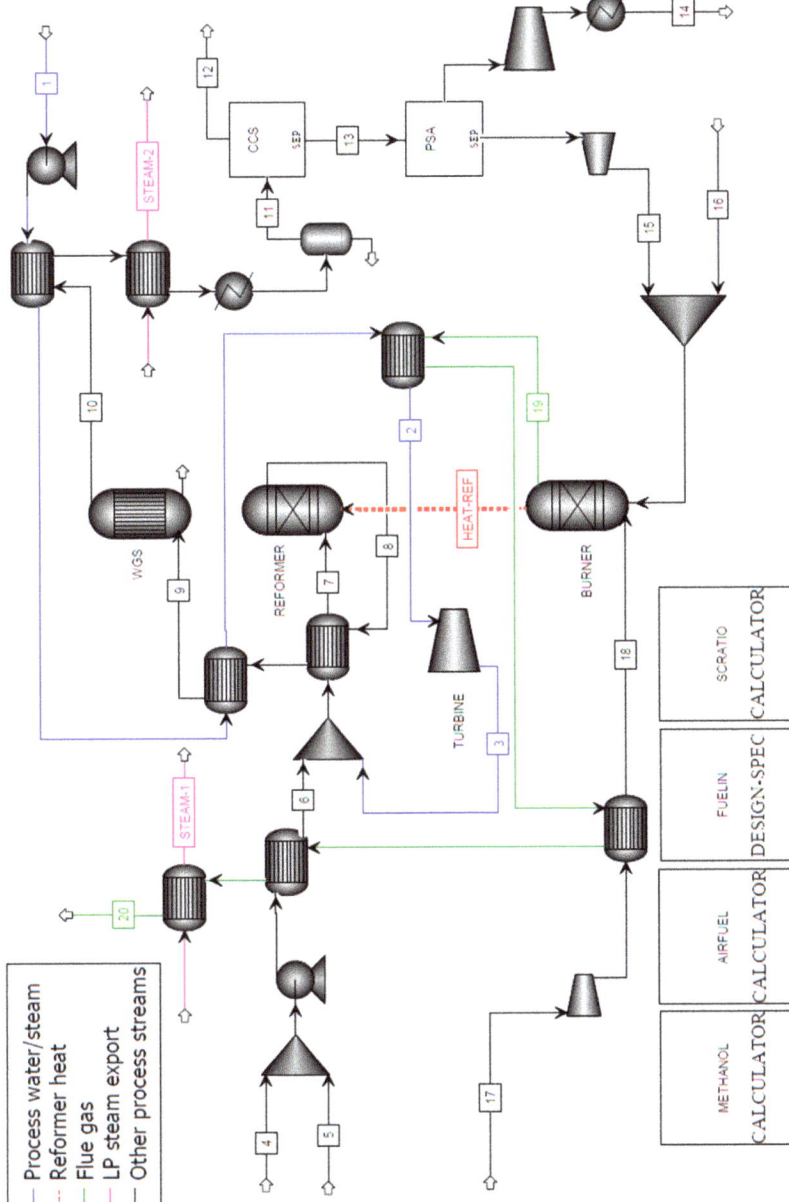

Figure 3. Aspen Plus flowsheet for C-SR of bio-oil with CO_2 capture.

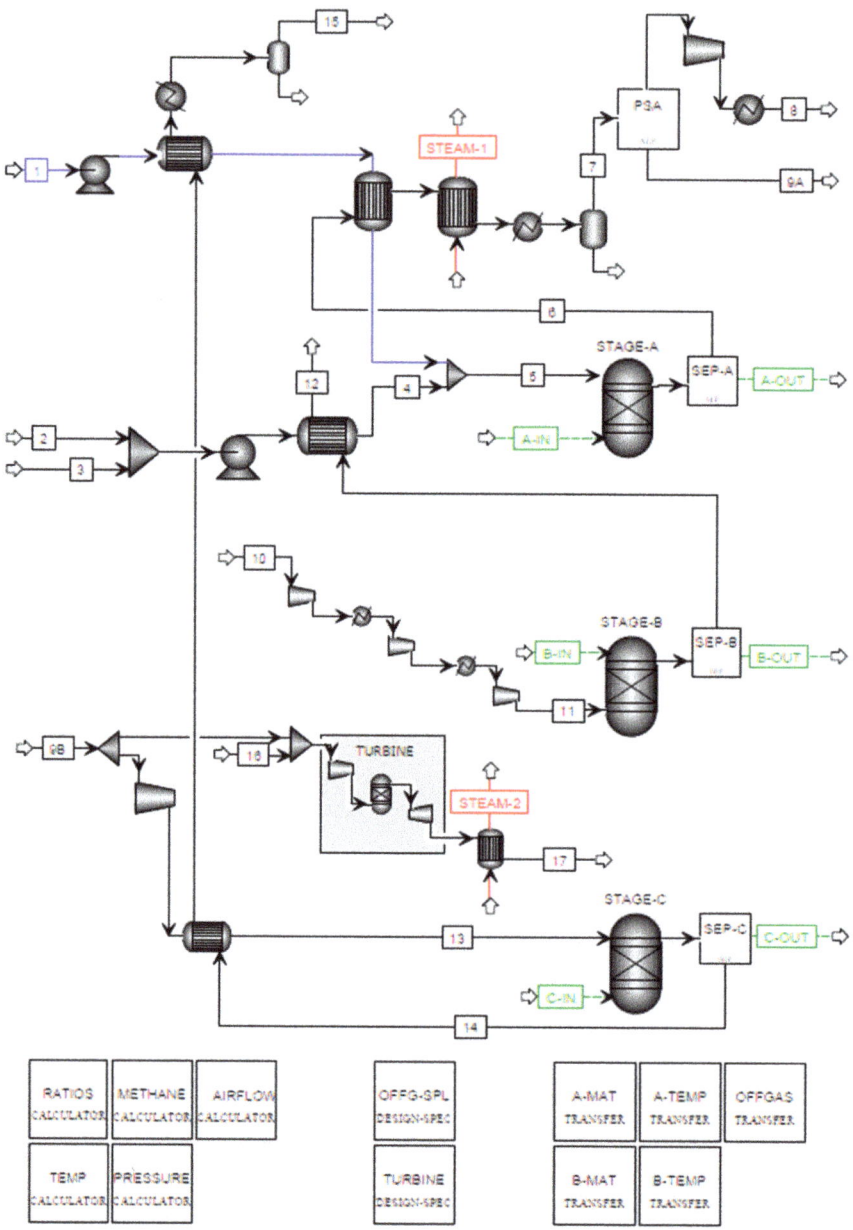

Figure 4. Aspen Plus flowsheet for SE-CLSR of bio-oil.

3. Results

3.1. Process Design Basis Selection

The process simulation performed on Aspen Plus provided sensitivity analysis for the impact of parameters such as temperature, pressure, steam:carbon (S/C) feed molar ratio, and steam export on the yield, thermal efficiency, and carbon emission potentials of each process. These results were utilised for the selection of a design basis for further economic analysis and comparison. The key conditions selected from the sensitivity analysis are provided in Table 5, with rationale provided in the following two sections.

Table 5. Design basis for economic comparisons.

	C-SR	C-SR-CCS	SE-CLSR
Reformer pressure (bar)	30	30	20
Reformer temperature (°C)	900	900	850
S/C ratio	5	5	2
NiO/C ratio	-	-	0.7

3.1.1. C-SR and C-SR-CCS

For the C-SR and C-SR-CCS processes, a reforming temperature of 900 °C was chosen, which provided close to maximum yields and thermal efficiencies. A pressure of 30 bar was also selected, which improved thermal efficiencies compared to lower pressures due to the benefits in operations such as hydrogen compression demand, despite providing lower reformer yields.

Under the C-SR scenario, increasing the S/C ratios between 3 and 7 boosted yields but decreased thermal efficiencies. The additional H_2 product was outweighed by the energy demand for supporting increased steam use. However, if LP steam is exported from the system, then increasing S/C between 3 and 5 provides marginally higher efficiencies as the exported steam counters the initial energy input for raising the steam. Exporting steam tends to increase thermal efficiencies by around 10%. However, not all plants will be able to export their steam, so this should be scrutinised on a case-by-case basis.

Due to the complex impact of S/C on operation, the effect of varying S/C on LCOH at varying scales has been analysed and is presented in Figure 5. It shows that there is only a marginal difference in LCOH between S/C 3 and 5 scenarios. This arises because the increased costs to raise the additional steam are offset by increased H_2 yields and furthered by steam export potential (Figure 5b). However, at S/C 7, the steam generation costs are considerably greater, causing the increase in LCOH illustrated in Figure 5. These results suggest an S/C of 5 may be optimal, especially when considering the added benefits in catalyst carbon deposition, which is not considered in the analysis. An S/C of 5 has, therefore, been selected for further economic evaluation.

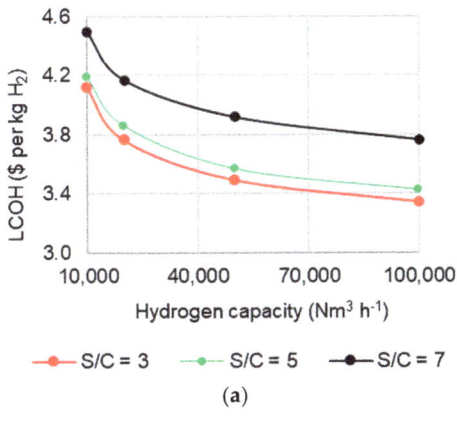
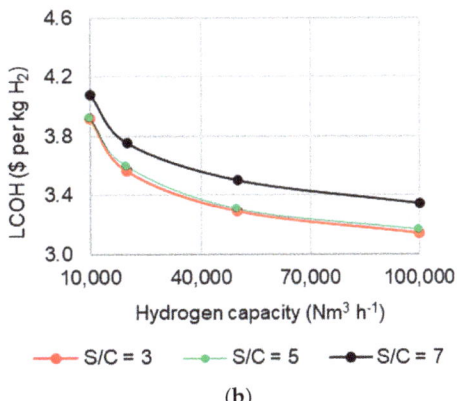

Figure 5. Effect of S/C ratio and capacity on levelised cost of hydrogen in C-SR at 30 bar and 900 °C (**a**) without steam export and (**b**) with steam export.

Figure 5 also shows the positive impact steam export has on the LCOH, reducing the value by around 10%. This further reinforces the worth of external heat integration. For C-SR-CCS, the process analysis showed the value of utilising the excess reforming heat within the CO_2 capture process and therefore holds inherent value.

3.1.2. SE-CLSR

Sensitivity analysis with the SE-CLSR process showed that increasing the temperature (T_1 as in Figure 1b) above 850 °C decreases the overall yield, as does increasing the S/C ratio. Therefore, 850 °C was selected for the base scenario and further analysis. The higher NiO/C ratio required with high S/C ratios to sustain the autothermal temperature cycle increases bed heat capacity, oxidation air requirement, and off-gas used for reduction. Therefore, both yield and thermal efficiency are reduced with the higher solids inventory associated with greater NiO/C ratios. For this reason, a low S/C ratio of 2 and NiO/C ratio of 0.7 were selected for further techno-economic analysis.

If steam export is not included in the SE-CLSR process, increasing pressure between 20 and 40 bar leads to an approximately 5% drop in thermal efficiency. However, if steam export is considered, thermal efficiency is not significantly affected by changes in pressure, as the ability to export spare heat compensates for the drop in hydrogen yield. A pressure of 20 bar was selected for further analysis, as the condition fits both scenarios.

3.2. Process Cost Comparison

The three processes applied with conditions selected in the design basis were compared for economic performance. Fixed capital, cost of manufacture, and LCOH results against capacity are presented in Figure 6. Here, the benefit of economies of scale is clear. For example, in the case of SE-CLSR, LCOH is reduced by 19%, from USD 4.63 to USD 3.76 per kg H_2 with an increase in process size from 10,000 to 100,000 $Nm^3\ h^{-1}$.

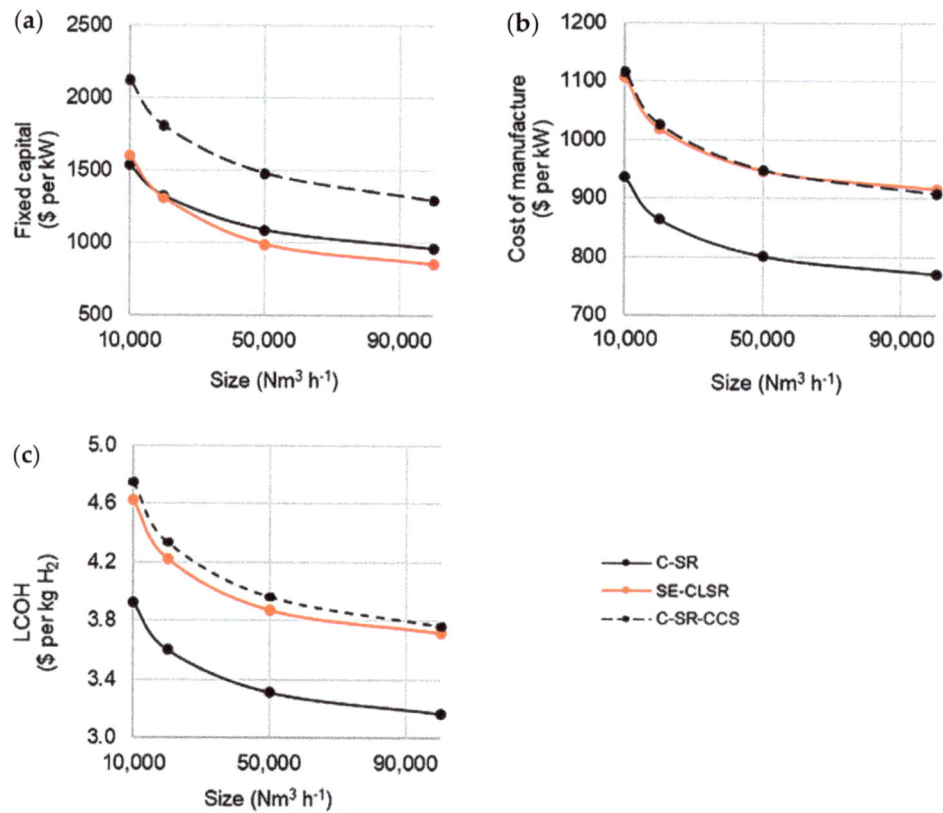

Figure 6. Cost analysis of base case C-SR and SE-CLSR processes with steam export: (**a**) fixed capital costs, (**b**) cost of manufacture, and (**c**) LCOH.

The LCOH for both SE-CLSR and C-SR-CCS is, as expected, greater than that of C-SR without CCS, reflecting the costs of CO_2 capture capabilities. SE-CLSR has comparable fixed capital requirements to C-SR, but that is countered by the high costs of manufacture forecasted for SE-CLSR. Nonetheless, the comparatively lower fixed capital range for SE-CLSR is reflected in a marginally lower LCOH than for C-SR-CCS. However, accounting for a level of uncertainty, the significance of this difference may be questioned.

Although the LCOH for each process option may be higher than that of other hydrogen production methods [65], these process options may still be competitive. Their low-carbon/carbon-negative status provides additional value. Additionally, the Hydrogen Council projects that H_2 costs at the pump of USD 6 per kg would still be cost-competitive for 15% of transport energy demand by 2030 [66].

The bare module and manufacturing costs have been broken down further for interpretation as displayed in Figure 7 for C-SR-CCS and SE-CLSR. For both processes, the PSA is the most costly module at 28% and 41% of totals for C-SR-CCS and SE-CLSR, respectively. The CO_2 capture unit for C-SR-CCS is a further 25%. This brings the total cost of gas separation in C-SR-CCS (PSA and CO_2 capture) to 53% of the total, whereas an equivalent CO_2 capture unit is not required for gas separation in SE-CLSR. The three-way valves incorporated in SE-CLSR also contribute significantly at 13% of the total cost.

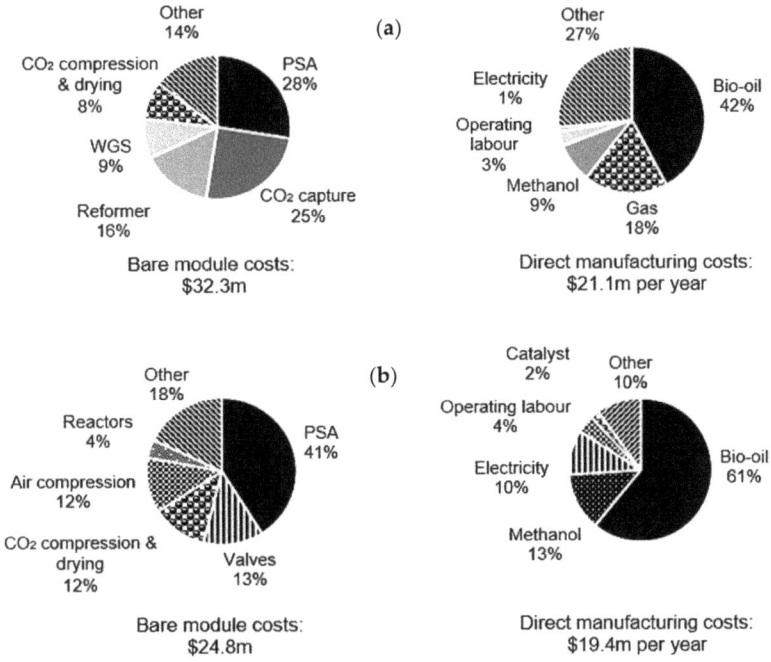

Figure 7. Breakdown of bare module costs and direct manufacturing costs in a 10,000 $Nm^3\ h^{-1}$ process for (**a**) C-SR-CCS and (**b**) SE-CLSR.

Under both scenarios, bio-oil purchasing presents the most substantial manufacturing cost at 42% and 61% of the total for C-SR-CCS and SE-CLSR, respectively. The greater influence of bio-oil on SE-CLSR manufacturing cost is the case, in part, because of the heat supplied from the bio-oil/methanol feeds rather than a cheaper fossil-fuel alternative, such as natural gas. Nonetheless, the emission reduction potential is highly attractive and should be factored into assessments. The electricity demand is also a noteworthy contributor to SE-CLSR operating costs, highlighting potential benefits from optimising towards greater self-sufficiency.

3.3. Carbon Emission Comparison

Table 6 details the emission balance for each process. The net CO_2 emissions are the biogenic CO_2 captured minus the fossil CO_2 emissions, whereas the avoided CO_2 is the biogenic CO_2 captured plus the difference in fossil emissions from the C-SR reference case. SE-CLSR has a superior emission outlook to C-SR-CCS because only 10% of the system CO_2 emissions are derived from electricity import and methanol use. The rest of the CO_2 is of biogenic origin, and all of the heat demand is met by the feedstock. For C-SR-CCS, around 15% of process emissions are fossil-derived, in part from the methane (natural gas) demand to top up the furnace requirements.

Table 6. Comparison of emissions from processes ($kg_{CO_2}\ kg_{H_2}^{-1}$).

Process	Fossil-Based CO_2 Emitted	Biogenic CO_2 Captured	Net CO_2 Emissions	CO_2 Avoided
C-SR	3.2	0	3.2	-
C-SR-CCS	0.46	8.7	−8.2	11.4
SE-CLSR	1.1	10.6	−9.5	12.7

The difference in the emissions is reflected in the cost of carbon, as presented in Figure 8 where the CCA for SE-CLSR is lower than that for C-SR-CCS at all scales. For the scale selection shown in Figure 8, the CCA of SE-CLSR ranges from 44 to 55 USD/teCO_2, whereas C-SR-CCS ranges from 52 to 72 USD/teCO_2.

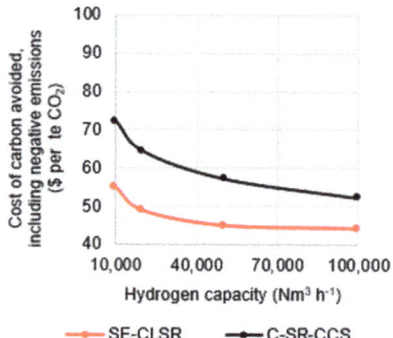

Figure 8. Cost of carbon avoided in C-SR-CCS and SE-CLSR, compared to bio-oil C-SR base case.

The avoided emissions and associated CCA analysis above is based on the use of bio-oil steam reforming as the reference case. If a conventional SMR process is used as a reference, the CCA for the same capacity range is between 94 and 144 USD/teCO_2 for SE-CLSR and between 103 and 163 USD/teCO_2 for C-SR-CCS. This is useful for comparison against other bioenergy with carbon capture and storage (BECCS) processes presented by Consoli [67]. The CCA of bio-oil C-SR-CCS and SE-CLSR is competitive against combustion and ethanol BECCS but 2–4 times greater than that of pulp/paper mills and biomass gasification.

3.4. Sensitivity

Figure 9 provides insight into the effect of key economic factors on the LCOH and CCA through sensitivity analysis. As expected, due to the weight of bio-oil purchasing on manufacturing costs (Figure 7), altering its price (+/−)20% has the largest impact on LCOH compared to other costs. For example, a 20% decrease in bio-oil price results in a 6.4% reduction in the LCOH for C-SR-CCS, whereas the same percentage reduction in natural gas price results in a 2.5% reduction in LCOH for C-SR-CCS. An alteration in PSA price has the greatest impact on LCOH of any of the bare module costs. This shows that future reductions or increases in the purchasing prices of these factors will significantly impact the financial outlook of process implementation.

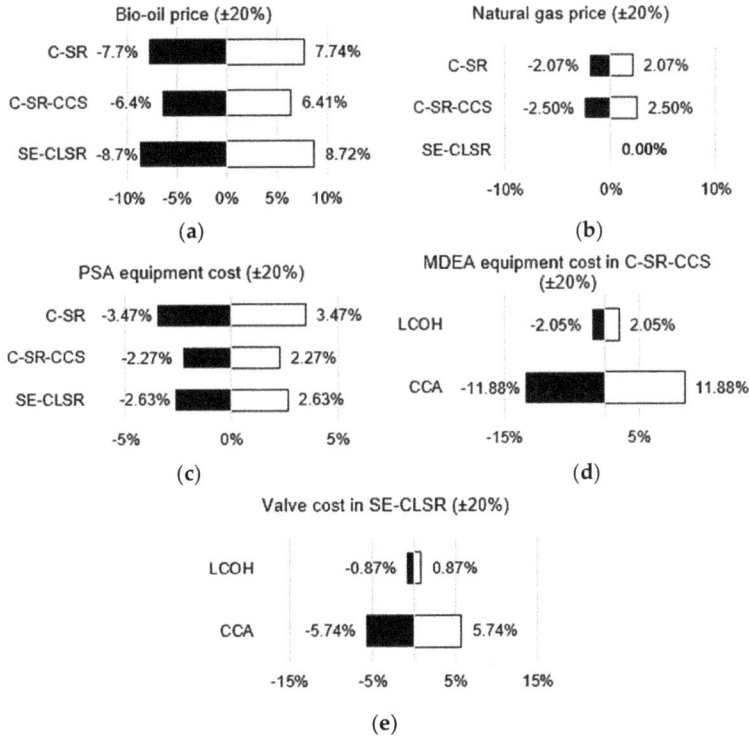

Figure 9. Various economic sensitivity analyses: (**a**) effect of bio-oil price on LCOH; (**b**) effect of natural gas price on LCOH; (**c**) effect of PSA equipment cost on LCOH; (**d**) effect of MDEA cost on LCOH and CCA for C-SR-CCS; (**e**) effect of valve cost on LCOH and CCA for SE-CLSR.

Variations in MDEA and valve cost are of interest because price variations have a limited impact on LCOH for C-SR-CCS and SE-CLSR but significantly influence CCA. For example, a 20% increase in MDEA equipment cost would increase the CCA of C-SR-CCS by almost 12%. Similarly, a 20% increase in the cost of three-way valves would boost the CCA by 5.74% for SE-CLSR. Therefore, if there are strong future incentives for BECCS/negative emissions, then plant operators need to be aware of factors such as this affecting the processing costs.

3.5. Desulphurisation Impacts

As mentioned, desulphurisation is often unaccounted for in bio-oil reforming studies due to the presumption of sufficiently low sulphur levels. Accordingly, a summary has been generated of the technical and economic impacts of HDS implementation based on a naphtha HDS system, reducing sulphur levels from 0.2 wt% to 1 ppmwt [31]. This is likely to far exceed the needs of a bio-oil plant and provides a conservative outlook for feasibility assessment. The analysis shows that despite the minimal impact on thermal efficiency and yields, the 11% addition bare module costs at a 10,000 $Nm^3_{H_2}$ h^{-1} scale would increase the LCOH by 5% from 3.93 to 4.13 USD $kg_{H_2}{}^{-1}$.

This analysis does not incorporate the positive impact desulphurisation may have on catalyst lifetime and, therefore, plant economics. As such, Figure 10 shows the effect of reforming catalyst lifetime on LCOH for C-SR at 10,000 Nm^3 h^{-1}. It shows that beyond a two-year catalyst lifetime there is not much difference in LCOH. However, below this and between 0 and 1 years particularly, there are definite benefits of prolonging the catalyst lifetime. In practice, the benefit of an improved lifetime would be superior to that shown in Figure 10 as fewer catalyst replacements would facilitate lower maintenance costs, downtime, and safety impacts. An improved understanding of catalyst lifetime in bio-oil reforming is required to further analyse and quantify the effects of improvement strategies.

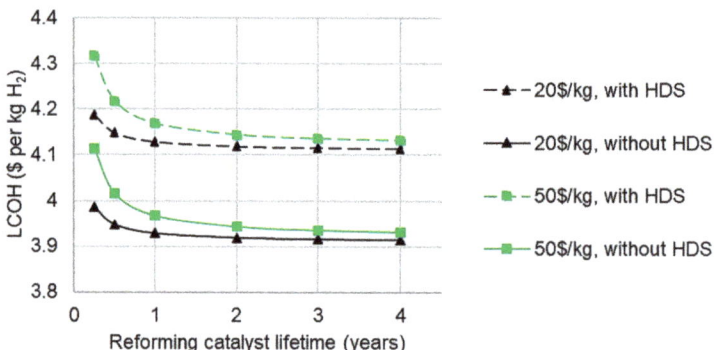

Figure 10. Effect of catalyst lifetime, cost, and hydrodesulphurisation on levelised cost of hydrogen from C-SR, 10,000 Nm^3 h^{-1}.

Considering the predicted lower sulphur content of bio-oil compared to naphtha, HDS may not be the most suitable. For example, metal oxide sulphur guard beds may be more cost-effective, as evidenced by ZnO sulphur guard beds for biogas and biomass syngas [68] or Matheson's Nanochem GuardBed used for bio-ethanol, bio-diesel, and biogas applications [69]. However, the limited available data restrict the current techno-economic impact analysis, and this area is suggested for future research.

4. Conclusions

The economic analysis showed that SE-CLSR is comparable to C-SR-CCS for the levelised cost of hydrogen (LCOH), with costs in the region of USD 3.8 to USD 4.6 per kg. These costs are similar to projected costs for the more conventional hydrogen with the BECCS biomass gasification route. As expected, they are higher than those of other H_2 production routes, but within the range that the Hydrogen Council has predicted will make H_2 cost-competitive at forecourts; thus, bio-oil might have a role in a diversified hydrogen production sector, especially if the potential value of negative emissions is considered.

The cost of carbon avoided (CCA) was shown to vary considerably depending on its method of calculation. If all emissions, both biogenic and fossil-based, were considered the same, the cost would range from USD 60 to USD 100 per ton equivalent (te) of CO_2. If biogenic emissions captured were considered as "negative emissions", the cost would be reduced to USD 40 to USD 70 per te CO_2. When a natural-gas-based SMR process was used as the reference to calculate CCA, the CCA increased to USD 90 to USD 160 per te CO_2 because methane is a considerably less expensive feedstock. However, for larger-scale plants (100,000 Nm^3 h^{-1}), the CCA of USD 95 to USD 105 per te CO_2 was within the range of BECCS in other industries.

Significant contributors to process cost were the PSA system, CO_2 capture (in the case of C-SR-CCS), and three-way valves (in the case of SE-CLSR). The high capital cost of hydrodesulphurisation could increase the LCOH by around 11%. However, this expense may be justified if required to extend catalyst lifetime, especially considering the potential costs associated with high process downtime.

Supplementary Materials: The following supporting information can be downloaded at: https://www.mdpi.com/article/10.3390/cleantechnol4020018/s1, S1: Bio-oil composition; S2: Equations for bare module cost of equipment; S3: reactor design.

Author Contributions: Conceptualisation, J.R., V.D. and T.M.; methodology, J.R.; software, J.R.; validation, V.D. and T.M.; formal analysis, J.R.; investigation, J.R.; resources, V.D. and T.M.; data curation, J.R.; writing—original draft preparation, O.G.; writing—review and editing, O.G. and V.D.; visualisation, O.G. and V.D.; supervision, V.D. and T.M.; project administration, J.R., V.D. and T.M.; funding acquisition, V.D. and T.M. All authors have read and agreed to the published version of the manuscript.

Funding: This research was funded by EPSRC Bioenergy Centre for Doctoral Training, grant number EP/L014912/1, and KCCSRC EPSRC consortium Call 2 grant "Novel Materials and Reforming Process Route for the Production of Ready-Separated $CO_2/N_2/H_2$ from Natural Gas Feedstocks", grant number EP/K000446/1.

Institutional Review Board Statement: Not applicable.

Informed Consent Statement: Not applicable.

Data Availability Statement: The data associated with this paper can be downloaded from https://doi.org/10.5518/1143 (accessed on 10 March 2022).

Conflicts of Interest: The authors declare no conflict of interest.

References

1. Abdin, Z.; Zafaranloo, A.; Rafiee, A.; Mérida, W.; Lipiński, W.; Khalilpour, K.R. Hydrogen as an Energy Vector. *Renew. Sustain. Energy Rev.* **2020**, *120*, 109620. [CrossRef]
2. Dawood, F.; Anda, M.; Shafiullah, G.M. Hydrogen Production for Energy: An Overview. *Int. J. Hydrogen Energy* **2020**, *45*, 3847–3869. [CrossRef]
3. Noussan, M.; Raimondi, P.P.; Scita, R.; Hafner, M. The Role of Green and Blue Hydrogen in the Energy Transition—A Technological and Geopolitical Perspective. *Sustainability* **2020**, *13*, 298. [CrossRef]
4. Intergovernmental Panel on Climate Change (IPCC) IPCC. *Climate Change 2021: The Physical Science Basis*; Cambridge University Press: Cambridge, UK; New York, NY, USA, 2021.
5. Wang, D.; Czernik, S.; Montané, D.; Mann, M.; Chornet, E. Biomass to Hydrogen via Fast Pyrolysis and Catalytic Steam Reforming of the Pyrolysis Oil or Its Fractions. *Ind. Eng. Chem. Res.* **1997**, *36*, 1507–1518. [CrossRef]
6. Sarkar, S.; Kumar, A. Large-Scale Biohydrogen Production from Bio-Oil. *Bioresour. Technol.* **2010**, *101*, 7350–7361. [CrossRef] [PubMed]
7. Dou, B.; Zhang, H.; Song, Y.; Zhao, L.; Jiang, B.; He, M.; Ruan, C.; Chen, H.; Xu, Y. Hydrogen Production from the Thermochemical Conversion of Biomass: Issues and Challenges. *Sustain. Energy Fuels* **2019**, *3*, 314–342. [CrossRef]
8. Tang, M.; Xu, L.; Fan, M. Progress in Oxygen Carrier Development of Methane-Based Chemical-Looping Reforming: A Review. *Appl. Energy* **2015**, *151*, 143–156. [CrossRef]
9. Adanez, J.; Abad, A.; Garcia-Labiano, F.; Gayan, P.; De Diego, L.F. Progress in Chemical-Looping Combustion and Reforming Technologies. *Prog. Energy Combust. Sci.* **2012**, *38*, 215–282. [CrossRef]
10. Ortiz, M.; De Diego, L.F.; Abad, A.; García-Labiano, F.; Gayán, P.; Adánez, J. Catalytic Activity of Ni-Based Oxygen-Carriers for Steam Methane Reforming in Chemical-Looping Processes. *Energy Fuels* **2012**, *26*, 791–800. [CrossRef]
11. Pröll, T.; Bolhàr-Nordenkampf, J.; Kolbitsch, P.; Hofbauer, H. Syngas and a Separate Nitrogen/Argon Stream via Chemical Looping Reforming—A 140 KW Pilot Plant Study. *Fuel* **2010**, *89*, 1249–1256. [CrossRef]
12. Yu, Z.; Yang, Y.; Yang, S.; Zhang, Q.; Zhao, J.; Fang, Y.; Hao, X.; Guan, G. Iron-Based Oxygen Carriers in Chemical Looping Conversions: A Review. *Carbon Resour. Convers.* **2019**, *2*, 23–34. [CrossRef]
13. Dou, B.; Zhang, H.; Cui, G.; Wang, Z.; Jiang, B.; Wang, K.; Chen, H.; Xu, Y. Hydrogen Production by Sorption-Enhanced Chemical Looping Steam Reforming of Ethanol in an Alternating Fixed-Bed Reactor: Sorbent to Catalyst Ratio Dependencies. *Energy Convers. Manag.* **2018**, *155*, 243–252. [CrossRef]
14. Pimenidou, P.; Rickett, G.; Dupont, V.; Twigg, M.V. Chemical Looping Reforming of Waste Cooking Oil in Packed Bed Reactor. *Bioresour. Technol.* **2010**, *101*, 6389–6397. [CrossRef]
15. Spragg, J.; Mahmud, T.; Dupont, V. Hydrogen Production from Bio-Oil: A Thermodynamic Analysis of Sorption-Enhanced Chemical Looping Steam Reforming. *Int. J. Hydrogen Energy* **2018**, *43*, 22032–22045. [CrossRef]
16. Brown, D.; Rowe, A.; Wild, P. Techno-Economic Comparisons of Hydrogen and Synthetic Fuel Production Using Forest Residue Feedstock. *Int. J. Hydrogen Energy* **2014**, *39*, 12551–12562. [CrossRef]
17. Collodi, G.; Azzaro, G.; Ferrari, N.; Santos, S.; Brown, J.; Cotton, B.; Lodge, S. *Techno-Economic Evaluation of SMR Based Standalone (Merchant) Hydrogen Plant with CCS*; IEAGHG Technical Report; IEAGHG: Cheltenham, UK, 2017.
18. Riva, L.; Martínez, I.; Martini, M.; Gallucci, F.; van Sint Annaland, M.; Romano, M.C. Techno-Economic Analysis of the Ca-Cu Process Integrated in Hydrogen Plants with CO_2 Capture. *Int. J. Hydrogen Energy* **2018**, *43*, 15720–15738. [CrossRef]
19. Romano, M.C.; Chiesa, P.; Lozza, G. Pre-Combustion CO_2 Capture from Natural Gas Power Plants, with ATR and MDEA Processes. *Int. J. Greenh. Gas Control.* **2010**, *4*, 785–797. [CrossRef]

20. Yang, F.; Meerman, J.C.; Faaij, A.P.C. Carbon Capture and Biomass in Industry: A Techno-Economic Analysis and Comparison of Negative Emission Options. *Renew. Sustain. Energy Rev.* **2021**, *144*, 111028. [CrossRef]
21. Northern Gas Networks. H21 Leeds City Gate. Available online: https://h21.green/projects/h21-leeds-city-gate/ (accessed on 24 May 2018).
22. Sutherland, F.; Duffy, L.; Ashby, D.; Legrand, P.; Jones, G.; Jude, E. Net Zero Teesside: Subsurface Evaluation of Endurance. In Proceedings of the 1st Geoscience & Engineering in Energy Transition Conference, Strasbourg, France, 16 November 2020; Volume 2020, pp. 1–5.
23. CO_2 Europipe Consortium. D3.1.2. Standards for CO_2: CO_2 Europipe Towards a Transport Infrastructure for Large-Scale CCS in Europe. 2009. Available online: http://www.co2europipe.eu/Publications/D2.2.1%20-%20CO2Europipe%20Report%20CCS%20infrastructure.pdf (accessed on 10 March 2022).
24. Pimenidou, P.; Dupont, V. Characterisation of Palm Empty Fruit Bunch (PEFB) and Pinewood Bio-Oils and Kinetics of Their Thermal Degradation. *Bioresour. Technol.* **2012**, *109*, 198–205. [CrossRef]
25. Abdul Halim Yun, H.; Dupont, V. Thermodynamic Analysis of Methanation of Palm Empty Fruit Bunch (PEFB) Pyrolysis Oil with and without in Situ CO_2 Sorption. *AIMS Energy* **2015**, *3*, 774–797. [CrossRef]
26. Dupont, V.; Abdul Halim Yun, H.; White, R.; Tande, L. High Methane Conversion Efficiency by Low Temperature Steam Reforming of Biofeedstock. In Proceedings of the REGATEC 2017 4th International Conference on Renewable Energy Gas Technology, Verona, Italy, 20 April 2017; pp. 22–23.
27. Garcia-Perez, M.; Chaala, A.; Pakdel, H.; Kretschmer, D.; Roy, C. Characterization of Bio-Oils in Chemical Families. *Biomass Bioenergy* **2007**, *31*, 222–242. [CrossRef]
28. Darmstadt, H.; Garcia-Perez, M.; Adnot, A.; Chaala, A.; Kretschmer, D.; Roy, C. Corrosion of Metals by Bio-Oil Obtained by Vacuum Pyrolysis of Softwood Bark Residues. An X-Ray Photoelectron Spectroscopy and Auger Electron Spectroscopy Study. *Energy Fuels* **2004**, *18*, 1291–13011. [CrossRef]
29. Heracleous, E. Well-to-Wheels Analysis of Hydrogen Production from Bio-Oil Reforming for Use in Internal Combustion Engines. *Int. J. Hydrogen Energy* **2011**, *36*, 11501–11511. [CrossRef]
30. Zhang, Y.; Brown, T.R.; Hu, G.; Brown, R.C. Comparative Techno-Economic Analysis of Biohydrogen Production via Bio-Oil Gasification and Bio-Oil Reforming. *Biomass Bioenergy* **2013**, *51*, 99–108. [CrossRef]
31. Maples, R.E. *Petroleum Refinery Process Economics*; PennWell Corporation: Nashville, TN, USA, 2000.
32. Srifa, A.; Chaiwat, W.; Pitakjakpipop, P.; Anutrasakda, W.; Faungnawakij, K. Advances in Bio-Oil Production and Upgrading Technologies. In *Sustainable Bioenergy: Advances and Impacts*; Elsevier: Amsterdam, The Netherlands; Oxford, UK; Cambridge, MA, USA, 2019.
33. Parkash, S. *Refining Processes Handbook*; Elsevier: Amsterdam, The Netherlands; Oxford, UK; Cambridge, MA, USA, 2003.
34. Davis, S.J.; Lewis, N.S.; Shaner, M.; Aggarwal, S.; Arent, D.; Azevedo, I.L.; Benson, S.M.; Bradley, T.; Brouwer, J.; Chiang, Y.M.; et al. Net-Zero Emissions Energy Systems. *Science* **2018**, *360*, eaas9793. [CrossRef] [PubMed]
35. Turton, R.; Bailie, R.C.; Whiting, W.B.; Shaeiwitz, J.A. *Analysis, Synthesis and Design of Chemical Processes*, 3rd ed.; Prentice Hall: Upper Saddle River, NJ, USA, 2009; ISBN 978-0-13-512966-1.
36. Turton, R.; Bailie, R.C.; Whiting, W.B.; Shaeiwitz, J.A. *Analysis, Synthesis, and Design of Chemical Processes*, 4th ed.; Prentice Hall: Upper Saddle River, NJ, USA, 2008; Volume 53.
37. Sadhukhan, J.; Ng, K.S.; Hernandez, E.M. *Biorefineries and Chemical Processes: Design, Integration and Sustainability Analysis*; John Wiley & Sons Ltd.: Chichester, UK, 2015.
38. Peters, M.S.; Timmerhaus, K.D.; West, R.E. *Plant Design and Economics for Chemical Engineers*, 5th ed.; McGraw-Hill: New York, NY, USA; London, UK, 2004.
39. Hamers, H.P.; Romano, M.C.; Spallina, V.; Chiesa, P.; Gallucci, F.; Annaland, M.V.S. Comparison on Process Efficiency for CLC of Syngas Operated in Packed Bed and Fluidized Bed Reactors. *Int. J. Greenh. Gas Control* **2014**, *28*, 65–78. [CrossRef]
40. Turner, J.; Sverdrup, G.; Mann, M.K.; Maness, P.-C.; Kroposki, B.; Ghirardi, M.; Evans, R.J.; Blake, D. Renewable Hydrogen Production. *Int. J. Energy Res.* **2008**, *32*, 379–407. [CrossRef]
41. Czernik, S.; French, R. Distributed Production of Hydrogen by Autothermal Reforming of Biomass Pyrolysis Oil. In Proceedings of the ACS National Meeting Book of Abstracts, Atlanta, GA, USA, 26–30 March 2006.
42. Campanario, F.J.; Gutiérrez Ortiz, F.J. Fischer-Tropsch Biofuels Production from Syngas Obtained by Supercritical Water Reforming of the Bio-Oil Aqueous Phase. *Energy Convers. Manag.* **2017**, *150*, 599–613. [CrossRef]
43. Do, T.X.; Lim, Y.I.; Yeo, H. Techno-Economic Analysis of Biooil Production Process from Palm Empty Fruit Bunches. *Energy Convers. Manag.* **2014**, *80*, 525–534. [CrossRef]
44. Methanex. Methanex Posts Regional Contract Methanol Prices for North America, Europe and Asia. Available online: https://www.methanex.com/our-business/pricing (accessed on 17 September 2019).
45. Song, H.; Ozkan, U.S. Economic Analysis of Hydrogen Production through a Bio-Ethanol Steam Reforming Process: Sensitivity Analyses and Cost Estimations. *Int. J. Hydrogen Energy* **2010**, *35*, 127–134. [CrossRef]
46. Swanson, R.M.; Platon, A.; Satrio, J.A.; Brown, R.C. Techno-Economic Analysis of Biomass-to-Liquids Production Based on Gasification. *Fuel* **2010**, *89*, S11–S19. [CrossRef]

47. Spallina, V.; Motamedi, G.; Gallucci, F.; van Sint Annaland, M. Techno-Economic Assessment of an Integrated High Pressure Chemical-Looping Process with Packed-Bed Reactors in Large Scale Hydrogen and Methanol Production. *Int. J. Greenh. Gas Control* **2019**, *88*, 71–84. [CrossRef]
48. Nazir, S.M.; Cloete, S.; Bolland, O.; Amini, S. Techno-Economic Assessment of the Novel Gas Switching Reforming (GSR) Concept for Gas-Fired Power Production with Integrated CO_2 Capture. *Int. J. Hydrogen Energy* **2018**, *43*, 8754–8769. [CrossRef]
49. Trane, R.; Dahl, S.; Skjøth-Rasmussen, M.S.; Jensen, A.D. Catalytic Steam Reforming of Bio-Oil. *Int. J. Hydrogen Energy* **2012**, *37*, 6447–6472. [CrossRef]
50. Kemper, J.; Sutherland, L.; Watt, J.; Santos, S. Evaluation and Analysis of the Performance of Dehydration Units for CO_2 Capture. *Energy Procedia* **2014**, *63*, 7568–7584. [CrossRef]
51. Gil, M.V.; Fermoso, J.; Pevida, C.; Chen, D.; Rubiera, F. Production of Fuel-Cell Grade H_2 by Sorption Enhanced Steam Reforming of Acetic Acid as a Model Compound of Biomass-Derived Bio-Oil. *Appl. Catal. B Environ.* **2016**, *184*, 64–76. [CrossRef]
52. Wright, M.M.; Román-Leshkov, Y.; Green, W.H. Investigating the Techno-Economic Trade-Offs of Hydrogen Source Using a Response Surface Model of Drop-in Biofuel Production via Bio-Oil Upgrading. *Biofuels Bioprod. Biorefin.* **2012**, *6*, 503–520. [CrossRef]
53. Arregi, A.; Amutio, M.; Lopez, G.; Bilbao, J.; Olazar, M. Evaluation of Thermochemical Routes for Hydrogen Production from Biomass: A Review. *Energy Convers. Manag.* **2018**, *165*, 696–719. [CrossRef]
54. Marda, J.R.; DiBenedetto, J.; McKibben, S.; Evans, R.J.; Czernik, S.; French, R.J.; Dean, A.M. Non-Catalytic Partial Oxidation of Bio-Oil to Synthesis Gas for Distributed Hydrogen Production. *Int. J. Hydrogen Energy* **2009**, *34*, 8519–8534. [CrossRef]
55. Spallina, V.; Pandolfo, D.; Battistella, A.; Romano, M.C.; van Sint Annaland, M.; Gallucci, F. Techno-Economic Assessment of Membrane Assisted Fluidized Bed Reactors for Pure H_2 Production with CO_2 Capture. *Energy Convers. Manag.* **2016**, *120*, 257–273. [CrossRef]
56. Dupont, V.; Ross, A.B.; Hanley, I.; Twigg, M.V. Unmixed Steam Reforming of Methane and Sunflower Oil: A Single-Reactor Process for H2-Rich Gas. *Int. J. Hydrogen Energy* **2007**, *32*, 67–69. [CrossRef]
57. Marquevich, M.; Farriol, X.; Medina, F.; Montané, D. Hydrogen Production by Steam Reforming of Vegetable Oils Using Nickel-Based Catalysts. *Ind. Eng. Chem. Res.* **2001**, *40*, 4757–4766. [CrossRef]
58. *Aspen Plus V.10*; Aspen Technology: Houston, TX, USA, 2015.
59. Chattanathan, S.A.; Adhikari, S.; McVey, M.; Fasina, O. Hydrogen Production from Biogas Reforming and the Effect of H2S on CH4 Conversion. *Int. J. Hydrogen Energy* **2014**, *39*, 19905–19911. [CrossRef]
60. Tian, X.; Wang, S.; Zhou, J.; Xiang, Y.; Zhang, F.; Lin, B.; Liu, S.; Luo, Z. Simulation and Exergetic Evaluation of Hydrogen Production from Sorption Enhanced and Conventional Steam Reforming of Acetic Acid. *Int. J. Hydrogen Energy* **2016**, *41*, 21099–21108. [CrossRef]
61. Baker, E.H. 87. The Calcium Oxide-Carbon Dioxide System in the Pressure Range 1-300 Atmospheres. *J. Chem. Soc. (Resumed)* **1962**, *87*, 464–470. [CrossRef]
62. Meerman, J.C.; Hamborg, E.S.; van Keulen, T.; Ramírez, A.; Turkenburg, W.C.; Faaij, A.P.C. Techno-Economic Assessment of CO_2 Capture at Steam Methane Reforming Facilities Using Commercially Available Technology. *Int. J. Greenh. Gas Control* **2012**, *9*, 160–171. [CrossRef]
63. Seider, W.D.; Seader, J.D.; Lwein, D.R. *Product and Process Design Principles: Synthesis, Analysis and Evaluation*; John Wiley & Sons, Ltd.: Hoboken, NJ, USA, 2009.
64. Schmidtsche Schack Flue Gas Convection Section in Steam Reformer Plants for Ammonia, Methanol, Hydrogen Production. Available online: https://www.schmidtsche-schack.com/products/flue-gas-convection-section-in-steam-reformer-plants/ (accessed on 25 April 2017).
65. Shahabuddin, M.; Krishna, B.B.; Bhaskar, T.; Perkins, G. Advances in the Thermo-Chemical Production of Hydrogen from Biomass and Residual Wastes: Summary of Recent Techno-Economic Analyses. *Bioresour. Technol.* **2020**, *299*, 122557. [CrossRef]
66. The Hydrogen Council. *Path to Hydrogen Competitiveness a Cost Perspective*; The Hydrogen Council: Brussels, Belgium, 2020.
67. Consoli, C. *Bioenergy and Carbon Capture and Storage-2019 Perspective*; Global CCS Institute: Melbourne, Australia, 2019.
68. Alptekin, G.O.; Jayataman, A.; Schaefer, M.; Ware, M.; Hunt, J.; Dobek, F. *Novel Sorbent to Clean Up Biogas for CHPs*; US Department of Energy: Washington, DC, USA, 2015. [CrossRef]
69. MATHESON Nanochem GuardBed for Sulfur Removal 2020. Available online: https://www.mathesongas.com/gas-equipment/ultrapurification/desulfurization/ (accessed on 9 February 2022).

MDPI
St. Alban-Anlage 66
4052 Basel
Switzerland
www.mdpi.com

Clean Technologies Editorial Office
E-mail: cleantechnol@mdpi.com
www.mdpi.com/journal/cleantechnol

Disclaimer/Publisher's Note: The statements, opinions and data contained in all publications are solely those of the individual author(s) and contributor(s) and not of MDPI and/or the editor(s). MDPI and/or the editor(s) disclaim responsibility for any injury to people or property resulting from any ideas, methods, instructions or products referred to in the content.

www.ingramcontent.com/pod-product-compliance
Lightning Source LLC
LaVergne TN
LVHW070420100526
838202LV00014B/1492